Concise College PHYSICS

Volume 2

Waves, Optics, Electricity, Electromagnetism, and Electronics

Concise College PHYSICS

Volume 2

Waves, Optics, Electricity, Electromagnetism, and Electronics

Samuel E. NONIE
B.Sc. (Hons.), M.Sc., Ph.D.
Senior Lecturer in Physics,
Papua New Guinea University of Technology
Formerly Lecturer in Physics, University of Sierra Leone (West Africa), and
Lecturer in Engineering Science and Mathematics, East Birmingham College, U.K.

CBS

CBS PUBLISHERS & DISTRIBUTORS PVT. LTD.

NEW DELHI • BENGALURU • PUNE • KOCHI • CHENNAI

ISBN : 81-239-1162-9

First Edition : 2005
Reprint : 2011

Published by Satish Kumar Jain and produced by V.K. Jain for
CBS Publishers & Distributors Pvt. Ltd.,
CBS Plaza, 4819/XI Prahlad Street, 24 Ansari Road, Daryaganj,
New Delhi - 110002, India. • Website: www.cbspd.com
e-mail: delhi@cbspd.com, cbspubs@airtelmail.in
Ph.: 23289259, 23266861, 23266867 • Fax: 011-23243014

Branches:
- ***Bengaluru:*** Seema House, 2975, 17th Cross, K.R. Road,
  Bansankari 2nd Stage, Bengaluru - 560070 Ph.: +91-80-26771678/79
  Fax: +91-80-26771680 • E-mail: cbsbng@gmail.com,
  bangalore@cbspd.com
- ***Pune:*** Bhuruk Prestige, Sr. No. 52/12/2+1+3/2,
  Narhe, Haveli (Near Katraj-Dehu Road By-pass), Pune - 411051
  Ph.: +91-20-64704058/59, 32342277 • E-mail: pune@cbspd.com
- ***Kochi:*** 36/14, Kalluvilakam, Lissie Hospital Road,
  Kochi - 682018, Kerala • Ph.: +91-484-4059061-65
  Fax: +91-484-4059065 • E-mail: cochin@cbspd.com
- ***Chennai:*** 20, West Park Road, Shenoy Nagar, Chennai - 600030
  Ph.: +91-44-26260666, 26208620 • Fax: +91-44-42032115
  E-mail: chennai@cbspd.com

Printed at :
Glorius Printers, Delhi

For my wife, Evelyn, and my boys, Edmond and Sammy, whose support and encouragement were beyond measure in the preparation of this work.

Preface

With the onward march toward technology, many undergraduate programs have adopted an instrumental approach, one that can only be achieved against a good background of the principles that underpin the operation of these instruments. Present day agricultural practices, for instance, include a wide range of instruments - use of centrifuges, automatic height adjustment harvesters, environmental data loggers, sensitive strain sensors, and many more. The trend is the same in other fields like surveying, architecture, forestry, chemical technology, and medicine - to name but only a few. Hence the need for a good level of physics for an expanding curriculum that caters for each of these disciplines.

With this need for physics, the role of the textbook to convey exactly the right message at the required level becomes of paramount importance. Standard physics texts, written with the major physics and engineering student in mind, can be overpowering to most other students. For these other students, the use of such texts tends to transform what should be a pleasant sojourn with the subject into a daunting time consuming task often at great expense to their major field of study.

Like the first volume, this second volume of the series attempts to focus attention to the wider first year physics student population. The first volume focused on the foundations of mechanics, and the building blocks of the physical world. It also introduced heat as a form of energy, and how this energy relates to other energy forms. Mechanisms of the thermal behaviour of substances were discussed, as were the thermal properties of materials. The volume ended with an introduction to the interesting and vital subject of thermodynamics.

This second volume focuses on the subjects of waves, optics, electricity and magnetism, and aspects of modern physics. It starts with a discussion of general wave motion, the nature of light and its propagation. Propagation phenomena like reflection and refraction are treated for all waves in general, and more specifically for light. The applications of these phenomena in the design and use of lenses and optical instruments are extensively discussed. The second part of the book deals with the subject of electricity. Although most people studying electricity tend to focus attention on current electricity, this book reveals the important fact that a clear understanding of electric current phenomena emanates from a study of charges at rest. This subject of electrostatics is presented in as simple a style as possible. Following electrostatic phenomena, the rest of the study is devoted to current electricity, and the interaction between electric and magnetic fields. The first chapter in this study focuses on direct current electricity. Useful electric quantities are introduced, together with the heating effect of a current and the power dissipated in an electric circuit. The magnetic effect of a steady current is introduced in the chapter that follows (chapter 10). This study is followed by electromagnetic induction - the interaction of electric and magnetic fields. Devices based on aspects of this interaction are studied, leading to the production of alternating currents. The behaviour of circuits to alternating voltages and currents is the topic of study of chapter 12. Chapter 13 draws attention, albeit in a simplistic manner, to the behaviour of circuits containing capacitances and resistances and those containing inductances and resistances, across a direct current supply voltage. In much the same simple manner chapter 14 presents an insight into the make-up and principles of operation of instruments designed for electrical measurements.

Modern technological world demands that studies of this kind are hardly complete without a treatment of electronics and computers. This call is addressed in chapters 15 and 16. These two final chapters give broad elementary introductions of modern electronics and computers. The two chapters draw on the experiences of the author both as a teacher of the subjects, and also as one whose job had involved design and operation of circuits and computer networking.

In the presentation of this second volume, a fair amount of mathematical points were found necessary to supplement the theoretical explanations of certain phenomena. The author has explained these mathematical points in Appendix 3. It is hoped that reading through this appendix will make it easy for the student to understand some of these concepts.

Between them, these two volumes provide a companion to the student reading physics both as a major course as well as a prerequisite to a main course. The down-to-earth explanations using common everyday examples where possible, coupled with further explanations in worked examples would, in the opinion of the authors, make this set of books an invaluable addition to the technology student's library.

The following are the highlights of these two volumes:

- Every chapter starts off with definitions of basic ideas, followed by applications. Each definition is highlighted and spelled out clearly, so the student reads the important parts.
- The material in each section follows a logical arrangement. Considering that most applications around us consist of the interplay of two or more principles, this logical presentation has not been easy (or possible) in a few instances. It is hoped the reader would appreciate this situation.
- Every chapter has a generous number of worked examples. In the solution of each example, the principles involved are explained to aid the students' reasoning. In many of these examples, certain lines have been omitted in the calculations on purpose, in the hope that by providing the missing links, the student can play a more active role in the solution, instead of just reading the worked example. There are exercises at the end of each chapter.

This project was completed in time, thanks mainly to the goodwill of some of my colleagues in the Department of Applied Physics, University of Technology, Papua New Guinea. These were Mr. Badira Varigini, and Mr. Ora Renagi. Their criticisms and invaluable suggestions helped map the details of the manuscript. Thanks also Mr. Peter Maclaurin, the Information Technology Manager of the University of Technology, for his valuable contribution to the contents of the final chapter on computers.

Apart from her normal encouragements in the project, my wife, Evelyn Nonie had the unenviable task of proofing the entire manuscript. Her quest for detail in every line made her role in the production of this book an exceptional one. I thank her very much for her efforts.

Finally, the author drew a lot of knowledge from the students taught at various levels over the past couple of decades. Thanks to all of them, for their input, albeit inadvertently, for most of them, in refining the material for this text.

S. E. NONIE

Contents

Part 1

Wave Motion and Optics

CONTENTS

CHAPTER 1

Waves and Wave Motion

1
CHAPTER

Waves and Wave Motion

1.1 WAVE MOTION

The ripples on the surface of a pond and the tsunamis are both examples of wave motion. A disturbance caused by a pebble dropped on to a pond excites water waves. When one end of a string is moved up and down at regular intervals, waves are sent along the string from the source of the vibration. A bob vibrating at regular intervals on the surface of water in a ripple tank creates water waves.

The above are all examples of wave motion. Two subtle differences are worth noting: When a pebble is dropped on to water in a pond a **disturbance** is created. A disturbance produces only one vibration, and only one crest is generated, which travels outward from the origin of the disturbance. A bob vibrating up and down at regular intervals on the surface of water in a ripple tank, and the regular movement of one end of a string, both create what is known as a **train of waves,** or **wave train**. A wave train consists of a regular set of identical disturbances. While we treat the two as waves, the analyses bear some differences.

Radiating outward from the source, the waves carry the signal to farther points within the medium. This signal is a packet of energy that gets transferred in the wave. It is this wave energy that makes a floating cork bob up and down on the surface of water, thereby changing its potential energy. While the energy is being transferred, however, there is no physical contact between the source and the object.

A wave is a means of transferring energy from source to object in a medium without physical contact between the source and the object.

There are many other types of wave motion, many of them not quite as obvious as those in the above examples. Sound is a kind of wave motion. Sound energy from musical instruments, for example, sets our eardrums into vibration in unison with the waves from the instrument. Light too, is a kind of wave motion, although we do not see the wave train quite as easily as we observe the water waves, or waves along a string. These less obvious waves nonetheless have a lot in common with the obvious ones, and their studies are very similar.

In this opening chapter of wave motion, we study the general characteristics of waves, the types of waves, and phenomena relating to the propagation of waves from one point to another. Our study here will focus mainly on mechanical waves. We shall leave the study other waves to later chapters.

1.2 GENERAL CHARACTERISTICS OF WAVES

Waves are classified in several ways. One of these is the medium supporting the transmission of the wave motion. Water waves, for example, denote waves that travel on the surface of water. Another classification is in the type of wave – e.g. sound waves, and light waves. Sound waves commonly travel in air, but also move through any medium. Light, we shall see later, is a form of wave motion that sends out waves capable of travelling through vacuum. Some waves do need a material medium to travel in, while for others, propagation does take place even in the absence of a material medium. Electromagnetic waves fall in this latter category. Radiant heat is an example of this wave motion. Irrespective of the type, waves have a set of properties in common.

1.2.1 Frequency of a wave

The frequency of a wave is defined as

the number of wave cycles generated per unit time.

A wave is always generated from a vibrating source. Since each vibration of the source generates a wave, the frequency of the wave is the same as the frequency of vibration of the source. For simplicity in analysing a wave motion, we start by assuming that the vibrations of the source are perfectly simple harmonic. Our study in Volume 1 defined simple harmonic motion as oscillatory motion in which the acceleration is proportional to displacement from the equilibrium position, and always directed towards that position. If a simple harmonic vibrating source is in contact with a medium, the vibrations from the source are communicated from one point to another within the medium. The frequency of the wave (i.e. how fast the crests are appearing) is only changed if the frequency of the source is changed. Changing the medium through which the wave travels has no effect on the frequency. Frequency is therefore a property of the source, and not of the medium.

The period of a wave is the time it takes to complete one wave cycle.

As in oscillatory motion, the frequency f is expressed in Hertz (or cycles per second), and the period T is expressed in seconds. The two quantities are related through the expression

$$f = \frac{1}{T}, \tag{1.1}$$

Example 1.1

A person standing in the ocean notices that after the first wave has passed by, twenty more waves pass in a time of 100 s. Calculate the frequency and period of the waves.

Solution

Since 20 waves pass through in 100 s, we have

Number of waves passing through per second = $\frac{20}{100}$ = 0.2 waves per second. Thus the frequency f of the source of vibration is 0.2 Hz, giving the period $T = \frac{1}{f} = \frac{1}{0.2} = 5.0$ s

1.2.2 Wavelength and Amplitude

Consider the wave set up along a horizontal string by vibrating one end vertically at regular intervals. This movement produces a train of waves in the string. **Fig. 1.1** is a representation of what the string would look like. The source of the vibration is at the left end O, of the string. The wave has crests C, and valleys or troughs V, equally spaced along the string. We define a property called the wavelength of the wave as

the distance between any two successive corresponding positions on the wave train.

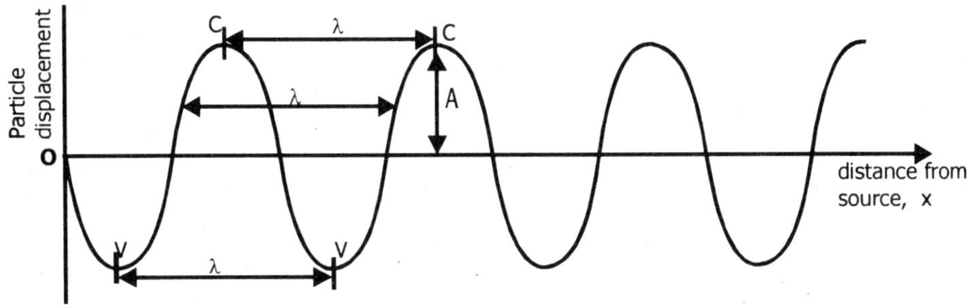

Fig. 1.1 Particle displacement vs. Position, showing the wavelength (λ), the crests (C), the troughs (V), and the amplitude (A).

In the diagram, the wavelength can be taken as the distance between any two successive crests, any two successive troughs, or indeed any two successive similar (or corresponding) positions. The Greek symbol λ (called lambda) is commonly used to represent wavelength. For mechanical waves, the unit of wavelength is m, although multiples and submultiples are used at times.

In our study of oscillatory motion in Volume 1, we came across the term amplitude of vibration, and defined it as the maximum displacement of the particle from its equilibrium position. In wave motion, too, we can talk of the amplitude of the particles of the medium constituting the wave motion. Recalling that each particle in the medium is oscillating, we define the amplitude as the maximum displacement from its equilibrium position. In **Fig. 1.1**, the amplitude A is shown as the distance from the baseline to the crest of the wave, the baseline representing the equilibrium position of the particle. Like wavelength, amplitude of mechanical waves is expressed in metres (m).

1.2.3 Intensity of a wave

Consider a cork floating on the surface of water. Waves generated on the water move the cork up and down, thereby giving potential energy to the cork. The maximum change in potential energy is a measure of the intensity of the wave. For an oscillating particle of mass m in the wave, having angular frequency ω, the maximum potential energy E is given by

$$E = \frac{1}{2} m\omega^2 A^2, \qquad (1.2)$$

Hence the maximum energy, or intensity I is given by

$$I = E = \frac{1}{2} m\omega^2 A^2. \qquad (1.3)$$

For an oscillating particle in the medium, and frequency of the source, we hence say that

the intensity of the wave is directly proportional to the square of the amplitude of vibration.

The intensity of a wave is a measure of the energy carried by the wave. The S.I. unit of wave intensity is $J/m^2/s$, or W/m^2. As the wave travels through a medium, the energy carried (or deposited) in the medium reduces with distance from the source of vibration. Studies show that

the intensity of a wave at any point in the medium is inversely proportional to the square of the distance from the source.

The above statement is known as the **inverse-square law** for the propagation of waves. Mathematically, we write the inverse-square law in symbols, as

$$I \propto \frac{1}{d^2}$$

or

$$I = \frac{\text{constant}}{d^2}, \qquad (1.4)$$

where d is the distance from the source. Since intensity at a given point depends on the amplitude of vibration, it follows that the amplitude of the wave decreases as we go farther away from the source. This explains why water waves created by a disturbance are felt only slightly, if at all, when you stand a long distance away from the source.

1.2.4 Speed of a wave

As the source vibrates, it sends out waves through the medium. At any point along the path, the crests are noticed to go past at a constant rate. The rate at which the wave progresses though the medium is the wave speed, or wave velocity.

Each time the source gets its maximum vibration, a crest is sent out. From our study of vibrations, the time interval between two successive positions of vibration of the source is the period. From the previous section, the distance between two successive crests is the wavelength. Thus, in the time the source vibrates through one period, the wave travels a distance of one wavelength. We hence write the equation for the speed v of the wave as

$$v = \frac{\lambda}{T} = f\lambda, \qquad (1.5)$$

where f is the frequency of the wave, λ the wavelength, and T is the period.

The speed with which a wave travels through a medium depends on two factors: the stiffness of the medium, and the density of the medium. As we studied in Volume 1, the stiffness depends on the elastic modulus of the medium. Using the method of dimensions, it is easy to show that

$$\text{speed of waves in a medium} = \sqrt{\frac{\text{Elastic modulus}}{\text{density of the medium}}} \qquad (1.6)$$

For a solid,

$$v = \sqrt{\frac{Y}{\rho}}, \qquad (1.7)$$

where Y is the Young's modulus of the solid, and ρ is its density. For a fluid of bulk modulus B, and density ρ,

$$v = \sqrt{\frac{B}{\rho}}. \qquad (1.8)$$

For a gas sample,

$$v = \sqrt{\frac{\gamma P}{\rho}}, \qquad (1.9)$$

where γ is the ratio of the heat capacities of the gas, and P is its pressure. For a string under tension T,

$$v = \sqrt{\frac{T}{\mu}}, \qquad (1.10)$$

where μ is the **mass per unit length**, or **linear density** of the string. Unlike other motions we have studied, waves travel at a constant speed in a given medium. The speed only changes when the wave travels from one medium to another, and when the properties of the medium change.

Example 1.2

A sound message is sent simultaneously in air and through the water between two sailing boats 2.7 km apart. It is found that there is a time difference of 6.1 s between the arrival of the two messages. Given that sound travels faster in water than through air, calculate the speed of the sound in water. Speed of sound in air = 343 m/s.

Solution

Time of travel $= \dfrac{\text{dis tan ce travelled}}{\text{speed of wave}}$. If t_1 and t_2 represent the times of travel of sound in air and in water, respectively, and v_1 and v_2 are the respective speeds in the two media, then $t_1 = \dfrac{D}{v_1}$, and $t_2 = \dfrac{D}{v_2}$, where D is the distance between the two boats. Thus, $t_1 - t_2 = D\left(\dfrac{1}{v_1} - \dfrac{1}{v_2}\right)$, or $\dfrac{1}{v_2} = \dfrac{1}{v_1} - \dfrac{(t_1 - t_2)}{D}$. From the data given in the question, $D = 2700$ m, $t_1 - t_2 = 6.1$ s. Substituting these, we get $v_2 = 1523\,\text{m/s}$.

Example 1.3

If a detonator is exploded on a steel fence, an observer standing along the fence 2.0 km away hears two reports. Given that the speed of sound in air is 343 m/s, calculate the time interval between the two reports. For steel, Young's modulus $= 2.0 \times 10^{11}\ \text{N/m}^2$; density $= 7.8 \times 10^3\ \text{kg/m}^3$.

Solution

This example is similar to **Example 1.2**. Here, we calculate the speed of sound in the steel using equation (1.4). If v_1 and v_2 are respectively the speeds in air and in steel,

$$v_2 = \sqrt{\frac{Y}{\rho}} = \sqrt{\frac{2.0 \times 10^{11}\ \text{N/m}^2}{7.8 \times 10^3\ \text{kg/m}^3}} = 5063.6\ \text{m/s}.$$

Using the ideas in the previous example, the time interval between the two reports is

$$t_1 - t_2 = D\left(\frac{1}{v_1} - \frac{1}{v_2}\right) = 2000\ \text{m} \times \left(\frac{1}{343\ \text{m/s}} - \frac{1}{5063.6\ \text{m/s}}\right) = 5.43\ \text{s}$$

1.3 TYPES OF WAVE MOTION

Whilst the medium can be used to classify a wave, there are other more common ways of describing waves. The usual method of classification uses the vibration of the particles of the medium in relation to the direction of travel of the wave.

1.3.1 Transverse waves

In a transverse wave, the particles of the medium vibrate at right angles to the direction of travel of the wave. Examples of transverse waves are water waves in a ripple tank, and waves set up along a string. **Fig. 1.2** shows the profile of a typical transverse wave.

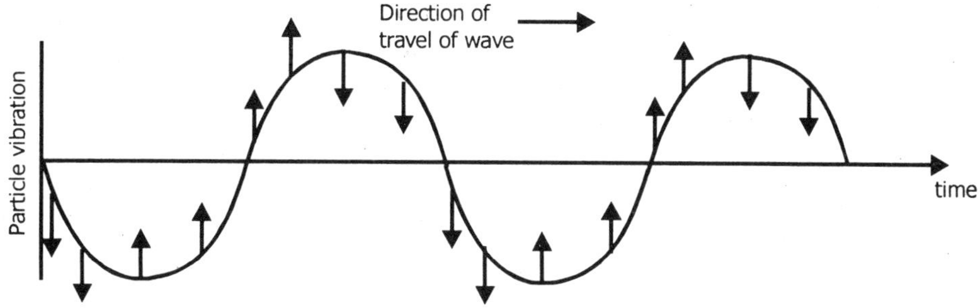

Fig. 1.2 Transverse wave motion. Direction of vibration of particle in medium is perpendicular to the travel of the wave.

1.3.2 Longitudinal waves

In a longitudinal wave, the particles in the medium vibrate in the same direction as the direction of travel of the wave. The wave travels in the medium by setting up areas of high pressure, or **compression**, and areas of low pressure, or **rarefaction**. **Fig. 1.3** depicts the propagation of a longitudinal wave in a medium. The figure shows the compressions C, and rarefactions R. As a result of the compressions and rarefactions, longitudinal waves are also known as pressure waves. These waves do not travel through vacuum. Examples of longitudinal waves are mechanical shock waves through a metal, heat waves through a solid, and sound waves in air.

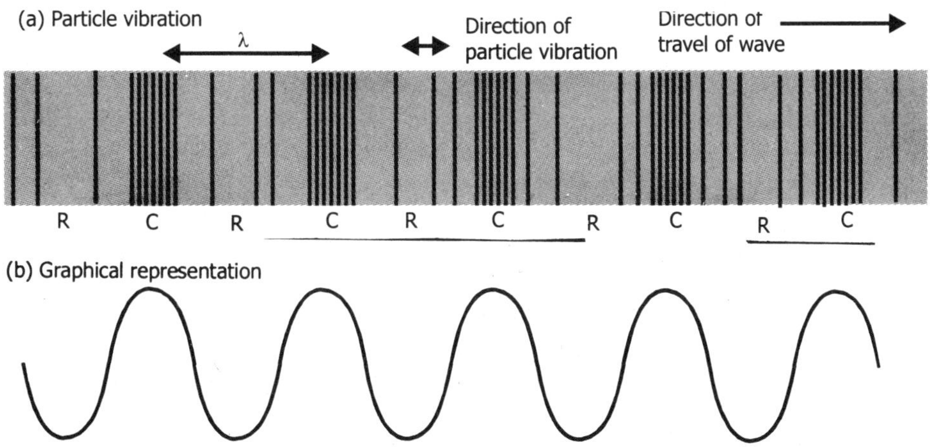

Fig. 1.3 Longitudinal wave motion, showing areas of compression C, and rarefaction, R.

In the graphical representation of longitudinal wave, the crests are made to correspond to the areas of compression, whilst the troughs correspond to rarefaction. These are shown in **Fig. 1.3** (b). The wavelength is the distance between two adjacent areas of compressions, or between two adjacent regions of rarefaction.

1.4 PROGRESSIVE WAVES

Whether the wave motion is transverse of longitudinal, it can either be a progressive wave, or a stationary wave. A progressive wave travels from one point to another in a medium, or from one medium to another, carrying energy along with it. Progressive waves can be reflected and refracted at the boundary between the two media.

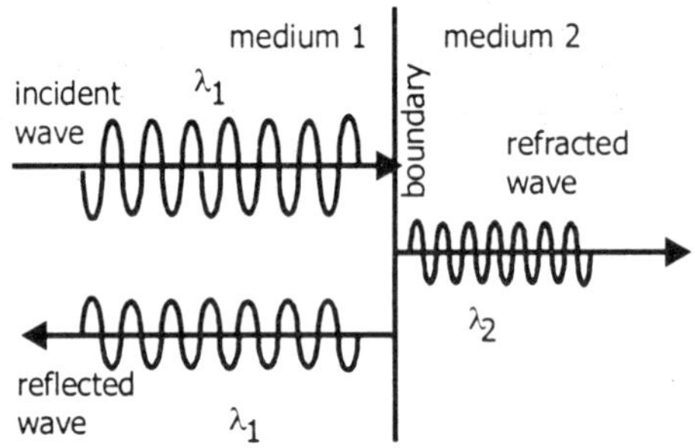

Fig. 1.4 Reflection and refraction of a travelling wave at a plane boundary, for normal incidence.

Fig. 1.4 shows a wave incident normally at the plane boundary between two media, labelled medium 1 and medium 2. The figure shows part of the incident wave at the boundary in medium 1, and the other part refracted through medium 2.

Note the following:

(a) There is a reversal in the form of the wave after reflection in medium 1. This is described by saying that the wave suffers a **phase change** of π on reflection. We discuss more on this phase change later in this, and in other chapters.

(b) In general, both the reflected and refracted waves are weaker than the incident wave. These intensities are represented in terms of the amplitudes of the waves. Considering that the wave carries energy with it, energy conservation principle can be used to help us understand why the intensities are different.

(c) Whilst the wavelength of the incident and reflected waves are the same, the wavelength of the refracted wave is different from that of the incident wave.

When a wave travels from one medium into another, the speed of the wave changes. A quantity known as the **relative refractive index** is used to express this change. If the speeds of the waves in the incident and refracted media are v_1, and v_2, respectively, the relative refractive index between the two media is written as ${}_1n_2$, where

$$ {}_1n_2 = \frac{v_1}{v_2} . \tag{1.11} $$

Since the frequency of the wave remains constant, equation (1.11) can be written in terms of the corresponding wavelengths λ_1 and λ_2 in the two media as

$$_1n_2 = \frac{\lambda_1}{\lambda_2}. \qquad (1.12)$$

In the example depicted in **Fig. 1.4**, the wave is traveling from a less dense medium to a denser medium. For this case, we have $v_1 > v_2$, and $\lambda_1 > \lambda_2$. The reverse obtains when the wave travels from a lighter medium to a denser medium.

If the wave is not incident at right angles to the boundary, the direction of the refracted wave is different from that of the incident wave. Drawing a normal to the surface at the points of incidence and refraction, we define the angles of incidence, reflection, and refraction as the angles made by the incidence, reflected, and refracted waves, respectively, with the normal at the points of incidence and refraction. Studies of the have revealed the following laws of reflection and refraction at plane surfaces:

1. ***The incident wave, the reflected wave, the refracted wave, and the normal all lie in the same plane,***
2. ***The angle of incidence is equal to the angle of reflection,***
3. ***The ratio of the sine of the angle of incidence to the sine of the angle of refraction is a constant.***

The third law is **Snell's law**, and the constant is the relative refractive index between the two media. **Fig. 1.5** shows the case of non-zero angle of incidence of a wave at a plane boundary.

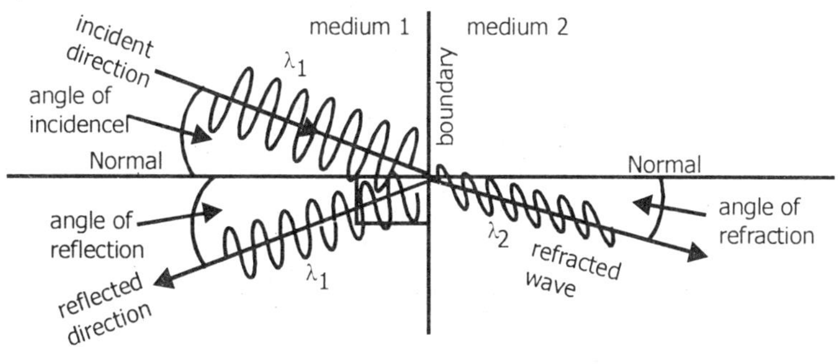

Fig. 1.5 Reflection and refraction at a plane boundary, for non-zero angle of incidence.

From law 3 above, we write

$$_1n_2 = \frac{\text{sine of angle of incidence}}{\text{sine of angle of refraction}} \qquad (1.13)$$

Reflection and refraction are studied in detail in geometrical optics.

Example 1.4

Surface water waves are produced in a large ripple tank by a small sphere vibrating with a frequency of 5 Hz. Near the sphere the wave crests are 0.05 m apart, and in some other region they are 0.02 m apart.

(a) Calculate the speed of the waves in these two regions.

(b) State the likely cause of the change in speed.

Solution

(a) From equation (1.4) connecting the speed of a wave with the frequency of its source and the wavelength, we have $v_1 = f\lambda_1$, and $v_2 = f\lambda_2$, where v_1 and v_2 are speeds of the water waves near the sphere, and far away from the sphere. λ_1 and λ_2 are the wavelengths. We hence calculate the speeds as follows:
$v_1 = f\lambda_1 = 5.0\text{ Hz} \times 0.05\text{ m} = 0.25\text{ m/s}$, and $v_2 = f\lambda_2 = 5.0\text{ Hz} \times 0.02\text{ m} = 0.10\text{ m/s}$,

(b) With the change in velocity, refraction has taken place. A likely cause for this is the water wave going from a shallow part of the tank to a deeper part.

Example 1.5

A fisherman at sea notices that his boat is moving up and down periodically due to waves on the surface of the sea. It takes 3.0 s for the boat to travel from the highest point to the lowest point, a total distance of 0.8 m. If the wave crests are 8.0 m apart, calculate the speed and amplitude of the water waves.

Solution

Fig. 1.6 depicts the pattern exhibited by the water waves. From the data given, the period T=6.0 s. Hence the frequency $f = \frac{1}{T} = \frac{1}{6.0\text{s}} = 0.167$ Hz.

The distance from the highest point C to the lowest point V is twice the amplitude A of the wave. From the data, we have $2A = 0.8\text{ m}$, or A=0.4 m. Since the wavelength is 8.0 m, the speed v of the water waves is $v = f\lambda = 0.167\text{ Hz} \times 8.0\text{ m} = 1.336\text{ m/s}$.

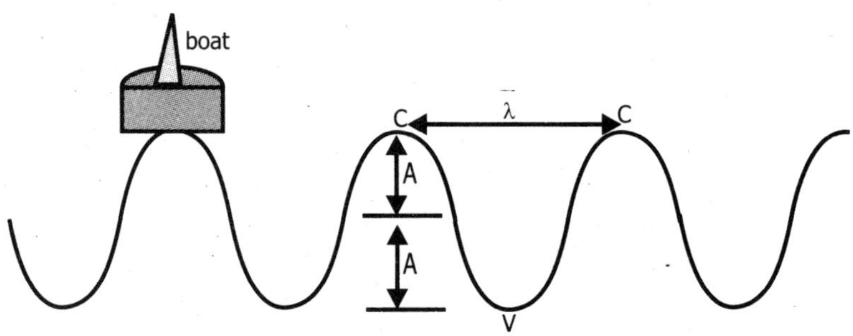

Fig. 1.6

1.5 DOPPLER EFFECT

It is noticed that as one approaches a blowing horn, the perceived frequency of the sound gets higher and higher until the horn is reached, and then gets lower as the horn is passed. Also, if a moving car sounds its horn continuously as it passes a stationary observer on the side of the road, the frequency of the sound is higher as the car approaches, and becomes lower as the car goes past. Similarly, the light from a distant star, observed from the earth shifts toward lower frequency when the star and the earth are going away from each other; the light shifts toward the high frequency if the star and the earth are moving toward each other.

There is an apparent change in frequency of a wave due to relative motion between the source of the wave and the observer. This phenomenon was first observed by the Austrian physicist, Christian Johann Doppler, and is today known as the **Doppler effect**. Doppler effect is used in modern astronomy to study the motions of stars, and also forms an integral part of modern theories of the universe.

We can formulate the observed frequency due to relative motion between source and observer by taking the simple case in which we assume the source and observer move along the same straight line. These formulae are derived in the following sections.

1.5.1 Observer moving, source stationary

Consider an observer moving toward a stationary source. The source gives out waves of frequency f, and wavelength $\lambda = \frac{V}{f}$, where V is the velocity of the waves. To the moving observer, the waves are observed to be travelling at a velocity $(V + v_0)$, where v_0 is the velocity of the observer. **Fig. 1.7** illustrates an observer moving relative to a stationary source. In (a), the observer is moving towards the source, and in (b), the observer is moving away from the source.

If we denote the observed frequency by f_0, then, $f_0 = \frac{(V + v_0)}{\lambda} = \frac{V + v_0}{\frac{V}{f}}$, or

$$f_0 = \left(\frac{V + v_0}{V}\right) f. \qquad (1.14)$$

A positive sign is used for v_0, when the observer moves toward the source, and a negative sign for the case in which the observer moves away from the source. With this notation, we see that the observed frequency is higher when the observer moves toward the source, and lower when observer recedes from the source. In this latter case,

$$f_0 = \left(\frac{V - v_0}{V}\right) f. \qquad (1.15)$$

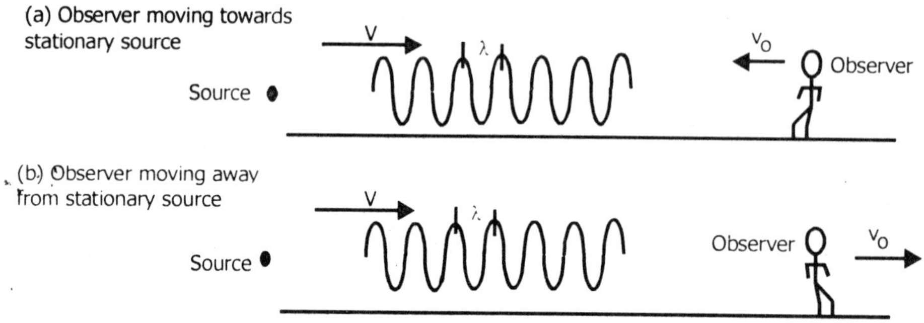

Fig. 1.7 Doppler effect. Source stationary, observer moving

1.5.2 Source moving, observer stationary

Consider that the source of the waves is moving with velocity v_s towards a stationary observer, and emitting waves with speed V, and frequency f. Since the medium has not changed, the wave speed relative to the medium is still V, as before. However, the wavelength is no longer the value it was. In one period, the wave travels a distance $VT = V/f$, where T is the period of the source. The wavelength is the distance between successive crests, and is determined by the relative displacement of the source and the wave. This displacement is different when source moves toward, and away from a stationary observer. Toward a stationary observer, the observed wavelength λ_1 is $\lambda_1 = \dfrac{V - v_s}{f}$, and

$$f_o = \frac{V}{\lambda_1} = \left(\frac{V}{V - v_s}\right) f. \qquad (1.16)$$

When the source moves away from the stationary observer,

$$f_o = \frac{V}{\lambda_1} = \left(\frac{V}{V + v_s}\right) f. \qquad (1.17)$$

1.5.3 Both source and observer moving

Equations (1.13) to (1.16) can be combined to give a general expression

$$f_o = \left(\frac{V \pm v_o}{V \mp v_s}\right) f. \qquad (1.18)$$

The upper signs are used when source and observer move towards each other, whilst the bottom signs are used when the source and observer move away from each other.

Example 1.6

Two vehicles, far apart, are travelling at the same speeds in opposite directions. One driver hears the horn of the other vehicle at a frequency that is 1.2 times the frequency he hears when the two vehicles are stationary. If the speed of sound is 343 m/s, calculate the speed of each vehicle.

Solution

In this example, both the observer and the source are traveling at the same. Thus $v_o = v_S$. Using equation (1.14), we write

$$\frac{f_0}{f} = \left(\frac{V + v_0}{V - v_S}\right) = 1.20 \qquad (1)$$

Substituting the numerical values, we get

$$\left(\frac{V + v_0}{V - v_0}\right) = 1.20 = \left(\frac{343 + v_0}{343 - v_0}\right) \qquad (2)$$

Rearranging (2), we have $v_0 = \dfrac{343 \text{ m/s} \times 0.2}{2.2} = 31.18 \text{ m/s}$ (or 31.2 m/s)

Example 1.7

A boy on a seashore hears the sounding of a ship's siren. When neither of them moves, the frequency he hears is 420 Hz. When the ship moves towards him, the frequency he hears is 430 Hz. However, when the ship moves away from him, the frequency he hears is 415 Hz. If the speed of sound is 343 m/s, find the speed of the ship in the two cases.

Solution

In both parts of this example, we use $f_0 = \left(\frac{V}{V+v_S}\right)f$, where v_S is the source speed, V the speed of the wave, and f the frequency of the wave. When the ship moves toward the observer, we use the negative sign for v_S. Thus,

$$f_0 = \left(\frac{V}{V - v_S}\right)f, \text{ or } v_S = V\left(1 - \frac{f}{f_0}\right) = 343 \text{ m/s} \times \left(1 - \frac{420}{430}\right) = 7.98 \text{ m/s}.$$

When the ship moves away from the shore, we have $f_0 = \left(\dfrac{V}{V + v_S}\right)f.$ Substituting values into this latter equation, we get $v_S = V\left(\frac{f}{f_0} - 1\right) = 343 \text{ m/s} \times \left(\frac{420}{415} - 1\right) = 4.13 \text{ m/s}.$

1.6 THE WAVE FUNCTION

The wave function is a mathematical expression for the wave. This expression contains information of the source of vibration, and the travel of the wave through the medium.

The wave function can be derived from our knowledge of the equation for the vibration of the source, and the speed of the waves in the medium.

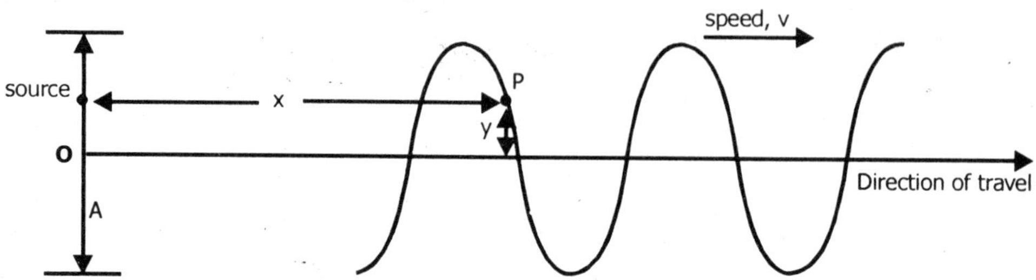

Fig. 1.8 Deducing the wave function.

Fig. 1.8 shows a vibrating source at O, sending out waves in a medium, the waves travelling with speed v, in the positive x-direction. Consider a point P, distance x from the source. A vibration arriving at P at time t was generated at the source at a time $(t - x/v)$. The displacement y, of the source from the equilibrium at this time is given by

$$y = A\sin\omega\left(t - \frac{x}{v}\right), \tag{1.19}$$

where A is the amplitude of vibration, and ω is the angular frequency of the source. Equation (1.19) gives the vibration of the medium at the point P, and hence represents the equation of the wave. Expressing ω in terms of f, and v in terms of f and λ, we get

$$y = A\sin 2\pi\left(ft - \frac{x}{\lambda}\right). \tag{1.20}$$

Equation (1.20) represents the wave function for a plane progressive wave travelling in a medium in the positive x-direction. At a glance, it is seen that from it we have information on the vibration of the source, the direction of travel of the wave, and the speed of the wave in the medium. The equation is more commonly written in the form

$$y = A\sin(\omega t - kx), \tag{1.21}$$

where $k\left(= \frac{2\pi}{\lambda}\right)$ is the **propagation constant.**

For a plane wave travelling in the negative x-direction, a similar reasoning gives the corresponding wave function as

$$y = A\sin(\omega t + kx). \tag{1.22}$$

The use of the sine function in equations (1.18) and (1.19), is based on the assumption that the initial particle displacement is zero at t = 0, and x= 0. This corresponds to the particle vibration starting on the baseline. In many cases, the wave is set into motion when the particle is at its maximum displacement. To describe such waves, we use the cosine function instead. The functions

$$y = A\cos(\omega t - kx), \tag{1.23}$$

and

$$y = A\cos(\omega t + kx), \tag{1.24}$$

are the wave functions for these cases. There are other alternatives to the wave function, depending on the starting conditions. Some of these are covered in the numerical examples. Whichever expression used, it is important to bear in mind that the angles should be expressed in radians, and not in degrees.

1.6.1 Wave properties from the wave function

Equations (1.19), through to (1.24) allow us to determine the important properties of a wave – the amplitude, the frequency, the wavelength of the wave, its direction of travel, and its speed. Alternatively, we can work out the wave function of the wave from a set of given properties - see Examples (1.8) and (1.9).

Example 1.8

A plane progressive wave is represented by the equation $y = 5\sin\left(500\pi t - \frac{\pi}{4}x\right)$, where x and y are in cm, and t is in seconds. Determine

(i) the direction of travel of the wave,
(ii) the frequency of vibration of the source,
(iii) the wavelength of the wave,
(iv) the speed of the wave,
(v) the amplitude of vibration of particles in the medium due to the wave.

Solution

(i) Comparing the given equation with the general wave function given by (1.14), we deduce that the wave is travelling in the positive x-direction.

(ii) In the usual notation, $\omega = 2\pi f = 500\pi$. This gives the frequency $f = 250$ Hz.

(iii) The propagation constant for this wave is given by $k = {}^{2\pi}/_{\lambda} = {}^{\lambda}/_{4}$. Thus, the wavelength of the wave is 8.0 cm.

(iv) With the frequency and wavelength determined, the speed v is given by $v = f\lambda = 250\text{ Hz} \times 0.08\text{ m} = 20\text{ m/s}$.

(v) From the wave function, the amplitude of vibration = 5.0 cm.

Example 1.9

Deduce the wave function from the following information: wavelength = 3.5 m, frequency = 420 Hz, amplitude 0.45 m, for a wave traveling in the positive x-direction.

Solution

For a wave travelling in the positive x-direction is given by $y = A\sin 2\pi\left(ft - x/\lambda\right)$, where A is the amplitude of the wave, f the frequency, and λ is the wavelength. Thus, the wave function is given by $y = 0.45\sin 2\pi\left(420t - x/3.5\right)$m.

1.6.2 Phase difference

In our study of oscillatory motion in Volume 1, we described the phase difference as the amount by which the two oscillatory motions are out of step with each other. We use the same idea for wave. Two points in a wave motion can be in phase, or out of phase with each other. If they are in phase, they vibrate in unison. This means both points are at crests at the same time, and similarly are at troughs at the same time. The difference in phase between two points on a wave can be expressed either in terms of the angle between their vibrations (as in oscillatory motion), or as a fraction of the wavelength. **Fig. 1.8** shows three points P, Q, and R, on a wave. The phase difference between P and Q is expressed as θ, where

$$\frac{x_2 - x_1}{\lambda} = \frac{\theta}{2\pi},$$

or

$$\theta = \frac{2\pi(x_2 - x_1)}{\lambda} = \frac{2\pi(\Delta x)}{\lambda}, \qquad (1.25)$$

where Δx is the distance between the two points. As the motions of the particles repeat themselves at regular intervals, it follows that two points exactly one wavelength apart are in phase. Thus, if two points in the medium have a phase difference of 3π, for instance, the two points are in fact only π radians $\left(\text{or } \lambda/2\right)$ out of phase, where λ is the wavelength. In **Fig. 1.8**, the points Q and R are in phase, since they are one wavelength apart. The phase difference between P and Q is the same as that between P and R.

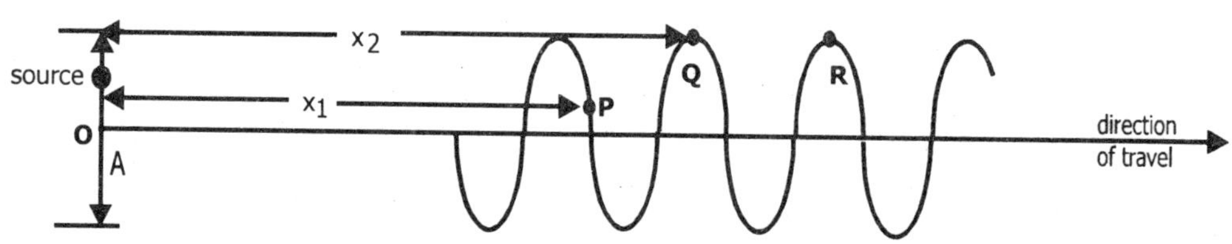

Fig. 1.8 Phase difference. Because Q and R are exactly one wavelength apart, the phase difference between P and R is the same as that between P and Q.

Example 1.10

A plane progressive wave is represented by the expression $y = 2\cos 200\pi(t + x/3000)$ cm, where x is in cm. Find

(i) the amplitude of the wave,
(ii) the speed of the wave,
(iii) the difference in phase between two particles in the medium 45 cm apart,

Solution

(i) From the given wave function, amplitude A = 2.0 cm.

(ii) The speed of the wave is determined from calculations of the frequency f of the wave, and the wavelength λ. From the wave function, $2\pi f = 200\pi$, or f = 100 Hz.

To find the wavelength, we set $\frac{2\pi}{\lambda} = \frac{200\pi}{3000}$, or λ = 30.0 cm

Hence the speed of the wave is given by $v = f\lambda$ = 100 Hz x 0.30 m =30 m/s.

(iii) The two particles are exactly 1.5 wavelengths apart. Since particles that are 1.0 wavelength apart vibrate in phase, particles that are 1.5 wavelengths apart will have a phase difference of π (or $\lambda/2$).

EXERCISES

1. From the wave function $y = 5\sin(2x + 5t)$, where y is in millimeters, x is in metres, and t is in milliseconds. Find the amplitude, wavelength, frequency, and speed of the wave.

2. Radio waves are transverse waves that travel at the speed of light, 3.0×10^8 m/s. An FM radio broadcasting station broadcasts at 91.9 MHz. Calculate the wavelength of this radio wave.

3. At a certain temperature, sound waves in air are found to travel at a speed of 345 m/s. If the wavelength of the waves is 1.30 m, calculate the period and frequency of the wave.

4. A wire of mass 200 g is stretched between two supports 15 m apart. The wire is under a tension of 100 N. Calculate the time it takes for a disturbance created at one end of the string to travel to the opposite end.

5. A wave is set up in a stretched string under a tension of 58 N. If the frequency of the wave is doubled, determine the new tension required to give the same wavelength.

6. An early morning jogger moving toward a stationary sounding siren observed that the frequency of the source is 1.04 times the frequency when he is stationary. Calculate the speed of the observer. Speed of sound in air = 343 m/s.

7. Deduce the wave function of a wave from the following information: wavelength 0.24 m, amplitude 5.1 cm, speed 8.0 m/s, travelling in the positive x-direction.

8. A motorist is travelling at a speed of 38 m/s. A Police car pulls up behind, and matches the same speed. The Police car sounds a siren demanding that the motorist stop. If the frequency of the siren is 860 Hz, when the Police car and an observer are stationary, and the speed of sound in air is 343 m/s, determine the frequency of the siren heard by the motorist while moving at the his speed.

9. The amplitude and frequency of a transverse wave on a string are 7.8 mm, and 6.0 Hz, respectively. Calculate the maximum speed of a particle of the string.

10. A radio transmitter uses a wave with frequency 1.20 MHz, and speed is 3.0×10^8 m/s. At what wavelength must a radio set be tuned to receive the broadcast signal?

11. The equation $y = A\sin(\omega t - kx)$ represents a wave function. If $A = 10^{-7}$ m, $\omega = 6600\ s^{-1}$, and $k = 20.0\ m^{-1}$, calculate the speed of the wave.

12. A train whistle emits sound at a frequency of 555 Hz. A person standing next to the train track hears the whistle at a frequency of 488 Hz. If the speed of sound is 343 m/s, calculate the speed of the train. Deduce whether the train is coming towards, or going away from the person.

13. A transverse wave is set up along a horizontal stretched string. The wavelength, amplitude, and speed of the waves are 1.60 m, 8.0 cm, and 100 m/s, respectively. Consider the left end of the undisturbed string as the origin. At time t = 0, the left end of the string is at the origin and moving downward.
    (i) Determine the frequency, angular frequency, and propagation constant of the wave.
    (ii) Work out the wave function for the wave.
    (iii) Deduce the equation of motion for the left end of the string.
    (iv) Find the equation of motion of a particle that is 1.2 m to the right of the origin.
    (v) Calculate the maximum speed of any particle in the string.

14. Transverse waves on a string are travelling in the positive x-direction, and have an amplitude of 2.50 mm, a wavelength of 30 cm, and a speed of 15.0 m/s. Deduce the equation that represents the displacement y as a function of time for a point located on the string a distance 40 cm from the source.

15. A wave incident in medium 1 is refracted at a plane boundary into medium 2. The

velocities of the waves in the two media are 2.0×10^8 m/s, and 1.5×10^8 m/s. Calculate the relative refractive index of the two media.

16. A plane progressive wave is represented by the function $y = 0.2\sin\left(200\pi t - \frac{20\pi x}{17}\right)$, where y is in millimeters, t is in seconds, and x is the distance from a fixed origin, in metres. Find
(i) the frequency of the wave,
(ii) the wavelength of the wave,
(iii) the speed of the wave,
(iv) the phase difference between a point 0.25 m, and another 1.10 m from the origin,

17. An ambulance travels at a speed of 120 km/h. Its siren emits sound waves at frequency 660 Hz. A passenger car is traveling at 88 km/h in opposite direction along the same road. Given that the speed of sound in air is 343 m/s, calculate the frequency of the siren as heard by an observer in the car
(i) as the car and ambulance are approaching each other,
(ii) as the car and ambulance go away from each other.

18. The displacement y, in metres, of the vibrating particles of a wave are represented by the equation $y = 0.30\cos 2\pi(t - 2x)$, where t is in seconds, and x is in metres.
(i) In which direction is the wave traveling?
(ii) Calculate the particle displacement when x = 10 m and t = 40 s.
(iii) Calculate the frequency, wavelength, and speed of the wave.
(iv) If this wave is totally reflected at a plane boundary, deduce the wave function for the wave formed when the reflected wave combines with the incident wave.

19. A particle initially displaced upwards by 5 cm from its equilibrium position, oscillates with frequency 50 Hz, and sends out a wave in a medium in the positive x-direction. It is found that the crests of the wave are 2.5 cm apart. Deduce the wave function of the waves generated.

20. The note emitted from the vibrating length of a wire of length 60 cm is in tune with a note from an audio oscillator. On using a different wire of linear density 10% more than before, the tuning is restored by adjusting its length to 62 cm. Determine the ratio of the tensions in the two wires, assuming the wires are of the same thickness.

21. A radio transmitter uses a carrier wave frequency of 104 MHz. The speed of the waves is 3.0×10^8 m/s. At what wavelength must a radio set be tuned to receive the broadcast ?

22. A plane wave is represented by the equation $y = 10\cos(4x + 5t)$ cm, where x in metres, and t in seconds. Determine
(i) the amplitude of the wave,
(ii) the frequency, wavelength and speed of the wave,

(iii) the particle displacement when $x = 0.5$ m, and $t = 1.0$ s,
(iv) the equation of the wave obtained when the above wave is totally reflected.

23. Waves of constant frequency are created on the surface of a lake. In one region A of the lake the wave crests are 2 cm apart, and in another region B they are 1.5 cm apart. Calculate the ratio of the speeds of the waves in regions A and B.

24. The equation of a travelling wave is given by $y = 2\cos 200\pi(t - 0.003x)$cm, where x is in metres, and t in seconds. Determine
(i) the period of the waves,
(ii) the wavelength of the waves,
(iii) the frequency of the waves,
(iv) the ratio of the maximum particle speed to the speed of the wave.

25. A transverse wave on a string is traveling in the negative x-direction, and has an amplitude of 2.50 mm, wavelength 30 cm, and frequency 50 Hz. Deduce the equation that represents the displacement y as a function of time for a point located on the string a distance 40 cm from the source.

26. A train travelling at a speed of 80 km/h sounds a whistle at a frequency of 555 Hz. A car driver travelling along a road parallel to the track travels at a speed of 60 km/h. Given that the speed of sound in air is 343 m/s, calculate the frequency of the whistle heard by the car driver when the car and the train travel
(i) in the same direction,
(ii) in opposite directions.

27. A marksman fires a rifle at a target. The bullet travels at an average speed of 600 m/s, and 4 seconds after he fires the gun, he hears the bullet strike the target. Given that the speed of sound is 343 m/s, calculate the distance between the marksman and the target.

28. Sounds with frequencies of 20 Hz and 20 kHz represent the usual frequency range for human audibility. Calculate the wavelength range of human audibility, given that the speed of sound in air is 343 m/s.

2

CHAPTER

Superposition of Waves

2
CHAPTER

Superposition of Waves

In the previous chapter we studied the basics of wave motion – the types of waves, their nature and properties, and the mode of propagation through a given medium. A single bob vibrating on the surface of water in a shallow tank, for instance, sends out waves. Quite a few interesting things happen when two bobs not far apart vibrate together, each sending out waves in the water in the same or opposite direction. What patterns do these waves form when taken together? How useful is this concept of combining waves? These questions, along with related ones, form the subject matter of this chapter.

2.1 SUPERPOSITION OF WAVES

The principle of superposition of waves helps us to find the effect of combining two or more waves travelling in a medium. This principle states that

The resultant vibration at a point due to two or more waves in a medium is the algebraic sum of the individual vibrations due to the waves.

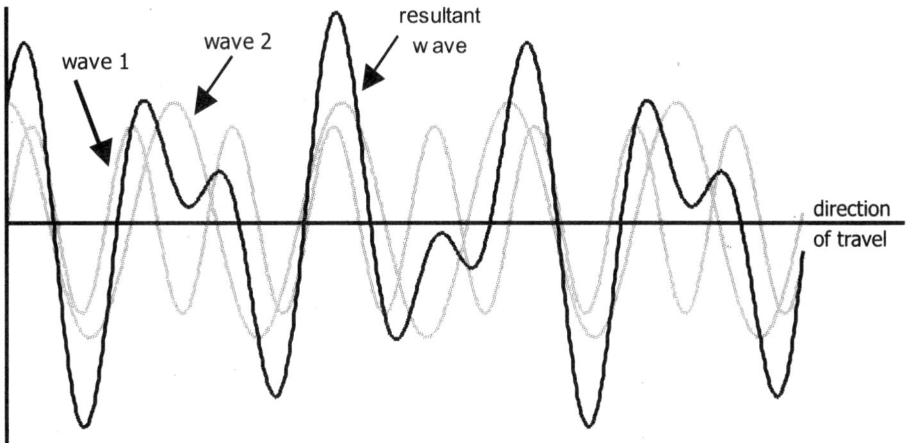

Fig. 2.1 Resultant of two waves obtained from the principle of superposition

The above statement can be understood with reference to **Fig. 2.1**, which shows two waves travelling in the same direction, for simplicity. At each point in the medium, the displacements of the waves 1 and 2 are added algebraically (i.e. taking their signs into consideration). This gives the resultant wave.

As can be seen from this example in which two simple waves are added together, the result of superposing waves can give complicated waves, especially if the waves have different frequencies, amplitudes and wavelengths. The process of graphically adding two completely different waves together in this manner can be very time-consuming, and often leads to a resultant wave that is of little practical value. Useful examples of superposition deal with adding waves that are very close in properties. In the sections that follow, we shall study some of these useful cases.

Using an analytical approach, we can determine the wave function for the resultant wave from the wave functions representing the individual waves. If y_1 and y_2 are the displacements due to the individual waves, we have

$$y_1 = A_1 \sin(\omega_1 t - k_1 x),$$

and

$$y_2 = A_2 \sin(\omega_2 t - k_2 x),$$

where the suffixes 1 and 2 refer to the respective parameters of each wave. For simplicity, the two waves are taken to be traveling in the same direction. When these two waves travel together through a medium, their combined effect produces the resultant wave. By the principle of superposition, the particle vibration y due to this resultant wave is

$$y = y_1 + y_2 = A_1 \sin(\omega_1 t - k_1 x) + A_2 \sin(\omega_2 t - k_2 x). \qquad (2.1)$$

If the two waves have very different parameters, equation (2.1) cannot be further simplified. The only approach possible in this case would be a point-by-point addition of the individual displacements – a process similar to the graphical approach. Simplified resultant wave functions result when the parameters of the two waves are related. The examples of superposition in this chapter are devoted to such simple cases.

The converse of superposition can also be done. Thus, it is possible to break a complex wave up into simpler waves, which, added together, give the complex wave. The simpler waves are known as the **harmonics** of the complex wave. Also called **overtones**, these harmonics often have simple frequency relationships with each other. We shall discuss more on harmonics in the next chapter on sound wave propagation.

2.2 STATIONARY WAVES

Stationary wave production is one of the special cases arising from simplifications that can be made to equation (2.1).

A stationary wave is produced by the superposition of two identical waves travelling in opposite directions.

Identical waves are waves with the same amplitude, same frequency, and the same wavelength. Thus stationary waves result from the superposition of two waves with the same amplitude, wavelength, and frequency, travelling in opposite directions.

A standard way of producing two identical waves travelling in opposite directions is to consider a wave totally reflected from a plane boundary. The reflected wave would have the same frequency and wavelength as the incident wave. Furthermore, assuming total reflection, the amplitude of the reflected wave is the same as that of the incident wave. This reflected wave then combines with the incident wave to form the stationary wave in the medium.

Fig. 2.2 illustrates the reflection of a wave at a plane boundary. For simplicity, the incident wave is drawn separated from the reflected wave. One important feature to note in the reflection is that on reflection, the wave is 'reversed', or turned upside-down. Thus, at every point in the medium, the particle displacement in the incident wave is opposite to the displacement in the reflected wave. In terms of phase, the reflected wave is out of phase by π (or $\lambda/2$) from the incident wave.

The equation for the stationary wave is obtained by adding together the particle displacements in the incident and reflected waves wave. To make calculations and equations simple, we assume the two waves are in the x-y plane. If we let the suffix 1 represent the incident wave, and 2 the reflected wave,

$$y_1 = A\sin(\omega t - kx), \tag{2.2}$$

and

$$y_2 = -A\sin(\omega t + kx) \tag{2.3}$$

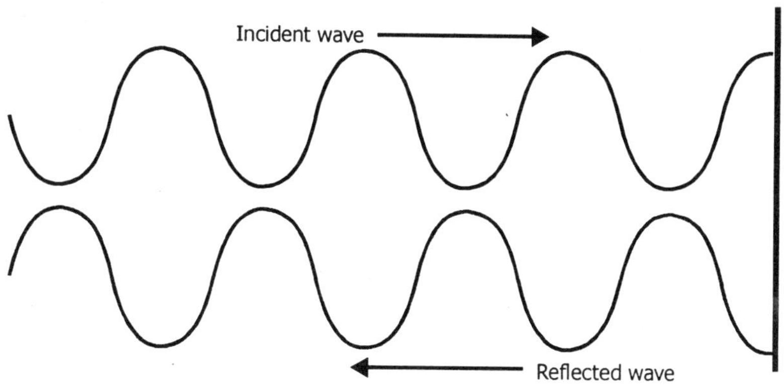

Fig. 2.2 Reflection of a wave at a plane boundary. At each point the particle displacement in the reflected wave is opposite to that in the incident wave.

In equations (2.2) and (2.3), we have used the same amplitude, angular frequency, and propagation constant for the two waves. Notice the sign differences between the two equations, representing the phase change, and directions of travel. By the superposition principle, the resultant displacement y of these two waves is given by

$$y = y_1 + y_2 = A\{\sin(\omega t - kx) - \sin(\omega t + kx)\}. \tag{2.4}$$

Applying trigonometric identities to (2.4), we get the stationary wave equation

$$y = 2A\sin kx\cos\omega t. \tag{2.5}$$

A close inspection of equation (2.5) would reveal the following:

(a) The stationary wave has the same frequency and propagation constant as the travelling waves giving rise to it.

(b) Depending on the wave function used to represent the incident and reflected waves, the equation for the stationary wave will turn out to be slightly different from (2.5).

(c) Unlike the travelling waves, the maximum particle vibration at points in the stationary wave depends on their positions. For the travelling waves, the amplitude of vibration of particles in the medium does not depend on the position of the particle. We say that the amplitude of the stationary wave is **modulated** by its position. This modulated amplitude M, is given by

$$M = 2A\sin kx, \tag{2.6}$$

and has a maximum value that is twice the amplitude of the travelling wave. Positions of this maximum particle vibration are called **antinodes**, whilst those of zero vibration are referred to as **nodes**.

Fig. 2.3 shows the two travelling waves combined to form the stationary wave. The antinodes and nodes on the stationary wave are shown at the positions marked A and N, respectively. In the sections that follow, we show how their positions relate to each other, and to the wavelength of the wave.

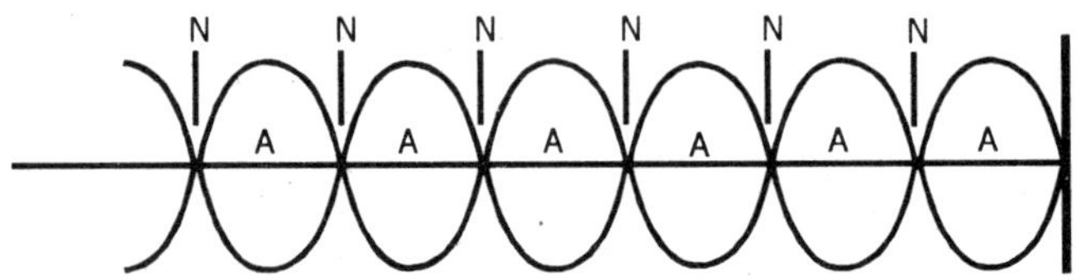

Fig. 2.3 Time profile of a stationary wave formed by the superposition of a wave and its reflection. Nodes and antinodes are shown marked N and A, respectively.

2.2.1 Positions of antinodes

As noted above, the modulated amplitude becomes a maximum at the antinodes. With the origin at the source, distances x that give antinodes are obtained from the expression

$$M = 2A\sin kx = 2A$$

giving

$$\sin kx = 1. \tag{2.7}$$

Equation (2.7) is satisfied for all values of x which are such that

$$kx_n = (2n+1)\frac{\pi}{2}, \tag{2.8}$$

where $n = 0, 1, 2$, etc. The suffix n is introduced on x to indicate the different positions. Substituting for the propagation constant in terms of the wavelength λ, we transform equation (2.8) to read

$$x_n = (2n+1)\frac{\lambda}{4}. \tag{2.9}$$

From (2.9), the position of the first antinode, for instance, is obtained by setting n equal to zero. Thus, the first antinode is a distance $\lambda/4$ from the source of vibration. Also, the distance between one antinode and the next (i.e. the separation of the antinodes) can be shown to be equal to $\lambda/2$.

2.2.2 Positions of nodes

Following the analysis for the antinodes, we have, for a node $M = 2A\sin kx = 0$, giving

$$\sin kx = 0, \tag{2.10}$$

from which,

$$kx_n = n\pi, \tag{2.11}$$

where $n = 0, 1, 2$, etc. This gives

$$x_n = n\frac{\lambda}{2}, \tag{2.12}$$

as the distances of the nodes from the source. It can also be shown that the separation of successive nodes is $\lambda/2$, the same as that for the antinodes. Thus the distance between a node and the next antinode is $\lambda/4$.

Example 2.1

A microwave transmitter is directed normally at a plane metal reflector. A small detector moving along the direction of travel of the wave is found to record successive positions of maximum and minimum intensity.

(a) Explain this observation.

(b) If the detector covers a distance of 12 mm in going from the first to the ninth successive position of maximum intensity, calculate the frequency of the microwave oscillator. Speed of microwaves = 3.0×10^8 m/s.

Solution

(a) The microwaves from the transmitter are reflected off the plane reflector. The reflected wave combines with the incident wave by the superposition principle to form a stationary wave. A small detector moving along the direction of travel of the wave would then pass through antinodes and nodes of the stationary wave formed. The antinodes represent positions of maximum intensity, whilst the nodes are those of minimum intensity.

(b) Construction of the stationary wave pattern as shown in **Fig. 2.3** would reveal that there are 4 wavelengths between the first to the ninth antinode. From the given data, we have $4\lambda = 0.012$ m, or $\lambda = 0.003$ m. We can hence calculate the frequency f of the microwaves from the speed v, and the wavelength λ. Thus,

$$f = \frac{v}{\lambda} = \frac{3.0 \times 10^8 \text{ m/s}}{0.003 \text{ m}} = 10^{11} \text{ Hz (or 100 GHz)}$$

Example 2.2

Standing waves in a medium are described by the equation $y = 2A\sin kx \cos \omega t$. If $k = 1.67\pi \text{ m}^{-1}$, calculate the distances from the source for the first 4 antinodes.

Solution

From the equation, the modulated amplitude M is given by $M = 2A\sin kx = 2A\sin(1.67\pi)x$. For antinodes, M = 2A, giving $\sin(1.67\pi)x = 1$.

Thus, $1.67\pi x = \frac{(2n+1)\pi}{2}$, or $x_n = \frac{(2n+1)}{2(1.67)}$, where n=0, 1, 2, 3, ...

For the first 4 antinodes, n=0, n=1, n=2, and n=3. Putting these values of n into the equation for x_n, we get the distances as 29.9 cm, 89.8 cm, 149.7 cm, 209.5 cm.

2.3 INTERFERENCE OF WAVES

Two wave motions travelling through a medium, will reinforce each other if they are in phase, and will reduce each other's effect if they are out of phase. An important phenomenon results when the two waves travel in the **same direction** with the same frequency, wavelength, and are either in phase or differ by a constant phase. They produce a phenomenon called **interference**. For two waves to produce an interference effect, they must be generated by **coherent sources**.

Two sources are said to be coherent if they produce waves of the same frequency and wavelength, and their vibrations are either in phase, or differ by a constant phase.

When two waves travelling in the same direction satisfy the above condition, they can interfere by reinforcing each other, by reducing each other. When the two waves reinforce each other, the interference is called **constructive interference**, and when they reduce each other, they produce **destructive interference**.

2.3.1 Constructive interference

When two waves interfere constructively at a point, the resultant particle displacement in the medium is greater than that produced by either wave separately. We can easily understand constructive interference with reference to **Fig. 2.4**. In this figure, waves A, B, C, D, and E travel in the same direction in a medium. All the waves are generated from coherent sources. Consider waves A, and B, for instance. These travel the same distances to get to the receiver, represented here by the plane vertical line. The crests of both these waves arrive at the same time. Hence the waves A and B would interfere constructively.

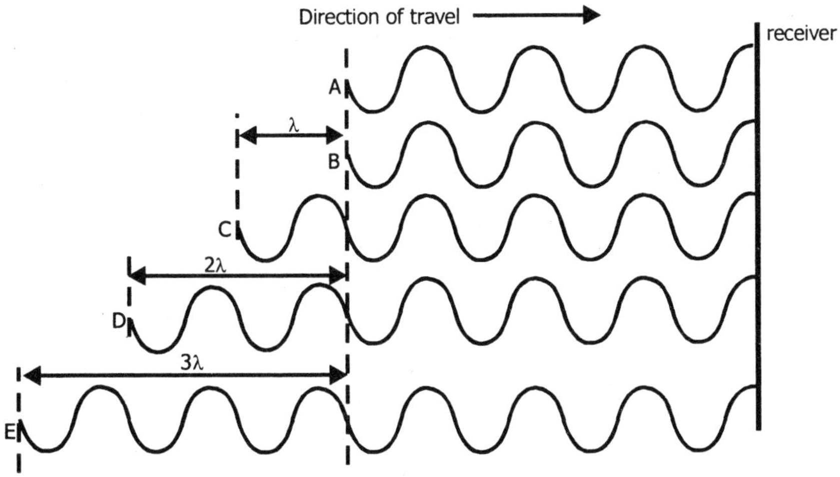

Fig. 2.4 Constructive interference of waves which differ in paths by integral multiples of the wavelength

Waves A and C travel different distances in getting to the receiver. The two waves have exactly one wavelength distance between them. Vibrations from B and C hence arrive in phase. They therefore interfere constructively. Similarly, waves A and D, being two wavelengths apart, interfere constructively, just as A and E will interfere constructively. In the latter case the two waves are three wavelengths apart.

In general, two waves interfere constructively if their difference in paths is an integral multiple of their wavelength. This difference in path is referred to as the **path difference**, normally represented by the Greek letter Δ. Thus,

constructive interference occurs when the path difference between the two waves is an integral multiple of the wavelength.

In symbols, we state the condition for constructive interference as

$$\Delta = m\lambda , \qquad (2.13)$$

where m = 0,1,2,3,... The quantity m is known as the **order** of the interference pattern. Thus we have the first, second, third, etc. order, signified by m = 0,1,2,3,..., respectively.

2.3.2 Destructive interference

In destructive interference, the resultant particle displacement at any point is the difference between the displacements due to the waves separately. In keeping with the method employed for constructive interference, we can relate the path difference for waves that interfere destructively. This is illustrated using **Fig. 2.5**.

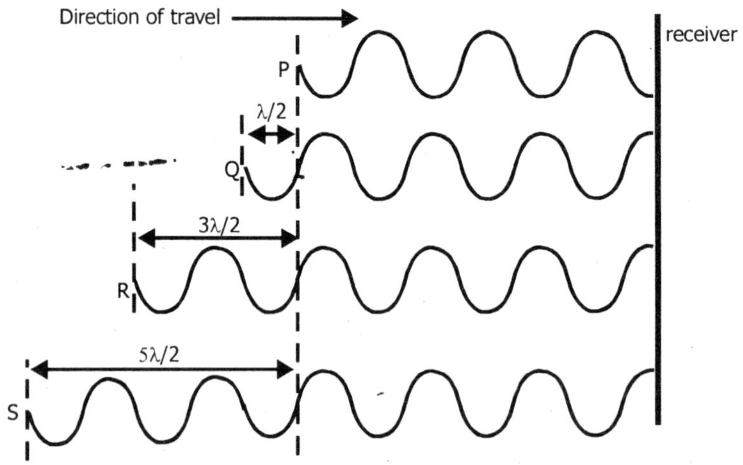

Fig. 2.5 Destructive interference of waves which differ in in paths by an odd multiple of half-wavelengths.

In this representation, two waves, P and Q from coherent sources, arrive at a receiver such that wave Q has travelled exactly one-half wavelength farther than wave P. The crest of one wave would coincide with the trough of the other, thereby producing a small resultant displacement. If the waves have the same amplitude, this resultant displacement would be zero. Similarly, waves P and R would interfere destructively, and so would P and S.

Destructive interference occurs when the path difference between the two waves is an odd multiple of half wavelengths.

In symbols, we have

$$\Delta = (2m+1)\frac{\lambda}{2} = \left(m+\frac{1}{2}\right)\lambda \qquad (2.14)$$

for destructive interference, where $m = 0,1,2,3,...$ In summary, we have the following:

(a) Interference pattern is observed for two waves if the waves are from coherent sources – i.e. waves having the same frequency and wavelength, and are either in phase, or differ by a constant phase.

(b) Constructive interference occurs if the path difference between the two waves is an integral multiple of the wavelength.

(c) Destructive interference occurs if the path difference between the two waves is an odd multiple of half-wavelengths.

At a glance, we might be tempted to confuse stationary wave production with interference. We bear in mind that stationary waves are produced when the waves travel in opposite directions, whilst interference pattern is observed when the waves travel in the same direction. Also, the amplitude of the individual waves contributing to the stationary wave must be equal, whilst interference pattern is observed even if the two waves do not have the same amplitude. Specific cases of stationary waves and interference are studied in the chapters of sound and physical optics.

Example 2.3

Fig. 2.6 shows the arrangement of two loudspeakers A and B, driven by the same amplifier, and emit sinusoidal sound waves which are in phase. Speaker B is 2.0 m to the right of speaker A. A small detector P is moved along the line joining the two speakers. It is found that second order interference maximum is obtained when P is 1.0 m to the right of B. Calculate

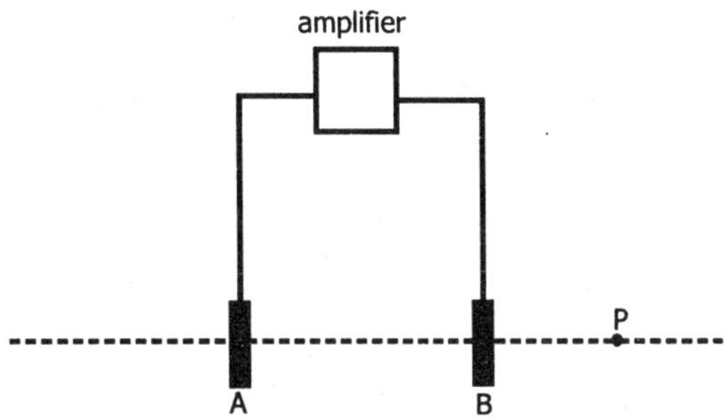

Fig. 2.6

(i) the wavelength of the waves,

(ii) the distance between successive points of maximum intensity when P is between A and B.

Solution

(i) The path difference Δ at P is given by $\Delta = 3.0 - 1.0 = 2.0$ m. For second order maximum, we have $\Delta = m\lambda$, where $m = 2$. Thus the wavelength of the wave is given by $2\lambda = 2.0$ m, or $\lambda = 1.0$ m. Hence the wavelength is 1.0 m.

(ii) Between the speakers, the waves are identical, and travel in opposite directions. Thus we have stationary wave formed between the speakers. The positions of P corresponding to maximum intensity are the antinodes. Recalling that successive antinodes are one-half wavelength apart, we deduce that the distance between points of maximum intensity between the speakers is 0.50 m, or 50 cm.

2.4 DIFFRACTION OF WAVES

When a wave meets an obstacle, or opening in its path, the waves have the ability to bend around the obstacle, or around the edges of the opening. It is observed that sound can be heard outside a room from a point that is not directly in line with the source of the sound.

The bending of waves around an obstacle or the edges of an opening is due to a phenomenon called **diffraction**, a property shown by all waves.

Diffraction effects are observed in situations where the size of the obstacle is of the same order of magnitude as the wavelength of the wave. As a wave travels through a medium, it sets the molecules in the medium in oscillation. Each point on the wavefront becomes a **secondary source** of waves, and generates waves that go out in all directions. The overall effect is that every point on the edge of the obstacle or the edge of the opening acts as a secondary source. Waves from any pair of these secondary sources would interfere either constructively or destructively to produce regions of maximum and minimum intensity. **Fig. 2.7** is a schematic representation of this phenomenon.

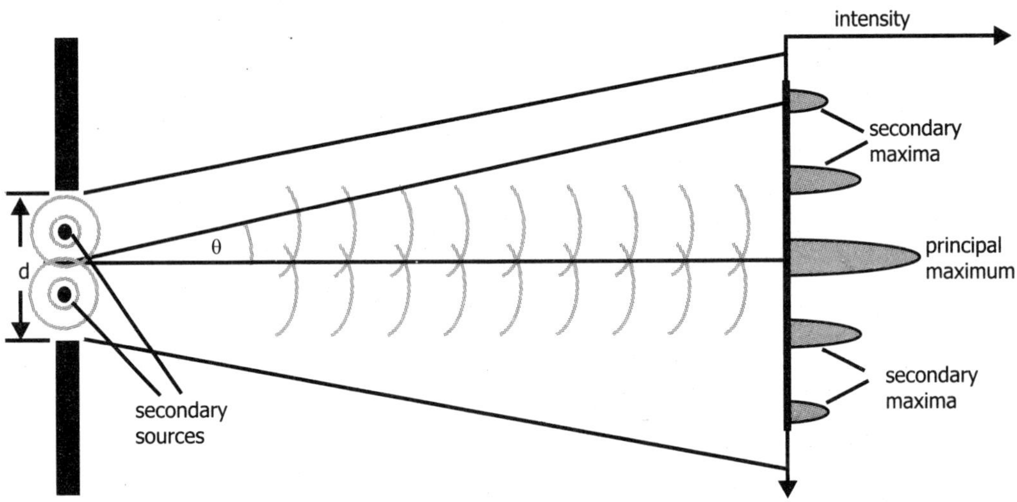

Fig. 2.7 Diffraction of waves through an opening. Each point of the opening acts as a secondary source of waves, as are indeed each point on the wave front as the wave travels through the medium. The graph shows how the intensity of the wave varies with distance along a line parallel to the plane of the slit. The intensity pattern shows a principal maximum followed by secondary maxima.

At large distances from the opening, the intensity is a maximum directly opposite the centre of the opening. On either side of this central maximum, the intensity decreases to zero, rises back to maximum, then to zero, and so on. The central maximum is the strongest one, whilst the secondary maxima are of much lower intensity. In the above figure, θ refers to the angular distance from the centre of the opening to the position of a secondary maximum of order m. For an opening of width d, we have

$$d\sin\theta = m\lambda \qquad (2.15)$$

From (2.15), we can determine the location of the first minimum of intensity on either side of the centre. For this position, m = 1. This gives

$$\sin\theta = \frac{\lambda}{d}, \qquad (2.16)$$

and if $d>>\lambda$, $\sin\theta \approx 0$. This case is interpreted by saying that the dimension of the opening is much larger than the wavelength of the waves. No diffraction pattern is then observed. This is the case in the well-known experiment you might have done in earlier years that demonstrates light travels in straight lines. In that experiment, the apertures through which the light passes is much larger than the wavelength of the light. Diffraction patterns would be observed if the apertures are the same order of magnitude as the wavelength of the light waves. Visible light waves have extremely small wavelengths, typically a few microns.

THE RIPPLE TANK

The ripple tank is a device for demonstrating the properties and propagation of transverse waves set up on the surface of water. While the investigations are done using transverse waves, the results are nonetheless true for longitudinal waves as well. The device consists of one or two bobs which are set into vibration on the surface of a shallow water in a flat-bottomed glass or perspex tank. The bobs are set into vibration by a small motor powered by a battery or some other power supply unit designed for the purpose. A block of wood carrying the bobs is suspended by means of rubber bands from a horizontal plank above. With the use of the rubber bands, the vibrations of the motor are communicated easily to the bobs. An overhead lamp throws the image of the waves on to white paper placed under the tank. In some versions of the setup, the tank is placed on an overhead projector, so the image is seen on a screen – a convenient method for large audiences. **Fig. 2.8** shows the features of the device. Using the ripple tank, we can study the properties of a single wave, and phenomena resulting from the superposition of two or more waves.

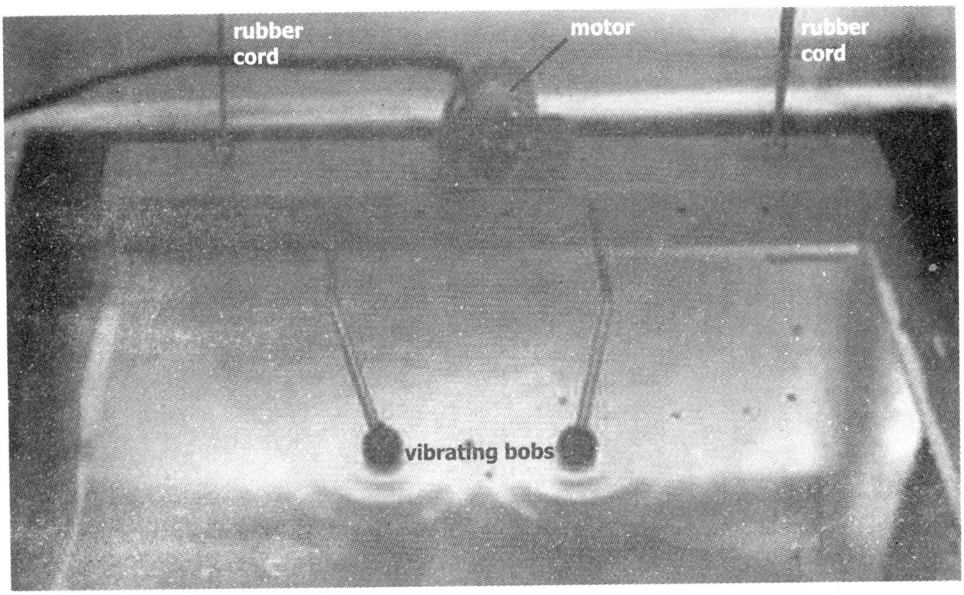

Fig. 2.8 The ripple tank

2.5.1 Measurement of wavelength

Using a single vibrating bob, circular waves emanate from the bob and travel outwards. **Fig. 2.9** shows the image of the travelling wave. Viewed from above the image will consist of bright and dark circles, corresponding to the wave crests and troughs. As the waves are constantly moving outward, this observation is often made easier through a hand-held stroboscope. The frequency of vibration of the bobs is tuned to be the same as the rotation of the stroboscope. With this adjustment, the waves then appear steady when viewed through the stroboscope.

The distance measured on the paper between the images of successive crests or troughs gives the wavelength of the wave. This method is often described as the direct method of finding the wavelength of the wave.

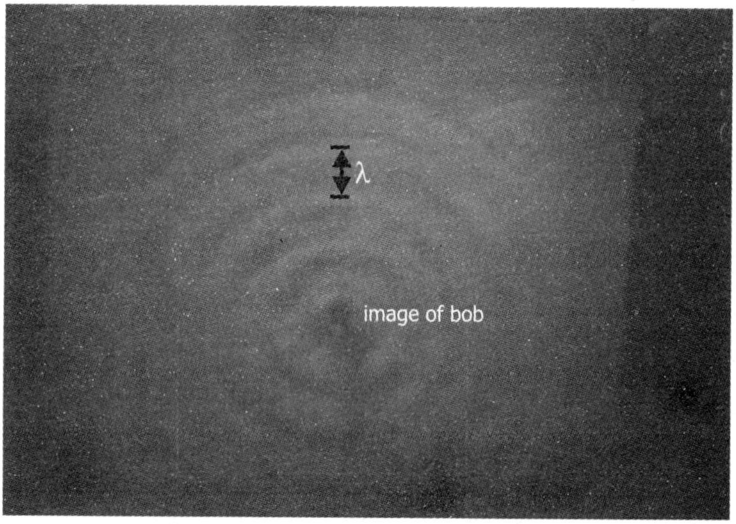

Fig. 2.9 Circular waves from a single vibrating bob

An alternative approach for finding the wavelength is through the formation of stationary waves. Stationary waves can be formed by placing a reflector, such as a metal strip, in the water at some distance away from the bob. Waves reflected from the strip combine with waves travelling outward to form a stationary wave. Steady nodes and antinodes are observed in the region where the stationary waves are formed. The distance apart of two successive nodes or antinodes of the stationary wave represents one-half wavelength. As these waves are not moving, they are more easily observed than the travelling waves. No stroboscope would be required for their observation.

2.5.2 Interference of water waves

To observe interference and diffraction phenomena in any wave, two sources are required, the two sources sending waves in the same direction. In the ripple tank demonstration, this is done by using two vibrating bobs. The experimental system is in line with the diagram of **Fig. 2.8**.

Each bob sends out circular waves in the same direction, and an interference pattern is observed in the region of overlap of the waves from the two sources. **Fig. 2.10** shows the wave profile that would be observed in this demonstration.

Along any line parallel to the line joining the bobs, the path difference can be determined. In this figure, P is a position of constructive interference between the waves generated by the two bobs, S_1 and S_2. O represents the position of the central maximum of interference. If the order of the interference at P is m, we write $S_2P - S_1P = m\lambda$, where λ is the wavelength of the waves.

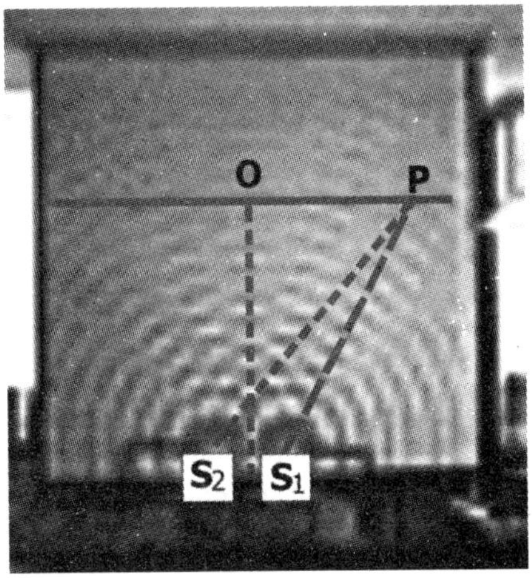

Fig. 2.10 Interference in water waves

EXERCISES

1. Two coherent sources produce waves with amplitudes of 1.0 cm and 3.0 cm. Calculate the resultant amplitude at points where the path difference from the two sources is
   (i) 1.5 wavelengths,
   (ii) 3.0 wavelengths

2. **Fig. 2.11** shows two speakers A and B driven by the same amplifier. The speakers are 3.20 m apart, along a horizontal straight line, and are both producing sound waves at 256 Hz. The speed of sound is 343 m/s. A listener C is 2.4 m directly in front of B, along the vertical line, as shown in the diagram. Determine if constructive or destructive interference occurs at C.

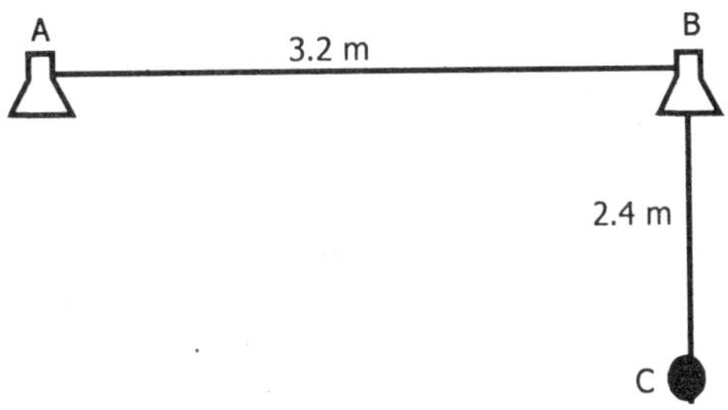

Fig. 2.11

3. An FM radio station transmits at a frequency of 100 MHz, and uses two identical antennas mounted at the same height, 15 m apart. The antennas are driven by the same amplifier, and send waves in the same direction. The resulting radiation pattern has a maximum intensity along the perpendicular bisector of the line joining the two antennas. Assume that the intensity is observed at distances far in excess of 15 m.

(a) At what angles (measured from the middle of the line joining the two antennas) is the observed intensity a maximum?

(b) For what angles is the intensity zero?

4. The equation $y = 4\sin(3x + 4t)$ mm, where x is in metres and t in seconds, represents a plane progressive wave.

(a) Find the amplitude, wavelength, period, and direction of travel of the wave.

(b) Find the equation of the stationary wave formed when this wave is superimposed on its reflection from a plane boundary.

(c) Calculate the distance between the first node and the seventh antinode of the stationary wave.

5. The sound produced by the loudspeaker in **Fig. 2.12** has a frequency of 5.7 kHz, and arrives at the small microphone alternatively via a path of fixed length, and another path

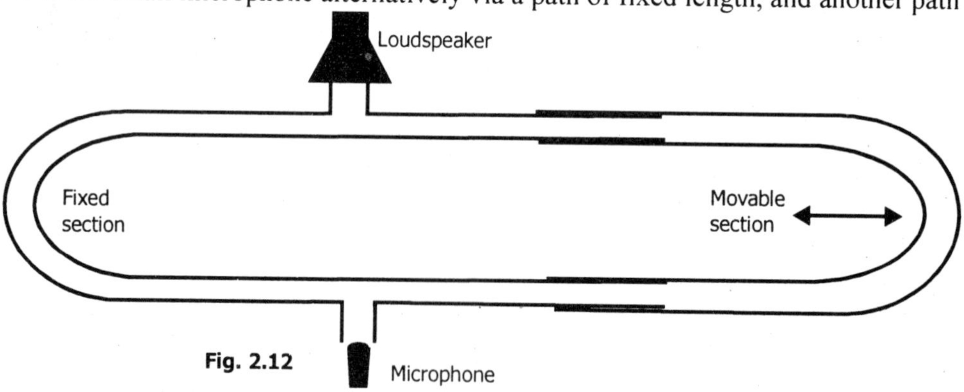

Fig. 2.12

whose length can be varied by pulling out, or in, the sliding section. By this pulling action, the intensity of the sound received changes from maximum to minimum. It is found that when the sliding section is pulled out by 2.0 cm, the intensity of the sound received by the microphone changes from maximum intensity to minimum. Determine the speed at which the sound travels through the gas in the tube.

6. The diagram of **Fig. 2.13** shows the geometrical arrangement of two bobs S_1 and S_2 vibrating in phase close to each other, sending waves on the surface of water in a shallow tank. On an observation parallel to the line of the bobs, an interference pattern is observed in the region of overlap of the waves from the two bobs. A second order constructive interference pattern is observed at the point P. Calculate

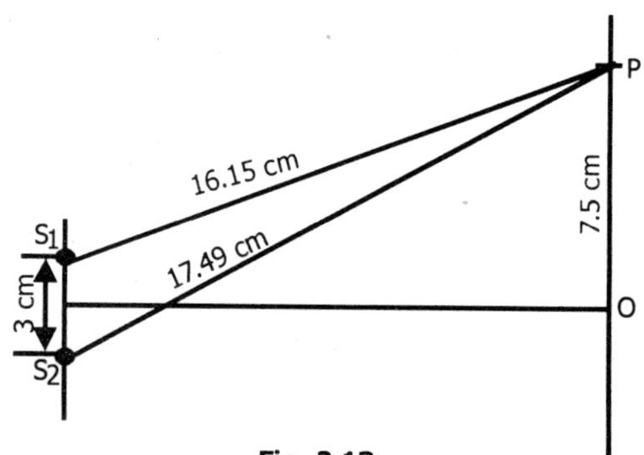

Fig. 2.13

(i) the wavelength of the waves,
(ii) the distance between the line of the bobs and the observation line,
(iii) the speed of the water waves, given that the bobs vibrate at 50 Hz.

7. Two microwave transmitters, driven by the same source, are mounted at the same height, and send out waves in opposite directions. Given that the speed of the microwaves is 3.0×10^8 m/s, calculate the frequency of the waves that would produce two successive positions of maximum intensity that are 0.05 cm apart.

8. The equations $y = 2\sin 2\pi\left(100t - \frac{x}{50}\right)$, and $y = 4\sin\frac{\pi}{25}x\cos 200\pi t$ represent two plane waves, where y is in centimeters, x in millimeters, and t in seconds.
(a) Find the frequency, wavelength, and speed of each wave in the medium.
(b) Show how the two waves are related to each other.
(c) For the second wave, determine the positions of the maximum and minimum particle displacements from the source.

9. The diagram of **Fig. 2.14** is a plot of three hypothetical waves travelling in a medium in the positive x-direction. Using 10 divisions in either direction to correspond to 2 cm, find
(i) the wavelength of the waves,
(ii) the amplitude of vibration of each wave,
(iii) the amplitude of the resultant wave obtained by superposing A and C,
(iv) the amplitude of the resultant wave obtained by superposing A and B.

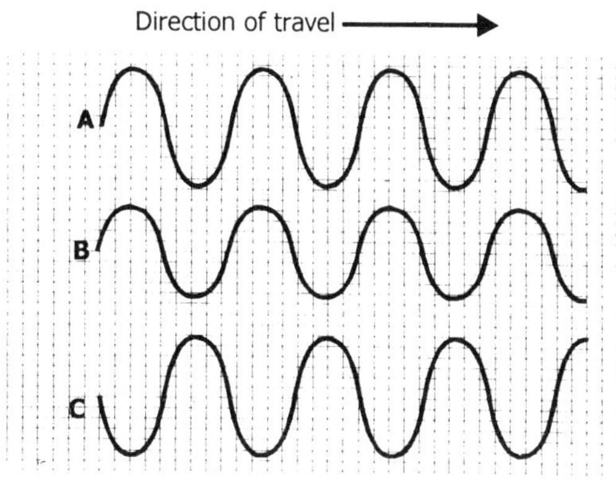

Fig. 2.14

10. Using **Fig. 2.14** above, draw a sketch showing the resultant of A and C if the two waves were travelling in opposite directions. Give the distance between successive points of zero displacement on this resultant.

11. The plane wave represented by the function $y = 0.01 \cos 2\pi(100t - 5x)$ m is totally reflected at a plane boundary.

(a) Give the maximum particle displacement of the stationary wave formed.

(b) Deduce expressions for

(i) the reflected wave,

(ii) the stationary wave formed.

(c) Determine the positions of the nodes and antinodes, taking $x = 0$ as the origin.

3

CHAPTER

Sound Waves

3
CHAPTER

Sound Waves

Through sound medium, we are able to communicate our thoughts and emotions with other people. Each of us is exposed to sound of various kinds – from blasts of whistles, the rustle of waves through the leaves of a tree, the running of a flowing river, the pleasant music of a household hi-fi system, the sound of barking dogs, and so on. Each of these sounds leaves a different impression on our sensations – some pleasant, and some not-so pleasant. Nonetheless, they are all classified as sound.

Sound has two different meanings, one psychological, and one physiological. On the psychological front, sound may refer to the hearing (or auditory) sensation. This starts with the sound exciting the eardrum, and the resulting excitation transmitted to the brain for interpretation. One effect of this is in the creation and recognition of languages, of which there is a near infinite number of them among the countries and races of the world. On the physiological or physical hand, we are concerned with the sound produced by a certain type of disturbance, the production and transmission of the disturbance, which gives the auditory sensation. It is obvious that of the two, the physical aspect is a far less complicated field of study than the psychological aspect. Furthermore, the physical aspect of sound lends itself to a more objective mode of study. In the study presented here, we concentrate primarily on the physical aspects of sound.

In the music industry, we describe the physical and psychological effects of sound using terms like **pitch**, **loudness**, and **tone quality**. In science, we describe sounds with such words as **frequency**, **intensity**, and **harmonics** (or **overtones**). While it is not always possible to get complete correspondence between these two sets of terms, we can discuss their relationships.

3.1 VELOCITY OF SOUND

It is a fact of common experience that the flash of a lightning spark is seen some time before the sound of the thunder is heard. This and quite a few other similar examples demonstrate the low speed at which sound travels compared to light. Because of this low speed, it is experimentally a lot easier to determine the speed of sound in air. One simple experiment involves the firing of a pistol a measured distance away, and measuring the time it takes between seeing the smoke of the gun and hearing the sound. Dividing the distance by the time allows one to determine the speed of sound. There are more accurate experiments that can be described for the determination of the speed of sound. We shall study these later, together with their underlying principles.

Measurements reveal that sound travels in air at a speed of about 343 m/s at a temperature of 20 °C. This speed depends on temperature. As air becomes less dense when heated, it follows that the speed of sound in air is higher at higher temperature. The speed of sound in air varies as the square-root of the Kelvin temperature. Thus, if v_o is the speed at temperature T_o, and v_1 the speed at temperature T_1, we have

$$\frac{v_o}{v_1} = \sqrt{\frac{T_o}{T_1}}. \qquad (3.1)$$

We note from chapter 1 that the speed of waves in a medium depends on the elastic property and density of the medium. Thus the speed of sound waves in media other than air will be different from the speed in air. **Table 3.1** shows values determined for the speed of sound in some common materials.

Table 3.1 Speed of sound in some common media

Medium	Speed (m/s)	Medium	Speed (m/s)
Aluminium	5,100	Alcohol	1,240
Brass	3,500	Water (at 4 °C)	1,420
Glass	5,010 – 6,010	Air (at 0 °C)	331.4
Iron	4,720 – 5,210	Air (at 20 °C)	343.0
Steel	5,103		
Brick	3,660		
Granite	6,010		
Oak	3,850		
Pine	3,320		

3.2 PROPAGATION OF SOUND

Like any other wave motion, the source of sound is always a vibrating system. Sometimes, it is not very easy to see the vibrations, but the vibrations nonetheless give the source of the sound waves. Sound is an example of longitudinal wave motion.

3.2.1 Sound needs material medium

Sound waves need a material medium for their propagation. This material medium can be a solid, liquid, or a sample of gas. That sound needs a material medium can be shown experimentally by having an electric bell placed in a bottle that is connected to a vacuum pump. The ringing bell may be clearly heard when air is present in the bottle. As the air is gradually drawn from the bottle, the sound becomes fainter and fainter. If it were possible to remove all the air, the sound would not be heard at all. As air is once again permitted to enter the bottle, the sound slowly increases until it finally regains its original intensity.

3.2.2 Reflection and refraction of sound

Sound waves can be reflected at all surfaces, and refracted as they go from one medium into another. Both these phenomena follow the laws covered earlier in chapter 1 on reflection and refraction of waves. Reflected sound is commonly referred to as echo. Methods based on echo sounding have become very useful in determining ocean depths, and the presence of objects beneath the surface of the ocean. The method is also used to determine the speed of sound in air.

Whenever sound passes from one medium to another of different density or elasticity, the velocity of the sound changes. All sound waves travelling in the open air are refracted to some extent. This is due to the fact that the air is seldom at rest, and varies throughout, in density and temperature. These variations in temperature and density cause changes of velocity, with consequent refraction of the waves. This phenomenon explains why sound does not appear to carry so far in the daytime as it does at night. In the daytime, the sun warms the surface of the earth, and the air immediately adjacent to the surface. Hence the air next to the surface is warmer that that at higher levels. Since the speed of sound increases with temperature, the part of the wavefront next to the earth's surface will accordingly be moving faster than that higher up. The result is the bending of the waves upward from the observer on the ground. The reverse is true at night; when the temperature of the ground is lower than that the air above. Often one can hear sound of voices or music more distinctly at night even though the source may be a considerable distance away.

As sound waves require a material medium to travel, it follows that the speed at which the medium or parts of it move can also have a bearing on the speed of sound. This explains why wind plays a role in the refraction of sound waves.

3.3 MUSICAL SOUND

Sound falls into two categories: noise, and musical sound. Most of the sound around us in speech and disturbances are classified as noise. Musical sound comes from instruments and/or singing. We distinguish the two sounds in terms of the effect each has on our human sensations. Musical sound is generally pleasing to the ear, whilst noise is not. We describe musical sounds in terms of **pitch**, **loudness**, and **quality**.

3.3.1 Pitch

The pitch of a musical sound depends on the frequency of vibration of the source. We think of pitch in terms of the "highness" or "lowness" of the tone as interpreted by our hearing sensations. Sound has a definite pitch when the source is vibrating at a fixed frequency. Increasing the frequency of vibration of the source increases the pitch.

The range of frequencies audible to the human ear varies with the individual, but generally falls between 20 Hz and 20 kHz, the range normally decreasing with the age of the person. Frequencies below 20 Hz are not normally heard, as they do not cause any meaningful vibration of the eardrum. Frequencies close to 20 kHz are painful to the ear, whilst frequencies in excess of 20 KHz are not heard at all. This latter category is referred to as **ultrasound**. Even though we do not detect them by the ear, ultrasound has turned out to have very useful technological applications. We shall study some of these applications later on in the chapter.

3.3.2 Loudness or Intensity

It is a fact of common experience that the harder we hit a piece of metal with a hammer, the louder the sound that will be produced. One important factor about human sensations is that they are limited when it comes to describing things quantitatively. Thus, statements like "twice as red", in light, and sound "twice as loud", and having "twice the pitch" are rather meaningless. It all depends on the interpretations put to the sensations by the brain. There is, however, a relationship between the sensation of loudness interpreted by the brain and certain physical properties of the sound reaching the ear. Like any wave, the intensity of sound depends on the amplitude of vibration of the source, and the distance from the source of the sound.

Early attempt to quantify the loudness or intensity of sound was done by two scientists, Weber, and Fechner. Accordingly known as the **Weber-Fechner law**, this approximate law gives the loudness L by the expression

$$L = \log_{10} \frac{I}{I_0}, \tag{3.2}$$

where I is the intensity of the sound, and I_o is the intensity at the threshold of hearing, determined to be $10^{-10}\,\mu\text{watt/cm}^2$. The unit of loudness is the **bel**, in honour of Alexander Graham Bell (1847-1922), inventor of the telephone. The definition of the bel appears to be a little technical. For practical purposes, we are more interested in sound intensity **differences**, for which the **decibel** (db) is used. One way to understand the meaning of decibel is to consider that each time the power delivered to the ear is multiplied by 10, the loudness sensation that results from it is increased by 1 bel, or 10 **decibels** (db). Thus,

$$\ell(\text{db}) = 10 \log_{10} \frac{I_1}{I_2}, \tag{3.3}$$

where I_1/I_2 represents the ratio of the intensities producing an intensity increase of ℓ (db).

Example 3.1

The speaker of a public address system produces sound with loudness rated at 90 db at a distance of 1.5 m from the speaker. Determine how far the sound travels in order to give a loudness of 40 db to a listener's ear.

Solution

Let d_1 and d_2 represent the distances of the speaker from the two positions of interest. The difference in loudness of the sound at the two locations is equal to 90 − 40 = 50 db (or 5 bels). Thus the intensity of the sound 1.5 m from the source is 10^5 times the intensity at the desired distance. Thus, $I_1 = 10^5 I_2$, and $d_1 = 1.5$ m. Using the inverse-square law, we write $\frac{I_1}{I_2} = \frac{d_2^2}{d_1^2}$, or $\frac{d_2^2}{1.5^2} = \frac{10^5 I_2}{I_2}$, giving the final distance as 474 m.

3.3.3 Quality

It is found that two different musical instruments, e.g. violin and piano, do not sound the same even when they are playing the same note. We describe this difference by saying that the **quality** of the note from the violin is different from the same note from the piano.

Most musical instruments do not produce a pure note – i.e. note of a single frequency. What is often heard from the instrument is the resultant of many frequencies put together, using the principle of superposition we studied in the last chapter. The quality of a musical note depends on the number of different frequencies, or harmonics put together to produce the resultant note.

3.4 STATIONARY WAVES IN STRINGS AND PIPES

In the last chapter, we defined a stationary wave as the wave resulting from the superposition of two identical waves travelling in opposite directions. We also saw that the simplest and most practical way of producing two identical waves in opposite directions is to have a wave reflected normally at a plane boundary. The incident and reflected waves then form the two identical waves that would combine with each other to form the stationary wave. Based on this, we derived the mathematical expression for the stationary wave, and obtained expressions for the positions of the displacement nodes and antinodes.

About the best practical use of stationary waves is in vibrating stretched strings and air columns in pipes. The importance of these lies in the operation of musical instruments. Notes from the guitar, for example, are obtained by plucking a string stretched between two pegs. Waves travelling along the string are reflected at the pegs, the reflected waves combining with the incident waves to produce standing waves. Wind instruments operate in similar fashion using stationary waves in vibrating air columns in pipes.

3.5 STATIONARY WAVES IN A STRETCHED STRING

Fig. 3.1 shows modes of vibration of a string stretched between two pegs, one at either end. Since the string is fixed at the ends, nodes appear at these points.

As we studied in chapter 1, the speed v of transverse waves along a stretched string is given in terms of the tension F in the string, and the linear density μ, by

$$v = \sqrt{\frac{F}{\mu}} = \sqrt{\frac{F}{m/L}},$$

where m is the mass of the string between the pegs, a distance L apart. In this analysis, we assume constant tension and linear density, hence the same speed.

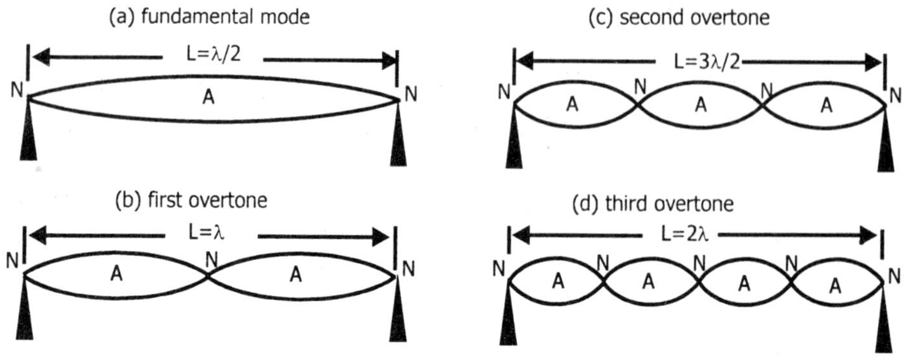

Fig. 3.1 Modes of vibration of a stretched string.

The simplest mode of vibration of the string would consist of two nodes N at the ends, and a displacement antinode A in the middle. This mode of vibration is called the **fundamental** mode, or **first harmonic**, shown in (a). For this mode of vibration, the wavelength is related to the length of the string as shown. The frequency of the fundamental mode f_o is given by

$$f_o = \frac{v}{\lambda} = \frac{1}{2L}\sqrt{\frac{F}{\mu}}. \tag{3.4}$$

The next higher mode of vibration is the **first overtone**, or **second harmonic**. This mode would show three nodes and two antinodes, as shown in (b).
The frequency f_1 for the first overtone is

$$f_1 = \frac{v}{\lambda} = \frac{1}{L}\sqrt{\frac{F}{\mu}} = 2f_o. \tag{3.5}$$

Similarly, the frequency f_2 of the **second overtone**, or **third harmonic** shown in (c), is given by

$$f_2 = \frac{v}{\lambda} = \frac{3}{2L}\sqrt{\frac{F}{\mu}} = 3f_o, \tag{3.6}$$

and the frequency f_3 of the **third overtone** (or fourth harmonic) is

$$f_3 = \frac{v}{\lambda} = \frac{2}{L}\sqrt{\frac{F}{\mu}} = 4f_o. \tag{3.7}$$

From (3.5) through (3.7), it follows that we can relate the frequency f_i of the ith overtone to the fundamental frequency f_o, by

$$f_i = (i+1)f_o. \tag{3.8}$$

Example 3.2

Standing waves along a stretched wire of length 2.4 m are described by the equation $y = 2A\sin kx\cos\omega t$, with $\omega = 314\ s^{-1}$, and $k = 1.67\ m^{-1}$, the left end of the wire being the origin. At what distances from the left end of the wire are

(i) the nodes of the standing wave,

(ii) the antinodes of the standing wave?

Solution

Substituting the numerical values of k and ω, the given equation for the stationary wave is written as $y = 2A\sin 1.67\pi x\cos 314t$.

(i) For nodes, $\sin kx = \sin 1.67\pi x = 0$, or $kx = n\pi$, where n = 0, 1, 2, 3, ...

But $k = \frac{2\pi}{\lambda}$, where λ is the wavelength of the wave. $\lambda = \frac{2\pi}{k} = \frac{2\pi}{1.67\pi} = 1.190$ m.

The nodes would hence occur at distances of 0, 0.595 m, 1.190 m, 1.785 m, and 2.380 m from the left end.

(ii) For the antinodes, $\sin 1.67\pi x = 1$, or $kx = (2n+1)\frac{\pi}{2}$, where n = 0, 1, 2, 3, ...

Using the relationship between k and λ as above, and substituting values, we find that the antinodes occur at distances of 0.292 m, 0.887 m, 1.482 m, and 2.077 m from the left end. As the wire is only 2.4 m long, we need not proceed further.

Example 3.3

A stretched string has a linear density of 7.8×10^{-4} kg/m. A wave travelling along the string has a frequency of 440 Hz. If the length of the string is 0.65 m, calculate the tensions in the string when it sounds notes of the first and second harmonics.

Solution

Here, the frequency of the note, the linear density, and length of the string all remain constant. The mode of vibration is controlled by the tension in the string. For the first harmonic (or fundamental mode), the tension F_o is given by $f = \frac{1}{2L}\sqrt{\frac{F_o}{\mu}}$. Thus $F_o = 4L^2 f\mu$.

Substitution gives the tension as 255 N.

For the second harmonic (or first overtone), the tension F_1 is given by $f = \frac{1}{L}\sqrt{\frac{F_1}{\mu}}$, giving $F_1 = L^2 f\mu$. Thus $F_1 = 63.75$ N.

3.6 THE SONOMETER

The sonometer is a laboratory setup for studying stationary waves in a stretched string.

The string passes over a pulley, and carries a mass at the end. The weight of this mass provides the tension in the string. **Fig. 3.2** is a section through the sonometer.

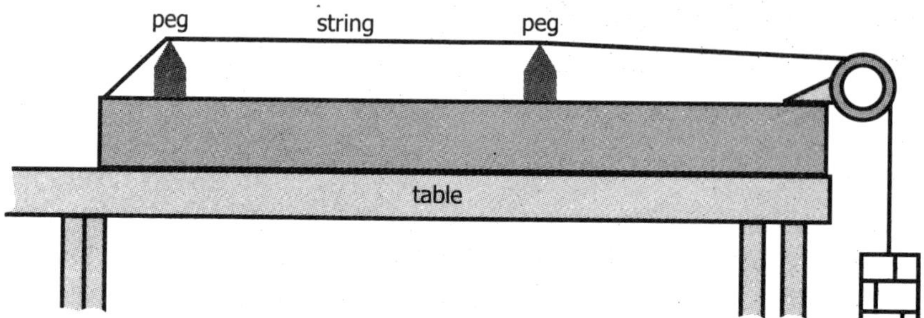

Fig. 3.2 The sonometer

The stretched string is set into vibration by means of **tuning forks**. The latter are metal forks made to give out a precise frequency when set into vibration. For a given tuning fork, the distance between the pegs is adjusted such that when plucked in the middle, the string gives out a note of the same frequency as that of the tuning fork. The distance between the pegs is then equal to one-half the wavelength. With the given frequency of the tuning fork, the speed of the wave can hence then be calculated. In addition to finding the speed of the waves, the sonometer is used to study the dependence of the frequency of the waves to the tension, and the linear density of the string.

3.7 STATIONARY WAVES IN PIPES

Many musical instruments operate on the principle of a stationary wave set up in a vibrating column of air. There are two categories of these systems. In one, the pipe is closed at one end and open at the other end. This is the closed pipe, and in it the vibrating air entering the open end is reflected at the closed end. The reflected and incident waves combine to form a stationary wave. The other pipe is the open pipe. This is open at both ends. Like the closed pipe, vibrating air forms a stationary wave.

3.7.1 The closed pipe

In a closed pipe the air particles in contact with the closed end are practically stationary. The closed end therefore will always be a node, no matter the mode of vibration. By contrast, the particles at the open end are very free to move, making this region an antinode. **Fig. 3.3** shows schematically the modes of vibration.

The fundamental mode has a node at the closed end, and one antinode at the open end – **Fig. 3.3** (a). The wavelength of the wave is hence equal to four times the length of the tube. The frequency f_0 of the fundamental mode is

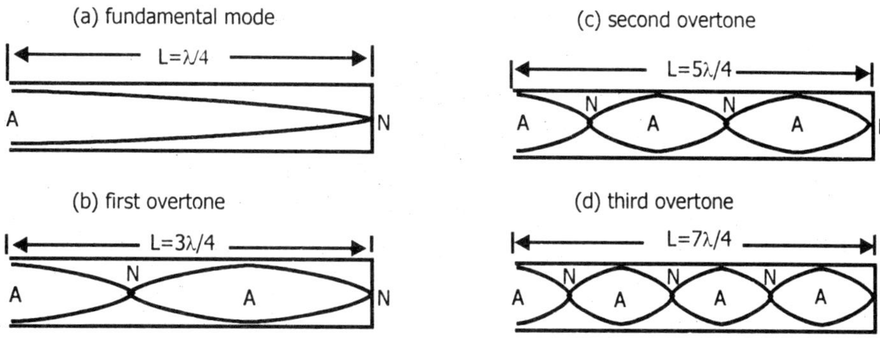

Fig. 3.3 Modes of vibration in a closed pipe

$$f_0 = \frac{v}{\lambda} = \frac{v}{4L}, \tag{3.9}$$

where v is the speed of the wave, and L is the length of the pipe. The schematic of the first overtone is shown in (b). The frequency f_1 is

$$f_1 = \frac{v}{\lambda} = \frac{3v}{4L} = 3f_0, \tag{3.10}$$

and for the second overtone (c), the frequency is f_2, where

$$f_2 = \frac{v}{\lambda} = \frac{5v}{4L} = 5f_0. \tag{3.11}$$

It is left as an exercise to show that in general, the frequency f_i of the ith overtone is

$$f_i = (2i+1)f_0 \tag{3.12}$$

3.7.2 The open pipe

Fig. 3.4 is a schematic diagram of the stationary modes in the open pipe. Since the pipe

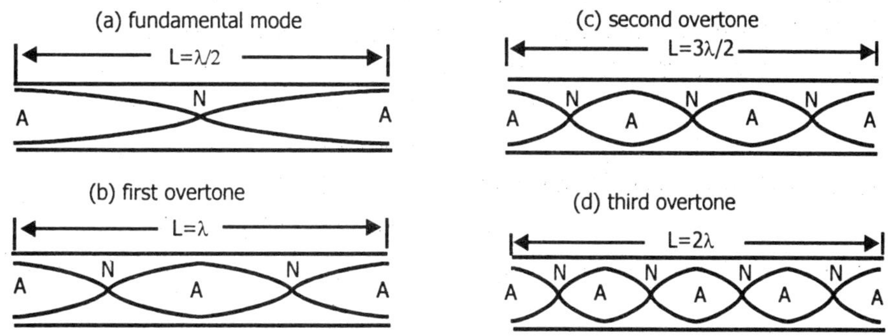

Fig. 3.4 Modes of vibration in an open pipe

is open at both ends, antinodes appear at these ends, no matter the mode of vibration. In the fundamental mode, we have $L=\lambda/2$, giving the frequency f_o as

$$f_o = \frac{v}{\lambda} = \frac{v}{2L}, \tag{3.13}$$

where v and L are the speed of the sound wave, and the length of the pipe, respectively. Using the figures, and in a manner similar to the closed pipe, we can derive expressions for the frequencies of the higher modes of vibration. It is left as an exercise to show that the frequency for overtone i is related to the fundamental frequency by

$$f_i = (i+1)f_o. \tag{3.14}$$

Example 3.4

The second overtones of an open pipe A and a closed pipe B have the same frequency at a given temperature. Calculate the ratios of

(i) the length of A to the length of B,

(ii) the fundamental frequency of A to the fundamental frequency of B.

Solution

Fig. 3.5 depicts the second overtone in both pipes, with lengths L_A and L_B.

(i) If f is the frequency, then $f = \frac{3v}{2L_A}$ for the open pipe, and $f = \frac{5v}{4L_B}$ for the closed pipe. Thus, $\frac{L_A}{L_B} = \frac{6}{5}$, giving $L_A : L_B = 6 : 5$.

(ii) If f_{oA} is the fundamental frequency of pipe A, and f_{oB} the fundamental frequency of pipe B, we have $f_{oA} = \frac{v}{2L_A}$, and $f_{oB} = \frac{v}{4L_B}$. This gives $\frac{f_{oA}}{f_{oB}} = \frac{4L_B}{2L_A}$. But $\frac{L_B}{L_A} = \frac{5}{6}$.

Thus $\frac{f_{oA}}{f_{oB}} = \frac{5}{3}$. The ratio of the fundamental frequencies is 5:3.

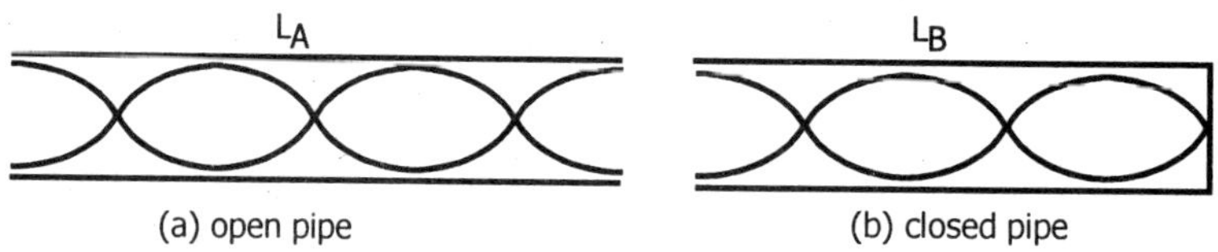

Fig. 3.5 Second overtones in (a) open pipe, and (b) closed pipe

3.7.3 End correction

In practice the wave profile extends beyond the open end of the pipe by an amount known as the **end correction**. For a fixed frequency of vibration, **Fig. 3.6** shows the profiles for the first two modes of vibration in (a) the open pipe, and (b) the closed pipe.

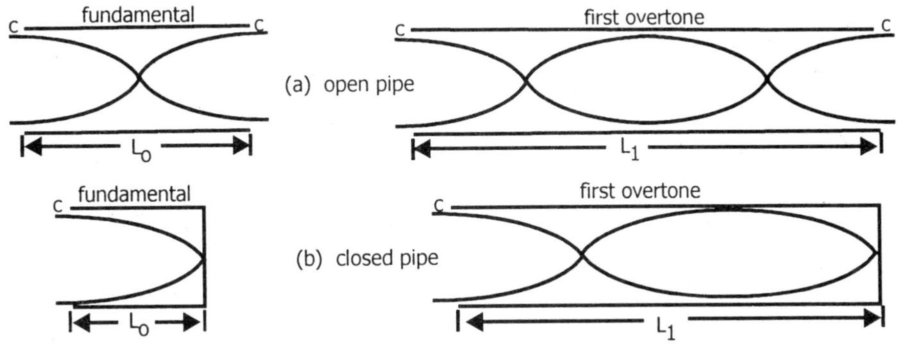

Fig. 3.6 End correction for open pipe (a), and closed pipe (b) for first two harmonics

The closed pipe has one end correction, whilst two end corrections must be added for the open pipe. For the closed pipe in the fundamental mode, we have $(L_o + c) = \lambda/4$; in the first harmonic, we have $(L_1 + c) = 3\lambda/4$, where L_o and L_1 are the lengths of the tube for the fundamental and first overtone, respectively, and c is the end correction. Similar analysis for the open pipe gives $(L_o + 2c) = \lambda/2$ in the first harmonic, and $(L_1 + 2c) = \lambda$, for the first overtone. It is found that the larger the diameter of a pipe the larger the end correction.

Example 3.5

Air in a closed pipe of variable length is excited by an unknown frequency. The air resonates with lengths of 11.5 cm, and 36.5 cm. If the speed of sound in air at the temperature is 340 m/s, calculate

(i) the wavelength of the wave,
(ii) the unknown frequency,
(iii) the end correction of the pipe.
(iv) the length of pipe required for the next overtone

Solution

Fig. 3.7 depicts three modes of vibration in the closed pipe. The three lengths of pipe are those for the fundamental mode, the first overtone, and the second overtone.

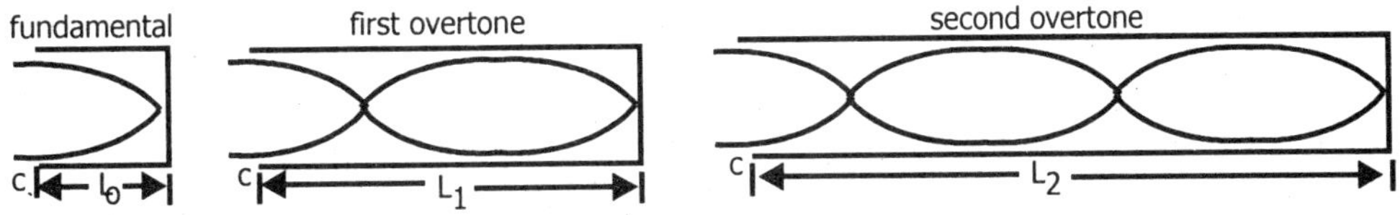

Fig. 3.7

Thus L_o = 0.115 m, and L_1 = 0.365 m,

(i) From the lengths of the fundamental mode and the first overtone, we have

$L_o + c = \lambda/4$, and $L_1 + c = 3\lambda/4$. Thus, $(L_1 - L_o) = \lambda/2$, or $\lambda = 2(L_1 - L_o)$.

Substituting the numerical values, we get the $\lambda = 0.50$ m (or 50 cm).

(ii) The unknown frequency f is given by $f = \frac{v}{\lambda} = \frac{340 \text{ m/s}}{0.50 \text{ m}} = 700$ Hz.

(iii) Using the fundamental mode, we have $L_o + c = \lambda/4$.
From the values obtained in (i) above, we get the end correction $c = 1.0$ cm.

(iv) The next higher mode of the pipe is the second overtone, shown in the figure. Comparing the wave profile for this mode and the fundamental mode, we can get the length L_2 of the pipe from the expression $(L_2 + c) - (L_o + c) = \lambda$. Substituting the values of λ and L_o from above, we get $L_2 = 0.615$ m (or 61.5 cm).

3.8 THE RESONANCE TUBE

The resonance tube is a device for finding the speed of sound in air, by using stationary waves set up in a column of air in a tube. **Fig. 3.8** is a diagram of the tube. A vibrating tuning fork is held over the open mouth of a tube. The length of the air column can be adjusted until sound of maximum intensity is heard from the tube. In this condition, the air column in the tube vibrates in resonance with the tuning fork. The lengths corresponding the different modes of vibration are measured. For simplicity, we here show the first two modes of vibration.

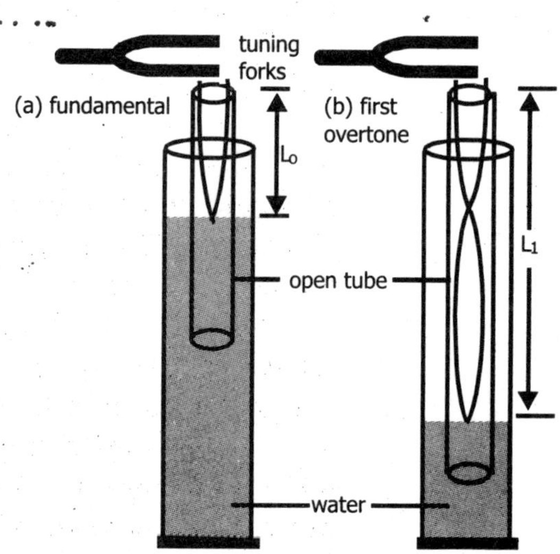

Fig. 3.8 Vibrating air columns in a resonance tube. (a) fundamental mode, (b) first overtone.

Using our earlier equations for the closed pipe, we have $L_o + c = \frac{\lambda}{4}$, for the fundamental mode, and $(L_1 + c) = \frac{3\lambda}{4}$ for the first overtone, where c is the end correction of the pipe. From the measurements, the wavelength of the sound can be determined, and hence the speed of sound in air. By performing the experiment at different air temperatures we deduce the variation of the speed of sound in air with temperature.

3.9 MUSICAL INSTRUMENTS

The principles thus far studied in connection with stationary waves are applied in the operation of musical instruments. Many of these instruments use vibrating columns of air, strings, or rods. We study here three of the most common instruments.

3.9.1 The Guitar and violin

In its older form, the guitar is a stringed instrument consisting of a hollow, waisted body with a flat, or slightly curved back, and a long neck crossed by a series of metal pieces called frets. The strings – usually six in number – are made of nylon, catgut or steel. They are all of the same length, but different thicknesses. A mechanism is provided for varying the tension in each string independently. As the frequency of a note in a stretched string depends on both its length and tension, it is possible to select a wide range of frequencies. In operation, all the strings vibrate in the fundamental mode. By pressing on strings behind the frets, the player is able to vary the frequencies in tune with the music being played.

The violin has a structure similar to the guitar. They are both stringed instruments, and operate on basically the same principles. Their major difference lies in the fact that whilst guitar strings are plucked, those on the violin are set into vibration by pressing with a bow, thus setting the string into continuous vibration. The schematic arrangements of the various parts of these two instruments are shown in **Fig. 3.9**.

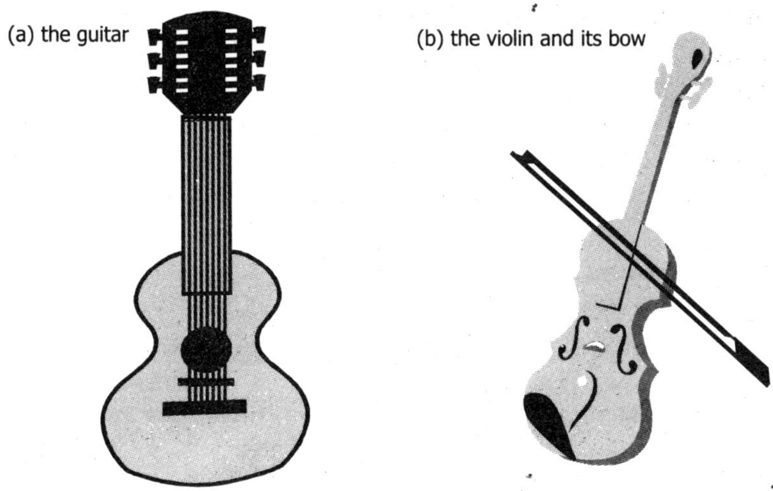

Fig. 3.9 Stringed instruments; (a) the guitar, (b) the violin and its bow. In both instruments, the string vibrates in its fundamental mode. Different masses and length of vibrating strings are used to vary the pitch of the note produced.

3.9.2 The flute

The flute is among a wide variety of wind instruments. The principles of operation of this class of instruments rely on vibrations of air columns in pipes. The flute has the form of cylindrical tube of metal, wood, or some other suitable material, with holes. The sound is produced by a stream of air directed against a sharp edge, setting into vibration the air enclosed in the instrument. The frequency of the note is varied by opening and closing the holes along the side of the flute. In the most professional brand, these holes are covered with spring-loaded keys – see **Fig. 3.10**.

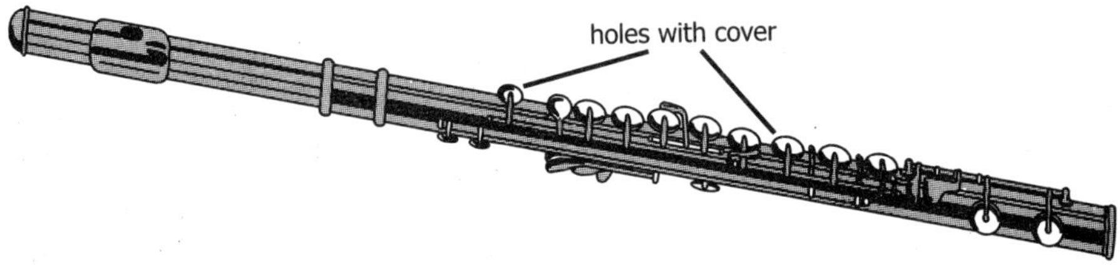

Fig. 3.10 The flute

3.10 ABSORPTION OF SOUND

When sound wave travels through a material medium, some of the sound energy is absorbed in the material and converted into heat. Most of the time, this heat energy cannot be measured in terms of a change in temperature; the change in temperature produced is very small to be measured accurately. The heat energy transferred to the material during the sound absorption is converted into increased motion of the molecules. Porous materials are effective sound absorbers because the many pockets of air they contain can readily be set into increased motion. A quantity known as the **absorption coefficient** is used to compare different materials in terms of their sound absorption.

The absorption coefficient of a material is the fraction of sound energy the material absorbs per unit length of travel within the material.

Some materials have low absorption coefficients, as sound waves are transmitted through them, with little loss of energy. Other materials such as sponge, foam, pressed plant fibres, are good sound absorbers. These materials have high absorption coefficients. They are used in applications where a sound barrier is to be created. An everyday use is in the manufacture of motor cars, where these materials are used to help reduce the sound of the engine and other external noise.

3.11 BEATS

We studied in the last chapter that waves can be superimposed on each other to yield interesting physical phenomena. Stationary waves and interference owe their origin to this superposition principle. Indeed the superposition principle is the single most important phenomenon used to explain the effects of combining two or more waves.

An interesting effect occurs when two waves differing only slightly in frequency travel together in a medium. The combination of such waves produces a phenomenon known as **beats**. This is the rise and fall of the resultant intensity at regular intervals. Although in principle, production of beats holds generally true for all wave motions, the phenomenon is most readily observed for sound waves.

Fig. 3.10 shows graphically the effect of superposing two sound waves that differ in frequency by a small amount – e.g. a wave of frequency 200 Hz, and one of frequency 204 Hz. The crests of the waves are coincident at regular intervals. The two waves tend to annul each other, also at regular intervals. The result is a periodic rise and fall in intensity, or beats.

The beat frequency is the difference between the two frequencies.

Thus, if f_1 and f_2 represent the frequencies of the two notes, with $f_2 > f_1$, then the beat frequency f is given by

$$f = f_2 - f_1. \qquad (3.13)$$

Beats are used by musicians to tune their instruments. In this process, a note of known frequency is sounded together with that from the instrument – e.g. the guitar – to produce beats. The number of beats heard per second is counted. Hence the frequency of the note from the instrument can be determined. This way, the frequency of the note from the instrument can be set to a predetermined value.

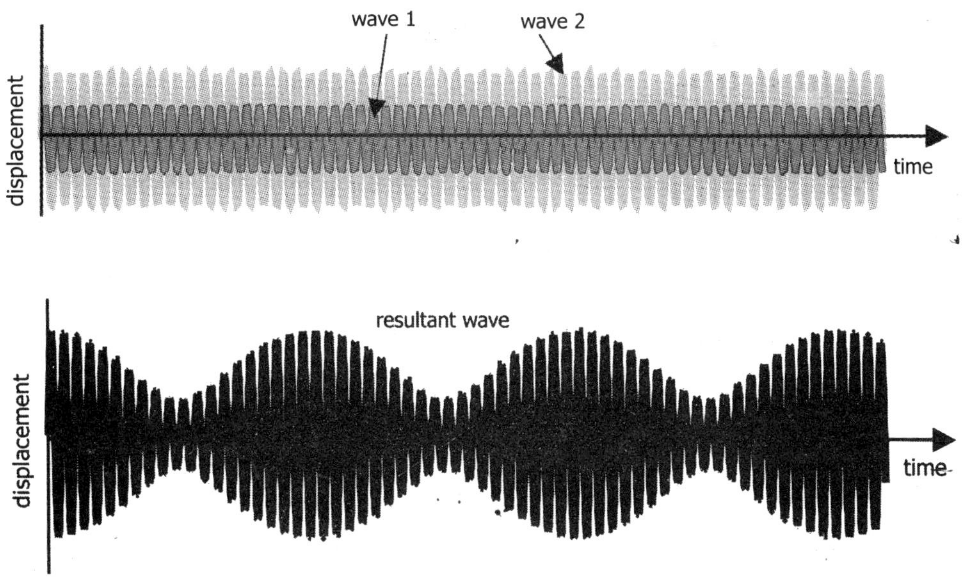

Fig. 3.11 Beats produced when sound waves of nearly equal frequencies are sounded together. The resultant wave goes through maximum and minimum intensity at regular intervals. In this example, waves of frequencies 200 Hz and 204 Hz are used.

3.12 THE EAR AND HEARING

Fig. 3.12 shows the important features of the human ear. On the outside is a curved structure of cartilage covered by the skin. This structure provides only an opening to the inner ear. The outer ear takes little or no part in the hearing; hearing takes place in the

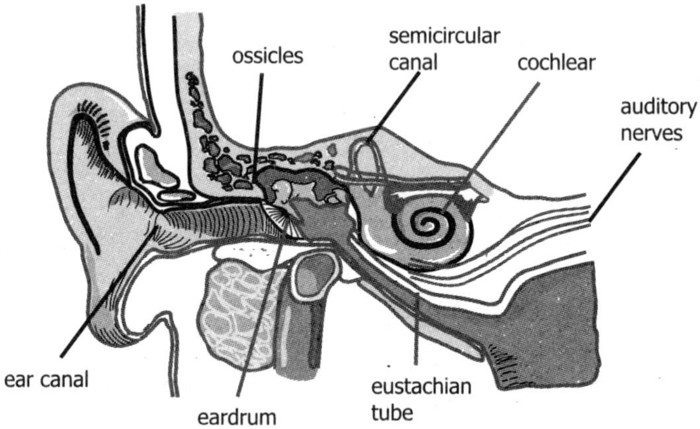

Fig. 3.12 Simplified diagram of the human ear

inner ear. The hearing process begins when the pressure wave in the ear enters the ear canal leading to the eardrum. The eardrum separates the ear canal from the middle ear, the latter containing the ear ossicles – the malus (or hammer), the incus (or anvil) and the stapes (or stirrup). The inner ear has a series of cavities, protected by the temporal bone. The two main parts of the inner ear are the semicircular canals that control balance of the body, and the cochlea. It is in the cochlear that the main process of hearing takes place.

The hearing mechanism of the ear essentially treats the organ as a pressure-sensing device, sensing systematic fluctuations in pressure. The air canal resonates slightly to sounds with frequencies in the range 3 kHz to 4 kHz. This resonance effect explains why the ear is most sensitive to frequencies in that range. The fluctuations in pressure at the eardrum are amplified in the middle ear. The amplification uses the fact that the oval window is some 20 to 30 times smaller than the eardrum. **Fig. 3.13** is a schematic view of the middle ear.

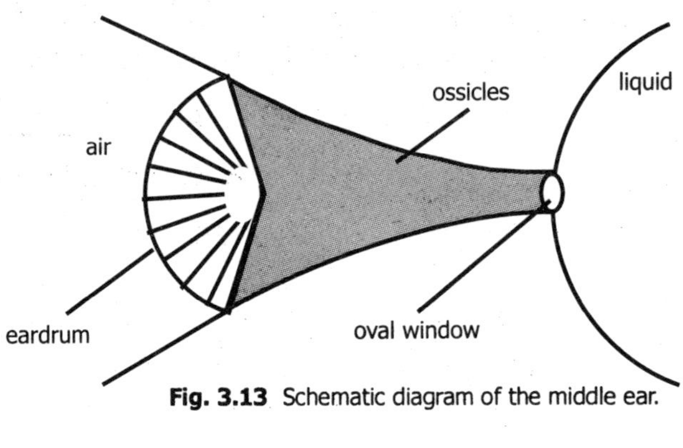

Fig. 3.13 Schematic diagram of the middle ear.

Inside the cochlear the original pressure wave in air becomes a pressure wave in a fluid. Changes in this fluid pressure are communicated to the basilar membrane in the cochlear. This membrane is such that it responds to changes in frequency of the sound wave entering the cochlear. The pressure waves entering the ear are finally converted into electrical pulses and sent along the auditory nerves to the brain.

3.13 ULTRASOUND

The properties and propagation of sound waves so far discussed are use increasingly in technical applications. In many of these applications, vibrations of frequency higher than those of normal sound waves – in excess of 20 kHz – are used. The waves may be longitudinal, or in solids, they may be transverse or shear waves.

There are many ways of producing and detecting ultrasonic vibrations. The most common are magnetorestriction, and piezoelectricity. In these, high frequency alternating magnetic fields or electric currents are converted into mechanical vibrations. Ultrasound applications are classified into high-power, and low-power. In high-power applications, the waves produce a distortion in the medium, whilst no distortion is noticed in low-power applications. We here discuss a few technical applications of ultrasound.

3.13.1 Low-power applications

Sound waves are now routinely used to find the depths of oceans and lakes. An instrument called the **fathometer** is used. In this measurement, high-frequency sound waves – known as supersonics – are sent from a source located at the keel of a ship. The sound travels to the ocean bed, gets reflected from the bed, and returns to the ship. An underwater microphone, or **hydrophone**, mounted on the ship detects the sound. The returning wave is amplified, and the time interval between the sending and returning wave is determined. Knowing the speed of sound in the water and this time interval, it is possible to calculate the depth of the seabed. With the advancement of technology, this depth sounding has been so improved that it is now possible to profile the seabed by continuous measurements made on a passing ship.

One of the powerful methods of nondestructive testing of materials and structures uses low-power ultrasound. Ultrasound pulse is sent through the material. Waves reflected at the boundaries of different structures within the specimen are detected. These reflected waves are used to detect the presence and size of cavities and cracks.

In medicine, low-power ultrasound is used in place of x-rays to produce a pictures of internal body structures. This method is especially useful for tissue examinations, and for cases in which the possible effect of x-radiation are to be avoided.

3.13.2 High-power applications

High-power ultrasound waves are used in welding and drilling, in which the direction of the vibration and thus of cutting is perpendicular to the surface of the material to be cut, so that holes may be drilled in any shape.

Ultrasonic irradiation of fluid suspensions is used in the clarification of wine by precipitation. High amplitude ultrasonic waves applied to a liquid creates air cavities within the liquid. This process is known as **cavitation,** and generates shock waves that produce vigorous streaming and shear stresses. These effects can be exploited in the production of emulsions, and the cleaning of surfaces. Thus ultrasound waves can be used for industrial cleaning of surfaces.

3.13.3 Other applications

The approximate locations of earthquakes can be determined using an instrument called the **seismograph**. Here, the earthquake is the source of the vibration, and the seismograph is used as the receiver. The principles of operation of this instrument rely on the propagation of sound waves through the earth. When an earthquake occurs, two distinct types of waves travel from the source of vibration. One of the waves is longitudinal, whilst the other is transverse. In the field of seismology, these waves are called the p-wave and the s-wave, respectively. These two waves travel with different velocities through the earth's crust. Hence it is possible to determine the distance to the source of the quake by recording with the seismograph the time interval between the arrival of the two types of waves. Similar observations carried out at many centres will enable geophysicists to determine the **epicentre** of the earthquake. The intensity of the measured vibration is used to estimate the size of the quake. The study of earthquakes has given us our most reliable information on the makeup of the earth's interior.

Seismographs are also used to determine underground geophysical structures in prospecting for petroleum. In this practice, an explosion is created underground using dynamite or TNT. In its travel through the earth's interior, this explosion is partly reflected at boundaries of layers with different densities. The reflections are detected by seismographs, and used to determine the difference between the times of arrival of the reflections. From the later a picture is formed about the structure of the subterranean profile. This is used in locating petroleum oil within the interior. The same data can be used to estimate the quantity of the find.

EXERCISES

You may use data in **Table 3.1** where necessary.

1. An ocean depth-sounding device emits a signal 38 kHz in water whose average temperature is 4 °C. The impulse is reflected from the ocean bed and returned to the same location 1.62 s after the signal is emitted. Calculate:
   (i) the depth of the water,
   (ii) the wavelength of the signal in water

2. Steam can be seen coming from a factory whistle on a day when the temperature is 24 °C, and 6.0 s later, the sound of the whistle is heard. How far is the factory away?

3. A workman strikes the steel rail of a railway track with a sledgehammer. The sound of the blow reaches an observer through the air and through the rail. The difference in time for the sound to reach the observer is measured to be 3.0 s. How far, in metres, is the observer from the workman?

4. In foggy weather, a lighthouse sends sound signals simultaneously under water (4 °C), and through the air (15 °C). A ship receives the signals 2.15 s apart. How far, in kilometers, is the ship from the lighthouse?

5. A listener doubles her distance from a source of sound that emits uniformly in all directions. By how many decibels does the sound level change?

6. Sound energy is radiated uniformly in all directions from a source at the rate of 40 watts.

(a) What is the sound intensity in watts per square centimeter at a point 35 m away from the source?

(b) What is the intensity level, in decibels, at that point?

7. A typical adult ear has a surface area of 21 cm^2. The sound intensity during a normal conversation is found to be about 3.2 $\mu W/m^2$ at the listener's ear. Assuming that the sound strikes the surface of the ear at right angles, determine the total power intercepted by the ear.

8. The signal-to-noise ratio of a cassette tape deck is the ratio of the sound intensity created when the tape is playing to the intensity created by the hissing noise of the tape winding on. This signal-to-noise ratio is normally expressed in decibels. If, in a typical case, the sound intensity due to the musical tones is 120 $\mu W/m^2$, while the intensity due to the tape hissing is $3.2 \times 10^{-11}\ W/m^2$, calculate the signal-to-noise ratio of the tape deck.

9. The lowest note on a piano has a fundamental frequency of 27.5 Hz, and is produced by a wire that has length 1.18 m. If the speed of sound in air is 343 m/s, calculate the ratio of the wavelength of the sound wave to the wavelength of the travelling waves on the string.

10. The G string on a guitar has a fundamental frequency of 196 Hz, and a length of 0.62 m. This string is pressed against the proper fret in order to produce a C note of frequency 262 Hz. Calculate the distance between the fret and the end of the string at the top end of the guitar.

11. A cylindrical pipe is closed at both ends. A wave is sent down the pipe by hitting one of the closed ends.

(a) Draw wave profiles for the standing waves in the pipe.

(b) For a fixed length of pipe, determine the ratio of the frequency of the second overtone to the frequency of the fundamental mode.

12. A tube, open only at one end, is cut into two shorter unequal lengths. The piece open at both ends has a fundamental frequency of 425 Hz, while the piece closed at one end has a fundamental frequency of 675 Hz. Find the fundamental frequency of the original tube.

13. An organ pipe, open at both ends, produces a note of frequency 262 Hz (i.e. middle C) in its third harmonic mode of vibration. If the speed of sound in air is 343 m/s, calculate the length of the pipe.

4 CHAPTER

Physical Optics

4

CHAPTER

Physical Optics

From early recorded times, man had wondered about the mysteries of light and imagined its exact nature and relationship to human vision. Today, much is known about the nature of light, and some are still being discovered. Each of these new discoveries is added to our growing knowledge on light. Based on these observations, laws have been formulated which predict the behaviour of light. Indeed one of the basic questions one needs to answer about light concerns its nature. Just what is light? When we scratch a match or switch on a light bulb in a dark room, we see not only the match or light bulb, but we see other objects too. Moreover, these objects may come in different colours. Colour is a word we use to describe certain physiological and psychological sensation of vision. Just what is it that makes the difference we have between colours? How do these colours combine with one another?

The observed behaviour of light as it travels from one place to another, and its interaction with matter has led to the development of a lot of technological tools. Tools have been developed which allow us to see objects as tiny as single blood cells and bacteria, and also see bodies in galaxies long distances away. We have developed high-speed cameras used for taking photographs of fast-moving objects, and also have tools for taking photos during the night and under the depths of the sea. All these instruments use the intricate behaviour of light as it travels from one place to another, and as it interacts with material objects.

We present the study of light in this book in two parts: the first part deals with the physical behaviour of light in its propagation. This aspect allows us to relate to the physical properties of light – mode of emission, colour, frequency, and energy. We call this branch **physical optics**, and it is the subject of this chapter. The second part of the study utilises the mapping of the path of light **rays**. We then use these rays in geometrical constructions and yield information on how these rays combine to give the images produced by the passage of light. This second part of the study is called **geometrical optics**. The two chapters following this chapter are devoted to geometrical optics.

4.1 THE NATURE OF LIGHT

We can understand the nature of light when we relate light to vision. An object becomes visible when light coming from it meets the eye. A body is visible either because it is luminous, or because it is illuminated. Luminous objects produce their own light. Examples include lamps, the stars, and the sun. Most common objects we find around us are non-luminous, and they include such things as wood, concrete, and writing paper. These non-luminous objects are made visible by the reflection of light from their surfaces.

Air, glass, and water readily transmit light. They are said to be **transparent**. Other substances tend to diffuse the light, and scatter it in many directions. The result is that it is barely possible to make out the profiles of objects when viewed through such materials. We are not able to discern details of the object being viewed. These materials are said to be **translucent**. Frosted light bulbs, paper soaked in paraffin or vegetable oil are examples of translucent materials. Substances which do not transmit light at all are said to be **opaque**.

Energy can be transmitted from one point to another by two methods. In one of these, matter may move from one point to the other. This is the **corpuscular theory**. In the other method, a wave disturbance can travel between the two points in the medium. This is the **wave theory**. Two theories have been put forward to help explain the behaviour of light as it travels from one point to another, and as it interacts with matter. These are the **corpuscular theory**, and the **wave theory**. Both theories have been advanced to account for the nature, propagation, and interaction of light with matter.

4.1.1 The corpuscular theory

The corpuscular theory assumes light to be made up of particles, or **corpuscles**, originating from the light source, and moving in straight lines through space or through a medium. One of the most famous champions of this theory was Sir Isaac Newton himself. The small corpuscles passing through the eye are thought to stimulate the sensory nerves and produce the sense of vision. The corpuscles were later named **photons**, the name they have even today.

The greatest success of the corpuscular theory lies in its account of the **photoelectric effect** – i.e. the production of electricity from certain metal surfaces when light of certain colour is shone on the surface. In this, it is thought that the energy of the photon is transferred to an electron in the metal surface, and the electron gets ejected. This leads to the electric current that is measured.

4.1.2 The wave theory

Christian Huygens (1629 – 1695) first conceived light as a wave motion. In his theory, he proposed light as a transverse wave motion. More than one hundred years later, this proposal was backed by the results of electromagnetic wave studies carried out by James Clarke Maxwell. In the wave theory, light is pictured as consisting of transverse waves of progressively changing electric and magnetic fields. The fields act at right angles to each other, and to the direction of propagation of the wave. In the propagation of light in a medium, Huygens proposed that each point on the wavefront acts as a secondary source of new disturbance that spreads out spherically. The new wavefront is the envelope of all the individual wavelets that originate from the old wavefront – shown in **Fig. 4.1**.

Many studies were carried out on the wave theory of light at the beginning of the nineteenth century as a result of discoveries made by Thomas Young of England, and Augustin Jean Fresnel of France. Young's theoretical work predicted the interference of light waves from two sources, just like surface waves in water. Fresnel carried out both theory and experiment much further, explaining diffraction by straight edges, obstacles and apertures, by a combination of Young's principle of interference, and Huygens' principle of constructing secondary wavefronts.

4.1.3 Wave-particle duality

There are two theories needed to explain all the properties and interactions of light. These are the wave theory, and the corpuscular theory. It is now known that light is a

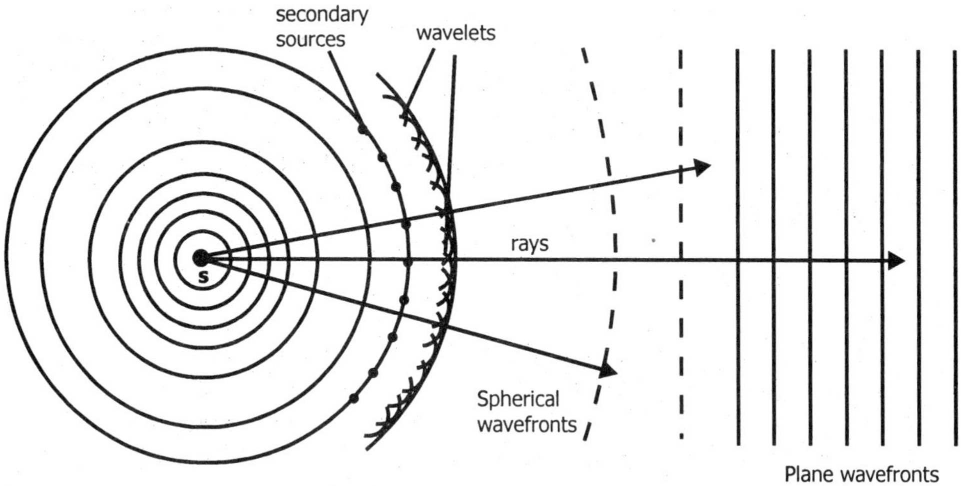

Fig. 4.1 Huygens' construction of secondary wavefronts. Spherical wavefronts from a point source become plane wavefronts at a great distance from the source.

form of transverse waves in which the electric and magnetic vibrations are perpendicular to the direction of travel of the wave. Thus light waves have wavelength and frequency. This wavelength controls the colour of the light. Light also exhibits other phenomena – e.g. interference, diffraction – that are explained only if we assume that light is a wave motion. On the other hand, there are other observed phenomena that can only be satisfactorily explained if light is assumed to be made up of particles, or corpuscles. About the most important of these is the photoelectric effect.

From the above, it seems that we thus have two seemingly contradictory ways of accounting for the behaviour of light. The existence of two theories to account for light is referred to as the **wave-particle duality**. In our approach to explaining the nature and behaviour of light, we sometimes assume wave properties, and at times we assume particle properties. In general the wave theory is used to explain phenomena associated with the propagation of light. Phenomena like interference and diffraction are explained using the wave theory. The particle or corpuscular theory comes into play when it is desired to explain phenomena dealing with the interaction of light with matter.

4.2 SPEED OF LIGHT

The speed of light is a very important quantity in physical calculations. Experience tells us that the speed of light far exceeds that of sound – the reason why we see the lightning flash before hearing the thunder, even though both occurred at the same time. Early measurements of the speed of light used astronomical methods, the details of which now belong to history. The most important and accurate terrestrial experiment to find the speed of light was done by Michelson, between the years of 1926 and 1929, over a distance of about 80 kilometres. **Fig. 4.2** is a schematic diagram of the setup used.

Light from a source S was made to emerge as a parallel beam and fall at an angle of 45^{0} on one face O of an octagonal mirror, the latter mounted on a rotating mechanism. The light reflected from the mirror travels to a parabolic mirror P. At the focus of P is a plane mirror M. The light reflected from P travels to M, to another point on P, reflected at another face O′ of the octagonal mirror into a viewing telescope V.

Without revolving, the initial alignment of the octagonal mirror was adjusted such that the light from S could be seen continuously through V. The octagonal mirror was then revolved with slowly increasing speed. At first no light could be seen. With increasing speed of revolution, however, a point was reached when the light was again seen continuously through the telescope. In this condition, the time taken for the light to travel the distance from O to O′ is equal to the time taken for the octagonal mirror to move from one face to the next. In the experiment, the speed of revolution of the mirror is set to be very accurate. The time τ it takes for the mirror to go from one face to the next is one-eight the period of revolution. Thus,

$$\tau = \frac{T}{8} = \frac{1}{8f}, \tag{4.1}$$

where T is the period of revolution of the octagonal mirror. If D is the total distance OO′ (i.e. O to O′ through M), and c is the speed of light, then

$$D = c\tau = \frac{c}{8f}, \tag{4.2}$$

giving

$$c = 8Df, \tag{4.3}$$

where f is the frequency of revolution of the mirror.

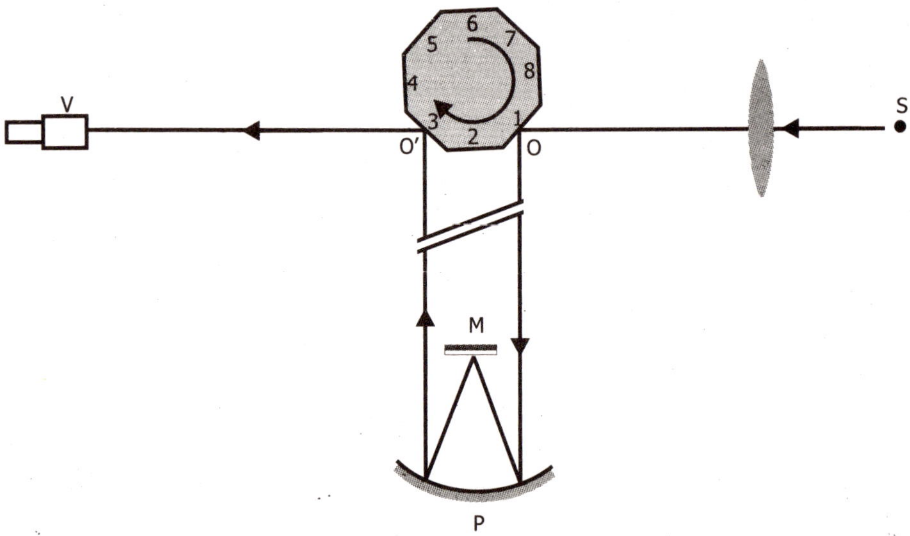

Fig. 4.2 Michelson's experiment for speed of light

Michelson's measurements gave the speed of light as 2.997921×10^8 m/s. Although many experiments have been performed using more sensitive procedures, these experiments have only served to confirm the value Mishelson obtained. For most practical purposes and calculations, we use $c = 3.0 \times 10^8$ m/s as the speed of light in vacuum. The speed in other media, of course depends on the refractive indices of these media. We often take it that the speed of light in air is the same as that in vacuum. This is only an approximation, as the refractive index of air is not exactly unity. As the refractive index of air is very nearly equal to 1, the above figure for the speed of light is used for most practical purposes.

4.3 PHOTOMETRIC QUANTITIES

Photometry is the branch of study dealing with the measurement of light. In this subject we define certain measurable quantities of light, and compare different light intensities on various surfaces. In the measurement of light, we have three quantities: the **luminous intensity** of the source, the **luminous flux** or flow of light from the source, the **illumination**, and **luminance** of the surface on which the light falls. In most applications, we are interested in the light actually falling on a surface. This amount, however, depends on the intensity of the source, and on the medium in which the light travels. The above quantities are very useful in the design and construction of houses, particularly in tropical countries. In these environments, it is desired that these houses receive the right amount of light. Construction engineers and architects make use of these photometric quantities in their designs.

4.3.1 Luminous intensity

The intensity of a light source is the amount of light energy coming from the source per unit time. It is measured in units of **candela**, (cd) where

One candela is the intensity of light from a perfect blackbody of surface area 1/60 cm² at the temperature of the melting point of platinum.

All other units of light intensity are derived from this basic unit.

4.3.2 Luminous flux

The **flux**, or flow of light is measured in lumens (lm).

One lumen is the flow of light emitted by a source of one candela per unit solid angle.

The solid angle is a three-dimensional angle in space subtended by a surface. We understand the meaning by using **Fig. 4.3**. Consider a point source S of light at a point in space. The locus of points in space the same distance from S generates a sphere. The solid angle Ω in three-dimensional space is defined by

$$\Omega = \frac{\delta A}{r^2}, \qquad (4.4)$$

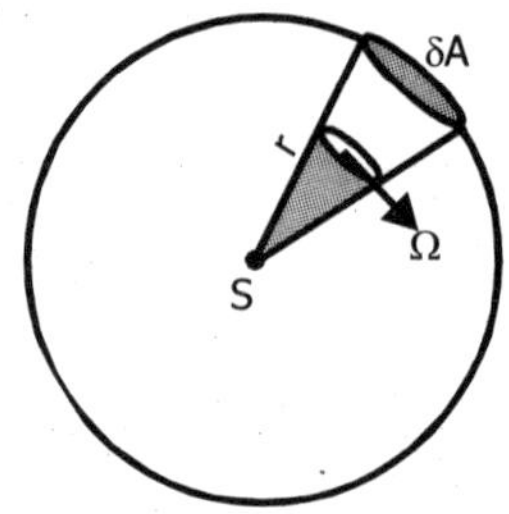

Fig. 4.3 Solid angle

where r is the radius of the sphere. Solid angle is given in **stearadians**. Like a two-dimensional angle, solid angle is a pure number. For the whole sphere of area $4\pi r^2$, the solid angle subtended at the centre is 4π stearadians. Luminous flux is expressed in cd/stearadian.

4.3.3 Illumination

Illumination (usual symbol E) is measured by the amount of flux falling on a unit area. The units of illumination is lumens per metre^2 (or lm/m^2).

4.3.4 Luminance

Luminance (symbol L) is a measure of the brightness of a surface. The SI unit for luminance is derived as follows: If a light source of 1 cd has a surface area of 1 m^2 – i.e. 1 cd distributed uniformly over an area of 1 m^2 – the luminance is 1 cd/m^2.

4.4 THE ELECTROMAGNETIC SPECTRUM

An orderly arrangement of radiation according to wavelength, frequency, or energy is known as a **spectrum**. The electromagnetic spectrum is such a spectrum. In it, the components of light are arranged in order of frequency and wavelength. The important parts of the electromagnetic spectrum are shown diagrammatically in **Fig. 4.4**.

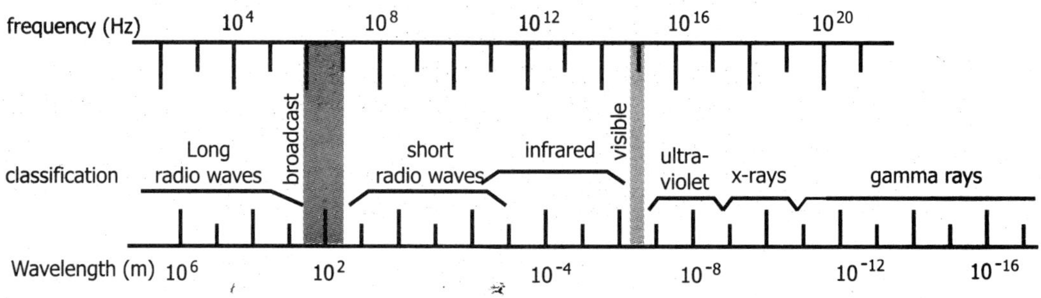

Fig. 4.4 The electromagnetic spectrum showing the component wavelengths and their frequencies.

All the radiations constituting the electromagnetic spectrum travel at the same speed of 3.0×10^8 m/s. This speed is one of the fundamental constants of nature. For each frequency of the radiation, the energy E of the wave is given by the expression

$$E = hf, \qquad (4.5)$$

where f is the frequency of the radiation, and h is a constant, known as Planck's constant, with numerical value of 6.7×10^{-34} Js.

The most familiar frequency range is visible light, the narrow band of electromagnetic radiation to which the human eye is sensitive. This extends from the violet on the high frequency (or short wavelength) side to red on the low frequency (long wavelength) side. X-rays are highly energetic type of atomic radiation with high penetrating power. Because of their high penetration, these rays have applications in medical imaging, in non-destructive testing of materials, and in studying the crystal structure of minerals. X-rays are produced from specialised tubes and also from radioactive substances. Closely related to x-rays are gamma rays. These are highly penetrating radiation produced from certain radioactive sources. Gammy rays in general have more energy than x-rays.

To the longer wavelength side of the visible region we start with the infrared radiation, known more commonly as heat radiation. Radio waves used for broadcast and other communication purposes have wavelengths ranging from about 1 mm (for short radio waves) to 10^6 m. The most useful region for normal broadcast has wavelength between 10 m and about 400 m.

Becoming increasingly important in technology are microwaves. These are electromagnetic radiation having frequency in the range 10^9 to 10^{12} Hz, and a wavelength between 1.0 mm to 1.0 m. The popular application of these waves nowadays is in cooking. A microwave oven is a relatively small, box-like oven that raises the temperature of an electrically non-conducting material (the food and its container) by subjecting the material to a high frequency electromagnetic field. The heating occurs throughout the food, rather than from the outside first – as in most other cookers. This greatly reduces cooking time and energy consumption.

4.5 INTERFERENCE OF LIGHT WAVES

One of the important properties of wave motion is interference. From our earlier study of this phenomenon in chapter 2, we have two types of interference: constructive interference and destructive interference. It was made clear that for interference to be observed between two wave motions, the sources must be coherent, and also the distances over which the interference pattern is observed must be of the same order of magnitude as the wavelength of the wave. Due to the general small wavelength of light, it took quite some ingenuity on the part of the early experimenters to observe interference in light. In this chapter, we study some of the experimental methods used for observing interference in visible light, and how the interference phenomenon is utilised in science and industry.

4.5.1 Coherent sources of light

If constructive or destructive interference is to continue occurring at a given point, the sources of the waves must be coherent. From our earlier study, coherent sources give out waves that maintain a fixed phase relationship. One difficulty encountered in producing two separate coherent light sources is due to the erratic nature of the processes that lead to light emission. Light is emitted when certain atoms are raised to higher energy levels by a process referred to as excitation. This excitation can be caused by heat. The release of this excess energy, called de-excitation, leads to the emission of light. Both the excitation and de-excitation processes are probability events. There is no way of telling which particular atom is going to be excited, or indeed when the excited atom will de-excite to produce the light. As a result, two separate sources of light give out light waves having erratic phase relationships, and hence will not be combined to bring about a stable interference pattern. Early attempts to create interference using two identically rated light sources (e.g. separate light bulbs) failed to produce interference.

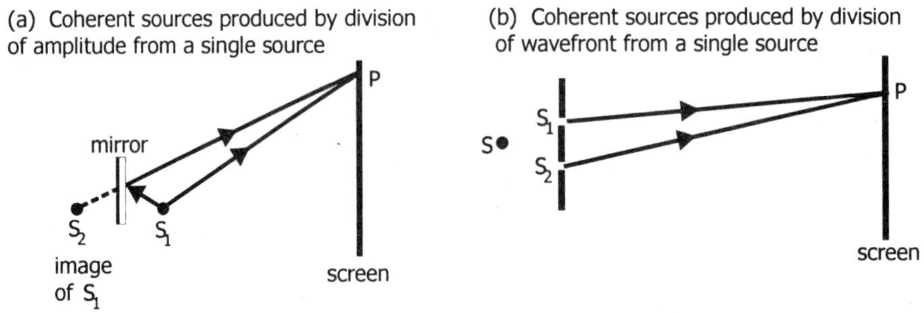

Fig. 4.5 Production of coherent light sources.

Coherent light sources are produced from a single light source by one of two methods, shown schematically in **Fig. 4.5**. In one of these, light can be reflected at a plane surface, the reflected wave having the same frequency and wavelength as the incident wave, and a constant phase difference of 180°. The light source and its image together would constitute two coherent sources. This method is known as **division of amplitude**. The other method involves dividing the light from a single slit into two secondary sources S_1 and S_2, by means of collimating slits. The secondary sources are always in phase. This method is known as **division of wavefront**. Using either method, the path difference Δ between the two waves at any point P on the screen is given by

$$\Delta = S_2P - S_1P \qquad (4.6)$$

4.5.2 Geometrical and optical paths

As we saw in chapter 2, calculations on interference make use of the difference in paths between two waves on arrival at the point of observation. For most cases, this path difference is taken as the difference between the geometrical distances between the two paths. This path difference is often called **geometrical path difference**. When a light wave goes through two media between two points, we have the concept of **optical path**. This is the distance travelled by the light that corrects for the fact that the light has two different speeds in the two media. If a material of thickness t, and refractive index n is introduced in the path of a ray of light, the material adds an extra optical path of $(n-1)t$.

4.5.3 Young's double slit

First demonstration of interference in light was done by Thomas Young. In this experiment, he observed the interference pattern produced when monochromatic light (i.e. light of one colour or wavelength) from two secondary sources overlap. **Fig. 4.6** is a schematic arrangement for demonstrating interference using Young's double slit method. The monochromatic source of light sends out waves, which fall on a single slit S. The wavefront spreading out from S is divided into two by the slits S_1 and S_2. The waves from the double slit interfere in the area of overlap, producing equally spaced bright and dark **fringes** on the viewing screen. In the experiment, these fringes are viewed through a telescope T mounted on the other side of a translucent screen.

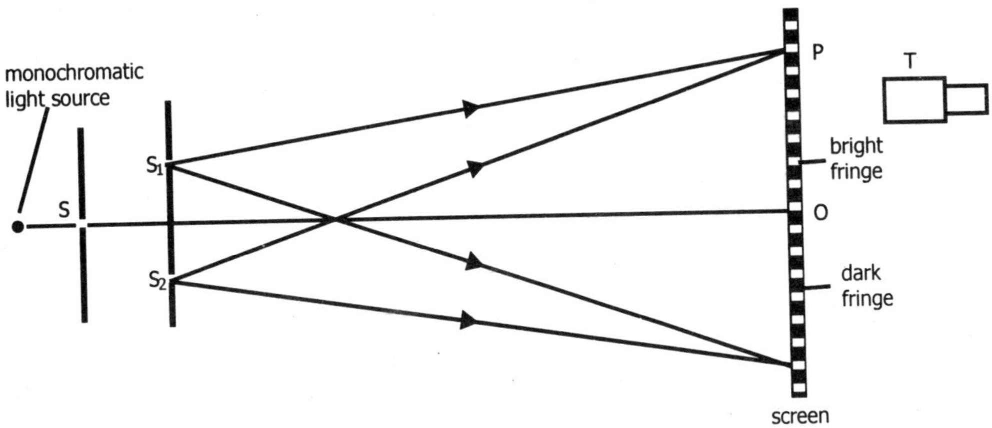

Fig. 4.6 Young's double-slit experimental setup. Interference observed in the area of overlap of the light beams from the two sources. Equally spaced bright and dark fringes are observer on the screen.

Using the geometry of the experimental setup, and measurements on the fringe system, Young was able to determine the wavelength of the light. We can better follow the analysis with reference to **Fig. 4.7**. The two slits S_1 and S_2 are a distance a apart, known as the **slit separation**. The viewing screen is a distance D away from the double slit. A perpendicular from S_1 meets S_2P at X. The distance S_2X is hence the difference in paths between the two waves. The perpendicular bisector of S_1S_2 meets the screen at O.

From the geometry, $\Delta = S_2P - S_1P = a\sin\theta$. The angle θ is very small indeed, as the wavelength of light is small. We hence write $\sin\theta \approx \tan\theta$, and $\Delta = a\sin\theta \approx a\tan\theta$. If the point P is a distance y_m away from O, then

$$\Delta = a\tan\theta = \frac{ay_m}{D}. \tag{4.7}$$

For constructive interference at P,

$$\Delta = \frac{ay_m}{D} = m\lambda; \tag{4.8}$$

and for destructive interference at P,

$$\Delta = \frac{ay_m}{D} = \left(m + \tfrac{1}{2}\right)\lambda, \tag{4.9}$$

where m is the order of the interference.

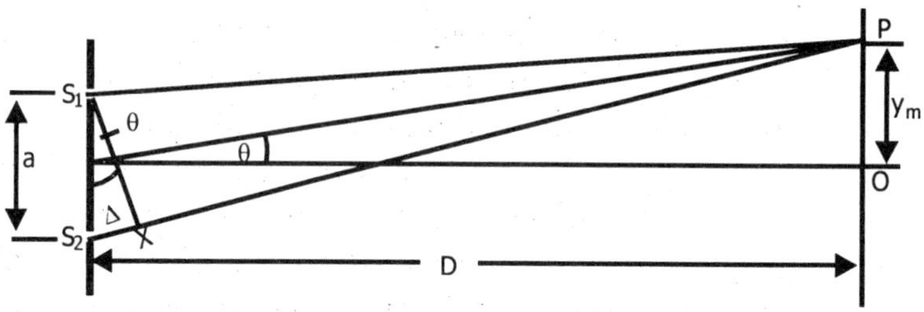

Fig. 4.7 Geometry of Young's double-slit experiment.

In the experimental, measurements of P from O for a given order of interference, the slit separation, and the distance from the slits to the screen allow a calculation of the wavelength of the light used. Alternatively, it is easier to measure the **fringe separation** – i.e. the distance between two adjacent bright or dark fringes. For two adjacent bright fringes corresponding to orders of interference m and (m+1), we have

$$\frac{ay_m}{D} = m\lambda, \quad \frac{ay_{m+1}}{D} = (m+1)\lambda,$$

and the fringe separation y therefore becomes

$$y = y_{m+1} - y_m = \frac{D\lambda}{a} \tag{4.10}$$

From equation (4.10), it follows that for a given wavelength, the separation of the fringes is greater, the smaller the slit separation, and the farther away the screen is from the slits. The interference pattern is influenced by the following:

(i) **The slit separation**. The fringes are farther apart when the slit separation is small. Conversely, less interference is observed when we increase the separation of the slits. In this latter case the fringes become very closely packed together, and appear almost as a continuous band.

(ii) **Slit-screen distance**. The fringe separation is greater the farther away the screen is from the slits. Of course we cannot move the screen too far away and still clearly observe the phenomenon, as the light intensity will become small to be visible beyond a certain distance.

When a thin transparent plate is put over one of the slits, the interference pattern shifts due to the additional optical path introduced by the plate. If a plate of thickness t produces a shift p in the fringe pattern, we write

$$p\lambda = (n-1)t. \qquad (4.11)$$

By measuring the shift in fringe pattern shift, we can find the refractive index of the plate.

Example 4.1

In a Young's double slit experiment, the slits are 0.5 mm apart, and are illuminated by monochromatic light of wavelength 0.60 μm. If the distance between the slits and the screen is 0.50 m, calculate

(i) the distance between the central maximum and the fourth bright fringe,

(ii) the fringe separation,

(iii) the angle subtended at the centre of the slits by the fourth maximum on either side.

Solution

Fig. 4.8 is a schematic of the experimental setup and the geometry of the system.

(i) The distance from central maximum to the fourth order maximum is y_4, where

$$\frac{ay_4}{D} = 4\lambda, \text{ or } y_4 = \frac{4D\lambda}{a}.$$

From the given data, $a = 5.0 \times 10^{-4}$ m, $D = 0.5$ m, and $\lambda = 6.0 \times 10^{-7}$ m. Thus,

$$y_4 = \frac{4D\lambda}{a} = \frac{4 \times 0.5\,\text{m} \times 6.0 \times 10^{-7}\,\text{m}}{5.0 \times 10^{-4}\,\text{m}} = 2.4\,\text{mm}$$

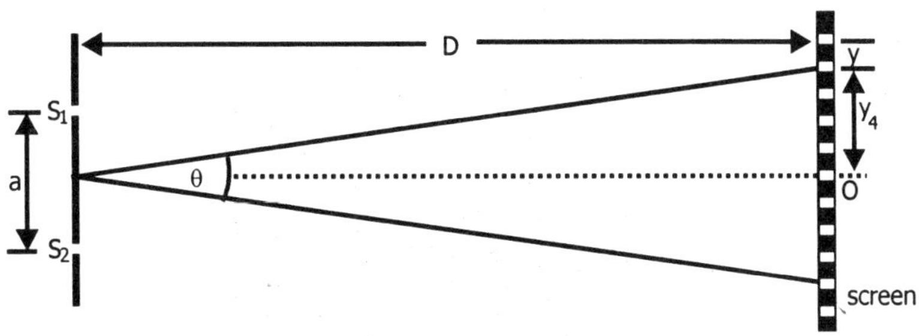

Fig. 4.8

(ii) The fringe separation y is given by $y = \frac{D\lambda}{a} = \frac{0.5\,\text{m} \times 6.0 \times 10^{-7}\,\text{m}}{5.0 \times 10^{-4}\,\text{m}} = 0.60\,\text{mm}.$

(iii) The angle θ subtended by the fourth bright fringe at the centre is given by

$$\tan\tfrac{1}{2}\theta = \frac{y_2}{D} = \frac{0.0024\,\text{m}}{0.5\,\text{m}} = 0.0048. \text{ Thus } \theta = 0.56^{\circ}$$

Example 4.2

In **Example 4.1**, it was found that if a plate of glass of refractive index 1.52 is placed over one of the slits, and white light is used, the central white fringe is displaced by a distance of 1.0 mm. Calculate the thickness of the glass plate.

Solution

White light is a combination of many colours – red, orange, yellow, green, blue, indigo, and violet. When white light is used as the source in Young's double slit setup, all the component colours will coincide at the principal maximum. On either side of this white maximum, the colours will be separated out according to the wavelengths of the components. A glass plate introduces an additional optical path difference, which results in the displacement of the central maximum on the screen. The amount x by which the central maximum is shifted is given by $x = (n-1)t$, where n is the refractive index of the plate, and t its thickness. Thus, $t = \frac{x}{n-1} = \frac{1.0\,\text{mm}}{1.5 - 1} = 2.0\,\text{mm}.$

4.5.4 Interference in thin wedges

A thin wedge consists of two glass plates (e.g. microscope slides) meeting at one edge, the opposite edges being separated by a small distance, usually by some material of small thickness – **Fig. 4.9**. A monochromatic ray of light incident on the top plate at A is partially reflected at the top plate, and partially refracted through the top plate on to the bottom plate. At the bottom plate, the ray is reflected, and the reflected ray travelling through the top plate, as shown. The two reflected rays combine to form an interference pattern.

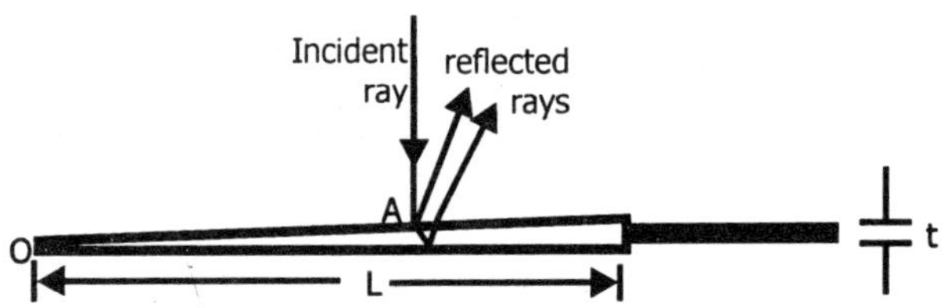

Fig. 4.9 Thin air wedge. Reflected rays combine with each other to produce interference fringes.

If y is the thickness of the air gap at A, and, with the ray incident vertically from above, the difference in paths Δ of the reflected rays is given by

$$\Delta = 2y\,. \tag{4.12}$$

At the position corresponding to the edge O, where the path difference is zero, we might expect to see a bright fringe. However, experiment shows that a dark fringe is seen instead. This is due to the fact that there is an extra path difference of $\lambda/2$ introduced by the reflection at the optically denser surface of the bottom plate. The equations for interference are hence modified to read

$$\Delta = 2y = \left(m + \tfrac{1}{2}\right)\lambda, \tag{4.13}$$

for constructive interference, or bright fringe, and

$$\Delta = 2y = m\lambda, \tag{4.14}$$

for destructive interference, or dark fringe. For two adjacent bright or dark fringes corresponding to wedge thicknesses y_1 and y_2, respectively, we have

$$2(y_2 - y_1) = \lambda. \tag{4.15}$$

If the angle of the wedge is θ, the fringe separation x, in the field of view is given by

$$\frac{y_2 - y_1}{x} = \tan\theta, \tag{4.16}$$

If t is the thickness of the spacer, and L the length of the wedge, then

$$\frac{\lambda}{2x} = \frac{t}{L}, \tag{4.17}$$

from which the thickness of the spacer can be determined. This method is useful in finding the thickness of thin foils.

Example 4.3

A wedge of air is formed by placing a thin aluminium foil between two glass plates at a distance of 75 mm from the line of contact of the plates. When the wedge is illuminated by light of wavelength 5.60×10^{-7} m from directly above, the fringe separation is found to be 1.20 mm. Calculate

(i) the thickness of the foil

(ii) the angle of the wedge,

Solution

From the given data, fringe separation, 1.20×10^{-3} m, L = 0.075 m, $\lambda = 5.60 \times 10^{-7}$ m.

(i) From equation (4.16), the thickness t of the foil is given by $t = \frac{\lambda L}{2x}$. Substituting the given numerical values, we get t = 0.0175 mm.

(ii) the angle θ of the wedge is given by the expression $\tan\theta = \frac{t}{L} = \frac{0.0175}{75}$. This gives the angle of the wedge as 0.013°

4.5.5 Newton's rings

Another demonstration of interference in light is shown in Newton's rings – in honour of its inventor. When a convex lens of large radius of curvature is placed on a flat plate of glass, a thin air wedge is formed between the two surfaces. Since the lens is circular, the interference pattern formed by this system of wedges is in the form of concentric rings. This is so because the thickness of the air film increases from zero as we go outward from the centre of the lens – shown schematically in **Fig. 4.10**.

For any point of the wedge between the lens and the plate, the path difference Δ between the emerging rays is given by

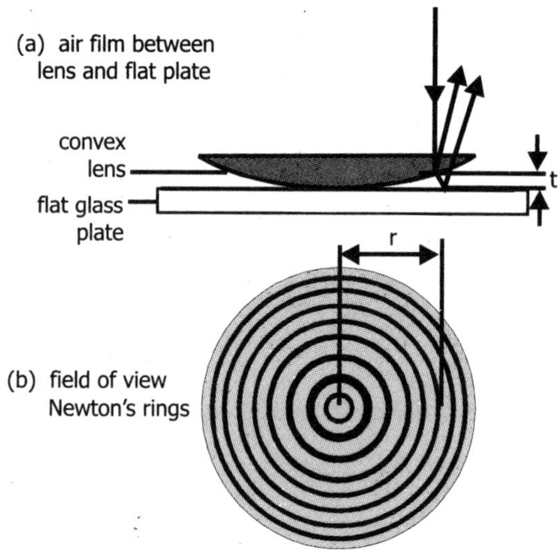

Fig. 4.10 Newton's rings. (a) Thickness of air film increases from centre, (b) concentric bright and dark rings observed in the field of view. Radius of ring depends on the thickness of the air film at the point.

$$\Delta = 2t,$$

Using a monochromatic source of light, we have for constructive interference,

$$\Delta = 2t = \left(m + \tfrac{1}{2}\right)\lambda, \qquad \text{(bright ring)}$$

and for destructive interference,

$$\Delta = 2t = m\lambda. \qquad \text{(dark ring)}$$

Using the geometry, and the theorem of intersecting chords, we can relate the thickness t of the air gap to the radius r of the ring. This gives

$$t(2R - t) = r^2, \text{ or } 2t = \frac{r^2}{R},$$

where R is the radius of curvature of the lens, and R>>t. Substituting for 2t into the equation for the path difference, we get

$$\frac{r^2}{R} = \left(m + \frac{1}{2}\right)\lambda, \tag{4.18}$$

for bright ring, and

$$\frac{r^2}{R} = m\lambda, \tag{4.19}$$

for a dark ring. Note that a dark ring appears at the centre, where the difference in paths is zero, due to the extra phase change on reflection.

4.5.6 Thin parallel-sided films

Fig. 4.11 is a magnified schematic showing the paths of rays incident on one face of a parallel-sided film.

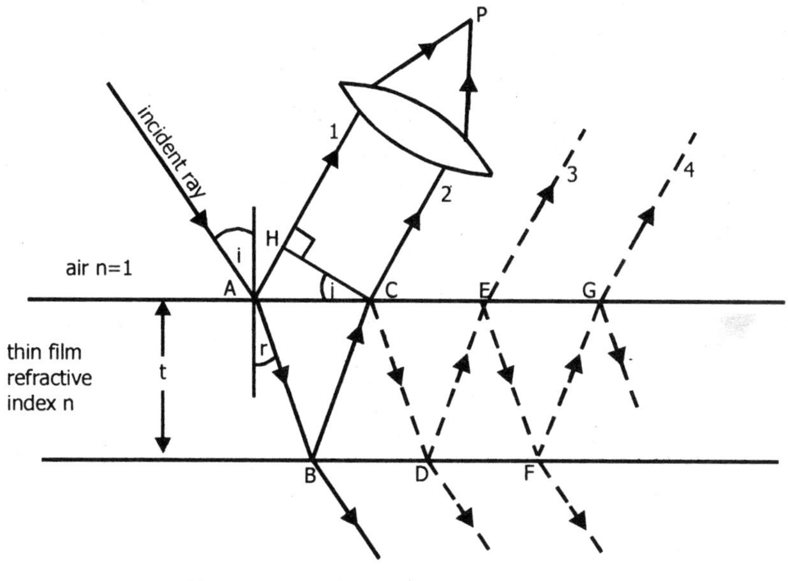

Fig. 4.11 Interference in a thin parallel-sided film.

When light meets one face of a thin film of refractive index n, most of the light is refracted through the film, and only a little is reflected. The refracted ray travels to the opposite face of the film where some reflection takes place, and the reflected ray travelling along the path BC. At C, reflection and refraction take place. The refracted ray passes into air. When the refracted ray at C and the reflected ray at A are combined through a suitable lens (e.g. the eye), an interference pattern is observed. Using monochromatic light we see a succession of bright and dark fringes in the field of view.

The geometric path difference between the rays 1 and 2 = (AB + BC) − AH. Bearing in mind that the refracted ray had to travel through a medium other than air, the optical path difference between the rays in question = $n(AB + BC) - AH$. From the figure,

$$AB = BC = \frac{t}{\cos r}, \text{ and } AH = AC\sin i = nAC\sin r = 2nt\tan r\sin r.$$

The optical path difference Δ between the two rays is therefore given by

$$\Delta = \frac{2nt}{\cos r} - 2nt\tan r\sin r = 2nt\cos r. \tag{4.20}$$

The conditions for constructive and destructive interference are then as follows:

$$\Delta = 2nt\cos r = \left(m + \tfrac{1}{2}\right)\lambda, \tag{4.21}$$

for constructive interference.

For destructive interference,

$$\Delta = 2nt\cos r = m\lambda \tag{4.22}$$

Again, there is an extra phase change introduced by reflection of one of the rays at the boundary of an optically denser medium.

When white light is used the component colours are refracted at different angles, thereby producing constructive and destructive interference at different positions depending on the angle of refraction. This explains the different colours seen on the surface of thin liquid films when illuminated by white light.

4.5.7 Reflecting and nonreflecting coatings

Thin film interference phenomena find applications in the lens industry in the fabrication of reflecting and nonreflecting coatings of lenses. A reflecting coating enables the lens surface to be highly reflective, whilst nonreflecting coating reduces the amount of light that is reflected from the surface of the lens. Both these coatings utilise the extra phase change brought about by reflection at an optically denser medium. In their fabrication, a thin layer of hard transparent material is deposited on the lens surface. A ray of light incident on the top face of the film suffers reflection and refraction. The refracted ray travels to the top surface of the lens, at which it is

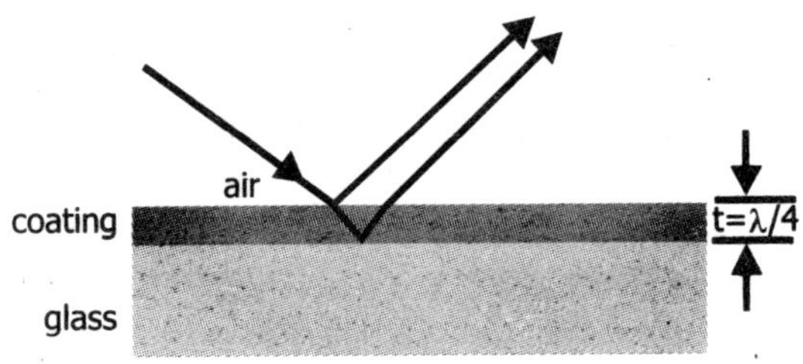

Fig. 4.12 Reflecting and nonreflecting film coatings

reflected - **Fig. 4.12**. The two reflected rays interfere constructively or destructively depending on the refractive indices of the film and the material of the lens.

If the coating thickness is one-quarter the wavelength of the light used, the surface becomes reflecting or nonreflecting, depending on the refractive index of the coating compared to that of the lens. If the coating material has refractive index greater than the lens material, the reflectivity is increased. There is a half-cycle phase shift of π (equivalent to $\lambda/2$) at the air-film interface, but none at the film-lens interface. The result is that reflected rays from the top and bottom of the film will interfere constructively, increasing the overall reflectivity.

In the nonreflecting coating, the refractive index of the coating material is smaller than that of the lens. The reflected rays will interfere destructively, thereby producing minimal reflection. If white light is used, the thickness of the coating can be one-quarter wavelength for only one particular wavelength, usually set to be in the centre of the yellow-green portion, at a wavelength of 550 nm, where the eye is most sensitive.

4.6 DIFFRACTION OF LIGHT WAVES

As we saw in chapter 2, diffraction is the ability of waves to bend round obstacles placed in their paths. This phenomenon can be explained using Huygens' construction of secondary wavelets. When a beam of light enters a small hole in an opaque screen, all the secondary wavelets are stopped except those arriving at the hole. The transmitted wavelets propagate the light into the geometrical shadow. Diffraction is considered to be the combination of interference effects due to many point sources of waves. Each infinitesimal part of the aperture acts as a source of waves, and waves from the entire secondary sources combine to produce the interference pattern.

There is no fundamental difference between interference and diffraction. The only slight difference between the two is that we use interference when only a small number of slits are involved, typically two. Diffraction is the interference of a large number of slits put together, thereby producing a continuous stream of Huygens' secondary wavelets. Thus we have a large number of continuous distribution of double-slit interference patterns.

There are two types of diffraction. In one, the source, the obstacle or aperture, and the observation screen are all relatively close to each other. This diffraction is known as **near-field diffraction**, or **Fresnel diffraction.** In the other kind of diffraction, the source, obstacle and the screen are all far enough away from each other that all rays can be considered parallel to each other. This diffraction is known as **far-field** diffraction, also referred to as **Fraunhofer diffraction**. In this section, we briefly study diffraction pattern due to single and multiple slits, using Fraunhofer diffraction. We end by studying the effect of large number of slits, in the **diffraction grating**.

4.6.1 Single slit diffraction

We assume here that the slit is very wide, so we neglect any effects at the edges, and make our observations in any plane perpendicular to the slit. This effectively reduces the problem to a two-dimensional observation. In **Fig. 4.13** (a) is shown a single slit, treated as a stack of narrow slits. To avoid making the diagram too complex, we have shown here only a few of these slits. An interference pattern would be produced on the screen when light waves from any pair of these sub-slits, A, and B, for example, combine with each other. The result is to produce consecutive bright and dark fringes on the screen, as shown in **Fig. 4.13** (b). Between every two neighbouring positions where darkness occurs, there is a position of maximum brightness. At $\theta = 0$, we have the central maximum. For a slit of width d, and using monochromatic light, the positions at which maximum intensity of the light occurs is given by

$$d \sin\theta = m\lambda, \tag{4.23}$$

and a dark fringe occurs when

$$d \sin\theta = \left(m + \tfrac{1}{2}\right)\lambda, \tag{4.24}$$

m being the order of the diffraction, and λ the wavelength of the light. The central maximum is the most pronounced, the other maxima (normally called secondary maxima) are relatively small in intensity.

The angular spread of the light due to diffraction is measured by the half-width of the central maximum. For a slit, this angular spread θ is given by

$$\theta \approx \sin\theta = \frac{\lambda}{d}. \tag{4.25}$$

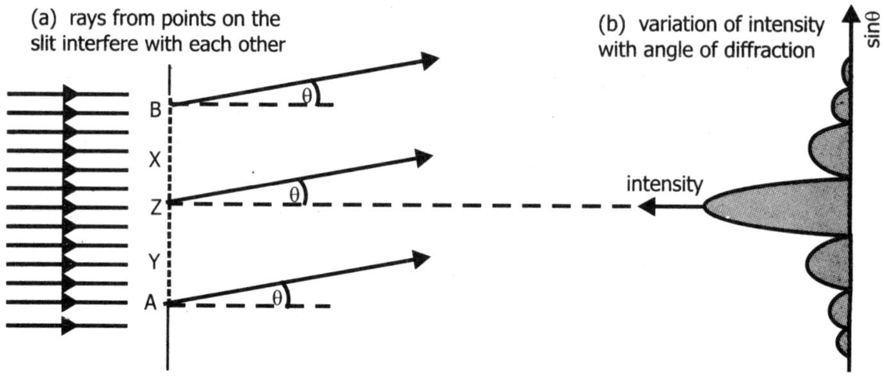

Fig. 4.13 Diffraction by single slit. Rays from points along the wavefront interfere with each other as in (a). The intensity of the diffracted beam varies with diffraction angle as shown in (b).

Using fairly complicated analysis outside our current scope, it is shown that if a circular hole of diameter d is used instead of the narrow slit, the angular spread of the principal maximum becomes

$$\theta \approx \sin\theta = \frac{1.22\lambda}{d}. \tag{4.26}$$

Example 4.4

A slit of length 0.1 mm is illuminated normally with light of wavelength 550 nm. At what angle to the straight-through beam would the first minimum occur?

Solution

Fig. 4.14 shows the intensity variation with angle in the field of view. The angle corresponding to the first minimum is θ. From our foregoing analysis, we have $\sin\theta = \frac{\lambda}{d}$, where λ is the wavelength of the light, and d is the width of the slit. Given that $\lambda = 5.50 \times 10^{-7}$ m, and $d = 10^{-4}$ m, $\sin\theta = \frac{\lambda}{d} = \frac{5.50 \times 10^{-7}\,\text{m}}{10^{-4}\,\text{m}} = 5.50 \times 10^{-3}$, from which $\theta = 0.315^{o}$.

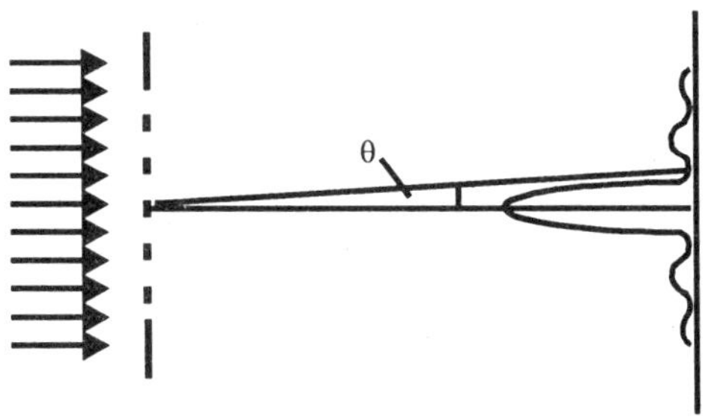

Fig. 4.14

4.6.2 Resolution

With diffraction, we are now able to study how close together two objects can be for them to be distinct when viewed through any optical instrument, or by the eye. This concept is described by the word **resolution**. For any optical instrument, the ability of the instrument to produce images of objects close to each other is known as the **resolving power**. Often, a good resolution means two objects can be very close together but we are still able to see two separate images instead of one.

When we have two sources of light or objects close to each other, the diffraction patterns from the sources overlap, and can be seen as two objects only if their separation is not smaller than some minimum distance. Through the efforts of many studies, a theoretical criterion was proposed by Rayleigh, that

two diffraction patterns will be resolved as long as the central maximum of one is no closer to the other than the position of the first minimum of the other.

The above statement is the "Rayleigh criterion". We can understand it better with reference to **Fig. 4.15**.

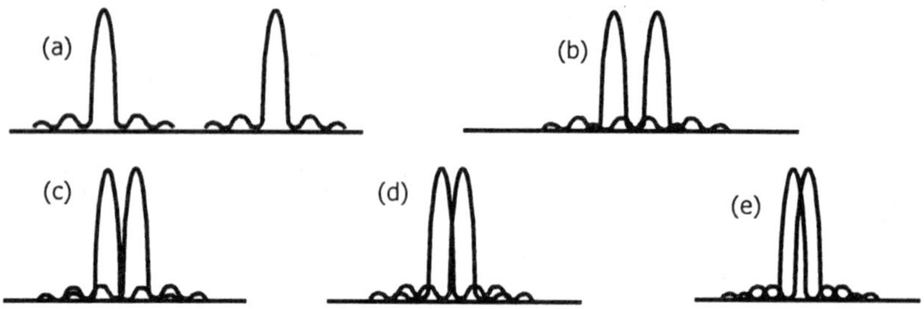

Fig. 4.15 Diffraction patterns produced due to two point sources or objects. In (a) and (b) the sources are far apart, and can be resolved quite easily. The limit of resolution is shown in (c), where the objects can just be seen as two, whilst in (d) and (e) the two objects cannot be resolved.

In both (a) and (b), the two objects or sources are far apart. The principal maxima are well separated from each other. In (c), the two principal maxima are just separated. It is seen that the principal maximum of one lies above the position of the first order minimum of the other. In (d) and (e) the two principal maxima cannot be resolved. At these distances, only one image of the two objects or sources is seen.

Example 4.5

Calculate the maximum separation of the two slits in a Young's double slit experiment when illuminated by light of wavelength 589 nm, if the interference pattern formed on a screen 1.50 m away is to be observed by the experimenter with a resolution of 0.166° (i.e. 1 minute of arc), at a distance of 25 cm.

Solution

Fig. 4.16 is a schematic representation of the observation of the fringes in question. The minimum distance is that for which the principal maximum for one slit coincides with the first minimum for the second slit. This is shown in the diagram. From the geometry, we have $y = d\tan\theta = 0.25\text{ m} \times \tan(0.166^\circ) = 7.243\times10^{-4}\text{ m}$. For the first order constructive interference, path difference $\Delta = a_{max}\dfrac{y}{D} = \lambda$, where a_{max} is the maximum slit separation.

Thus, $a_{max} = \dfrac{D\lambda}{y} = \dfrac{1.50\text{ m} \times 5.89\times10^{-7}\text{ m}}{0.0007243\text{ m}} = 1.22\text{ mm}.$

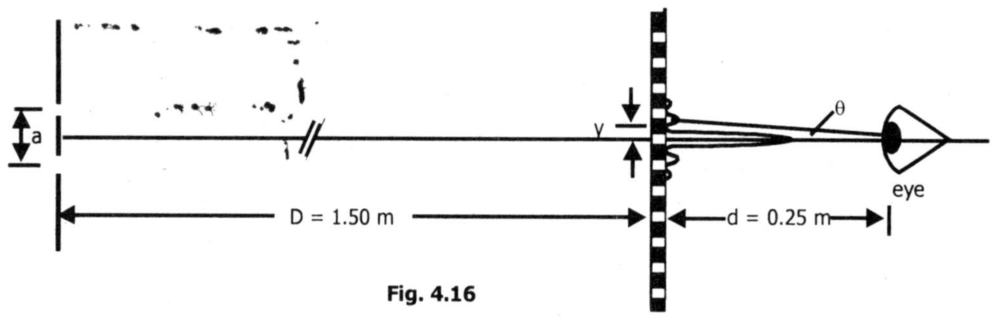

Fig. 4.16

4.6.3 The diffraction grating

The diffraction grating consists of a large number of equally spaced slits very close to each other. In practice, gratings are produced by ruling lines close together on a plane glass plate. Often a diamond head is used for the ruling. The lines scatter the light in all directions, effectively stopping the light from passing straight through. The space between the lines act as the apertures. Thus the diffraction grating can be considered as a large number of double slits close together.

Fig. 4.17 (a) shows the passage of a parallel beam of monochromatic light incident normally on a diffraction grating whose grating spacing (i.e. the distance from the centre of one slit to the centre of the next) is d. Using light of wavelength λ, for any two adjacent slits, the light scattered by the pair will have maximum intensity if

$$d\sin\theta = m\lambda, \tag{4.27}$$

where m is the order of the interference. By extending this idea, we can see that any pair of slits will produce either constructive or destructive interference depending on their relative positions on the grating. These will all combine to produce an overall variation in intensity shown in **Fig. 4.17** (b).

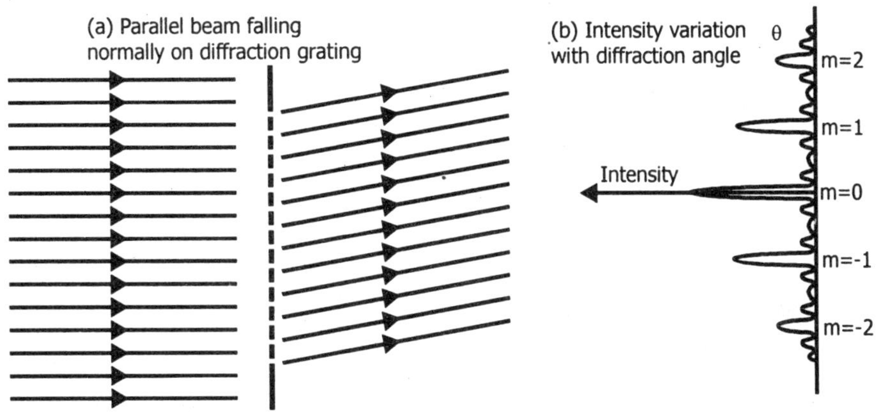

Fig. 4.17 The diffraction gating

The grating manufacturers usually specify the number of lines per unit length of the grating – e.g. 500 lines per cm. From this specification, we calculate the spacing of the grating. If N is the number of lines per cm of the grating, then we have

$$d = \frac{1}{N}, \tag{4.28}$$

and equation (4.26) is modified to read

$$\frac{1}{N}\sin\theta = m\lambda, \quad m = 0, 1, 2, 3, \ldots \tag{4.29}$$

Example 4.6

A diffraction grating has 4000 lines per centimetre, and is illuminated by light of wavelength 500 nm. Determine the directions in which maxima will occur.

Solution

Using the diffraction, equation, maxima will occur if $d\sin\theta = m\lambda$. From this, the directions of the maxima will be given by $\sin\theta = m\lambda / d = Nm\lambda$, where N is the number of lines per unit length of the grating. This gives $\sin\theta = 0.20m$, m being the order of the diffraction pattern. As $\sin\theta$ can only lie in the range $-1 \leq \sin\theta \leq 1$, the only possible values of $\sin\theta$ are thus 0, ±0.20, ±0.40, ±0.60, and ±0.80. The corresponding values of θ are 0, ±11.53°, ±23.58°, ±36.87°, and ±53.13°.

The diffraction grating is used in the laboratory for two purposes. One of these is to determine the wavelength of a monochromatic source of light. In this use, a monochromatic source of light is made to fall normally on the grating. Using a telescopic eyepiece that moves over an angular scale, the angles for the diffraction maxima can be measured. Knowing the grating spacing, the wavelength of the light can be determined. The other use is in splitting a light beam into its component colours. The beam is made to fall normally on the grating. Since the different colours have different wavelengths, the secondary maxima will fall at different angles. By measuring these angles, we can determine the wavelengths of the different components.

4.7 POLARISATION OF LIGHT

Interference and diffraction are not the only phenomena exhibited by waves. A polarised wave is one that has vibrations in only one direction. Unlike interference, polarisation only takes place for transverse waves. As vibrations in longitudinal waves are in the same direction as the direction of travel of the waves, longitudinal waves are in principle already polarised by their very nature.

In light waves, the electric and magnetic vibrations take place in all directions at right angles to the direction of travel of the wave. It is possible to obtain light in only one direction. Such light is said to be **plane-polarised.** Several methods exist for producing plane-polarised light. The most important of these make use of refraction through certain types of crystal. A crystal such as calcite has the property that a beam of light falling upon it is split up into two beams, each of which contains light that is plane-polarised, the two polarised beams producing vibrations at right angles to each other. This property of producing two beams is known as **double refraction**. In some crystals, tourmaline, for example, in addition to the double refraction, the crystal absorbs both the refracted beams, but one beam is absorbed more than the other. If the crystal is thick enough, only one beam will emerge from the other side of the crystal. Such crystals are said to be **dichroic**, and this phenomenon is called **dichroism**.

Polarisation is used in the manufacture of Polaroid sunglasses. These glasses produce polarised light, thereby reducing the intensity of the transmitted light. Many of these glasses use dichroic crystals, or sheets of plastic packed with dichroic crystals.

EXERCISES

1. Young's double slit experiment is performed using light of wavelength 589 nm. Fringes are measured on a screen 1.20 m away. The distance between the perpendicular bisector and the twentieth bright fringe is measured to be 11.8 mm. Calculate

   (i) the separation of the slits,
   (ii) the separation of the fringes on the screen

2. The slits in a Young's double slit experiment are spaced 0.30 mm apart, and the observation screen is 60 cm from the slits. Calculate the distance between the seconds and fourth dark fringes in the interference pattern when light of wavelength 589 nm is used as the illuminating source.

3. A strip of foil of thickness 0.10 mm thick is used to separate two glass plates 10.0 cm long, the other ends of the plates being in contact. The plates are illuminated from directly above by light of wavelength 540 nm, and interference fringes are formed. Determine the number of interference fringes that would be observed per centimetre.

4. Parallel rays of wavelength 589 nm fall on a Young's double slit. On an observation screen placed 1.50 m away from the double slit, it is found that the separation of the fringes is 2.0 mm. Calculate the slit separation.

5. Two radio antennas radiating in phase are placed at points A and B, 200 m apart, as shown in **Fig. 4.18**. The radio waves are coherent, and have a frequency of 5.00 MHz. A radio receiver is moved along the line BC, perpendicular to the line joining the antennas. If the receiver is not to be moved too far away, at what distances from B will there be constructive interference from the two sources?

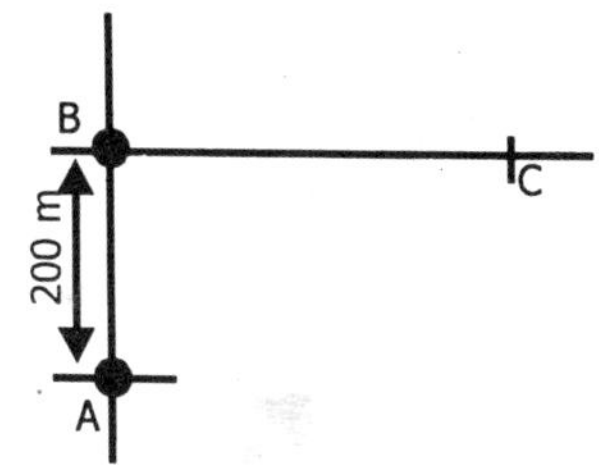

Fig. 4.18

6. Two rectangular flat glass plates touch at one edge, and are inclined to one another at a small angle. The plates are illuminated normally by monochromatic light of wavelength 5.89×10^{-7} m. Bright and dark fringes are produced in the field of view above the plates, and the fringe separation is measured. The light source is now changed to monochromatic light of wavelength 5.46×10^{-7} m. Determine the percentage change in the fringe separation.

7. Two equal glass plates of length 10 cm touch at one end whilst their other ends are separated by a piece of paper 0.02 mm thick. The system is illuminated by monochromatic light of wavelength 500 nm directly from above. Calculate the separation of bright and dark fringes seen in the field of view.

8. Young's interference fringes are produced using monochromatic light. If the slit separation is 1.0 mm, and the fringes are 0.59 mm apart on a screen placed 1.0 m from the slits, calculate
   (i) the wavelength of the light used,
   (ii) the angle subtended by the fourth bright fringe at the centre of the slits, using light of the wavelength in (i).

9. A car is approaching on a straight road on a pitch-dark night. The headlights of the car are 100 cm apart. How far from an observer will the car be when the observer can be sure that two lights are seen instead of one? Consider that the aperture of the iris of the eye is 0.5 cm, and the effective wavelength of the light as 550 nm.

10. A diffraction grating with 4000 lines per centimeter is illuminated normally by light that has two wavelengths, 484 nm, and 605 nm. At what angles to the incident beam will maxima due to the two wavelengths coincide?

11. The visible region of the electromagnetic spectrum contains wavelengths from 380 nm to 700 nm. Find the approximate number of lines per centimetre of the diffraction grating that will spread this spectrum over 15° in the second order.

12. A diffraction grating with 4000 lines per centimetre is illuminated normally with light of wavelength 500 nm. Determine the directions in which the maxima will occur.

13. Light of frequency 5.20×10^{14} Hz passes normally through a parallel-sided block of thickness 5 mm, and refractive index 1.50. Calculate the number of wavelengths in the glass.

14. Two identical wave trains of wavelength 550 nm in vacuum travel identical paths in reaching a point, except that one has travelled through 5 cm of vacuum while the other was travelling through 5.0 cm of air of refractive index 1.0003. Calculate the phase difference between the two wave trains when they arrive at the final point.

15. Two slits 0.5 mm apart are illuminated normally by a parallel beam of light of wavelength 589 nm, and the interference pattern is observed on a screen 1.0 m away from the slits. Calculate
   (i) the separation of the fringes of the interference pattern,
   (ii) the distance from the central fringe on the screen to the sixth dark fringe.

17. When blue light of wavelength 426 nm in vacuum passes through a particular biological specimen, the light splits into two beams plane-polarised at right angles. The refractive indices for these beams are 1.316 and 1.332. Calculate the velocities and wavelengths of the two beams in the specimen.

5
CHAPTER

Geometrical Optics 1
(Reflection and Refraction at Plane Surfaces)

5
CHAPTER

Geometrical Optics 1
(Reflection and Refraction at Plane Surfaces)

From our study of physical optics in the last chapter, we saw that light is a form of wave motion. We also studied the properties peculiar to this wave motion, as distinct from, for instance, sound waves. Using Huygens' construction of secondary wavelets, we constructed the wavefronts from a point source propagating in space, and indicated the relationship between a ray of light and the wavefront propagating in the medium. In this chapter and the next, we trace the paths of rays as the light travels from one medium to another, and through certain types of media. This study is called geometrical optics. In geometrical optics, we deal with phenomena that occur when the size of the object or apparatus is many orders of magnitude greater than the wavelength of the light. Thus we do not expect interference and diffraction effects.

When light travels within a medium, or from one medium to another, there are two important phenomena: reflection, and refraction. In both of these, the direction of the light ray is altered. Surfaces encountered by the light can be straight or curved. Accordingly, we study the reflection and refraction of light in two parts: at plane surfaces, and at curved surfaces. The present chapter focuses on plane surfaces, whilst the next is devoted to curved surfaces.

5.1 REFLECTION AT PLANE SURFACES

If a ray of light strikes a plane surface, some of the light is absorbed and some reflected. We extend our initial study on the reflection of waves to cover the reflection of light rays. The laws governing reflection are as follows:

1. ***the incident ray, reflected ray, and the normal to the surface at the point of incidence all lie in the same plane;***

2. ***the angle of incidence (i.e. the angle between the incident ray and the normal) is equal to the angle of reflection.***

5.2 REFRACTION AT PLANE BOUNDARIES

A ray of light passing from one medium to another is bent from its original direction at the interface of the two media. This phenomenon is known as **refraction**. **Fig. 5.1** shows a ray of light passing from one medium to another, the two media being separated by a plane boundary. At the plane interface, some of the incident light is reflected whilst some of it is refracted into the second medium.

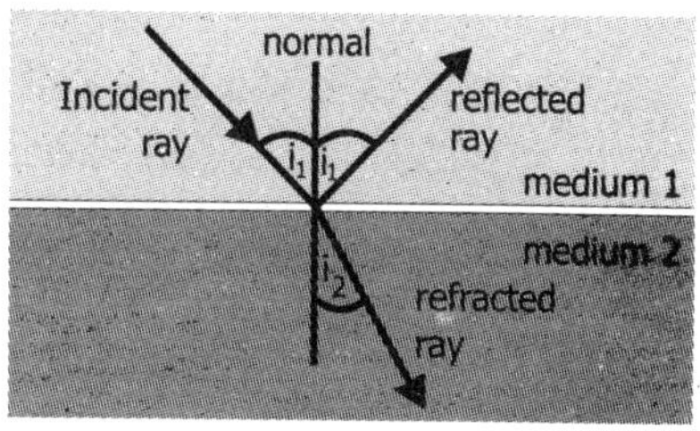

Fig. 5.1 Refraction at a plane boundary between two media.

The laws governing the refraction are as follows:

1. ***the incident ray, refracted ray, and the normal to the interface at the point of incidence all lie in the same plane;***

2. ***the ratio of the sine of the angle of incidence to the sine of the angle of refraction is a constant.***

The second law is called Snell's law, in honour of its inventor. The constant is called the **relative refractive index** between the two media. In symbols, we have

$$\frac{\sin i_1}{\sin i_2} = {}_1n_2, \qquad (5.1)$$

where ${}_1n_2$ is the relative refractive index of the two media, for a ray of light travelling from medium 1 into medium 2. Often, if the prefix and suffix are missing, this signifies that the incident ray is in air (or vacuo), and the refractive index is known as the **absolute** refractive index of the medium. The absolute refractive index (or simply the refractive index) is a constant, and a property of the material. This property depends on the colour of the light used, and on the material's internal structure. **Table 5.1** lists the absolute refractive indices of some common substances for white light.

Table 5.1 Absolute refractive indices of some common substances for white light

Substance (Solids)	Refractive index, n	Substance (Glasses)	Refractive index	Substance (Liquids at 20 °C)	Refractive index
Ice	1.309	Crown glass	1.52	Methanol	1.329
Polystyrene	1.49	Light flint	1.58	Water	1.333
Rocksalt	1.544	Medium flint	1.62	Glycerine	1.473
Quartz	1.544	Dense flint	1.66	Benzene	1.501
Zircon	1.923				
Diamond	2.417				
Rutile	2.62				

As we saw in chapter 1, we can relate the refractive index to the speed of the wave (in this case light) in the medium. Equation (5.1) can then be written in the form

$$\frac{\sin i_1}{\sin i_2} = {}_1n_2 = \frac{\text{speed of light in medium 1}}{\text{speed of light in medium 2}}$$

$$= \frac{\text{speed of light in medium 1}}{\text{speed of light in vacuo}} \div \frac{\text{speed of light in medium 2}}{\text{speed of light in vacuo}}$$

$$= \frac{n_2}{n_1},$$

where n_1 and n_2 are the absolute refractive indices of the two media – i.e. the refractive index with ray incident in vacuo, or air. The general form of Snell's law is as follows:

$$n_1 \sin i_1 = n_2 \sin i_2, \tag{5.2}$$

also written as

$$n \sin i = \text{constant}. \tag{5.3}$$

Example 5.1

An illuminated object is located at the bottom of a swimming pool containing water of refractive index n, to a depth D.

(a) Draw the paths of rays from the object entering the eye of the observer from directly above.

(b) Show that the object is displaced by an amount t, given by the expression $t = D\left(1 - \frac{1}{n}\right)$

(c) Given that the refractive index of water is 1.33, and that the pool is 2.50 m deep, determine the apparent depth of the pool.

Solution

(a) **Fig. 5.2** is a ray diagram showing rays from the illuminated object to an observer directly above. Rays from the object refracted into the air are bent away from the normal, and appear to come from an image displaced by distance t.

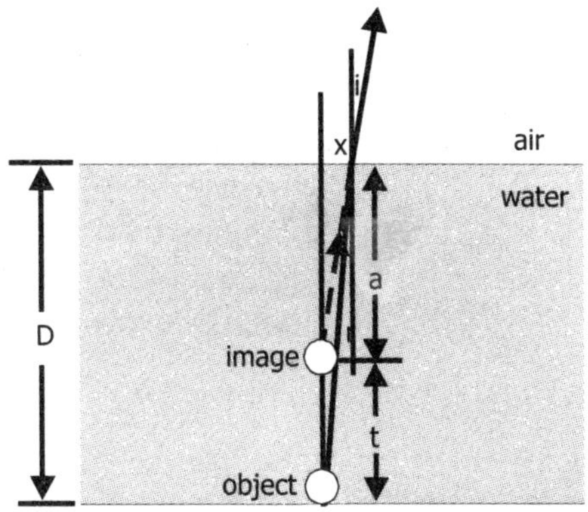

Fig. 5.2 Displacement of object.

(b) The rays entering the eye directly above are so close to the normal that the angles of incidence and refraction at the water-air surface are very small. From the geometry, we have

$$x = D \tan r = a \tan i \tag{1}$$

For the small angles considered here, we can write $\tan r = \sin r$, and $\tan i = \sin i$, i and r being the angles of incidence and refraction, respectively. Equation (1) becomes $D \sin r = a \sin i$, giving $a = \frac{D}{n}$, where n is the refractive index of the medium (water). The image is displaced from the object by t = D − a. Thus, we have $t = D\left(1 - \frac{1}{n}\right)$

(d) The apparent depth $a = D - t = D\left[1 - \left(1 - \frac{1}{n}\right)\right] = \frac{D}{n} = \frac{2.50 \text{ m}}{1.33} = 1.88 \text{ m}$.

5.3 DEVIATION

The term **deviation** is used as a measure of how far a ray of light had been taken off its original direction as a result of reflection or refraction. Deviation is expressed in degrees. The deviation angle is the angle between the final direction of the ray and the incident direction. In **Fig. 5.3** are shown deviations produced on reflection and refraction. It is seen from these diagrams, that for reflection, the angle of deviation D is given by

$$D = 180 - 2i, \qquad (5.4)$$

and for refraction

$$D = i - r, \qquad (5.5)$$

all angles being expressed in degrees.

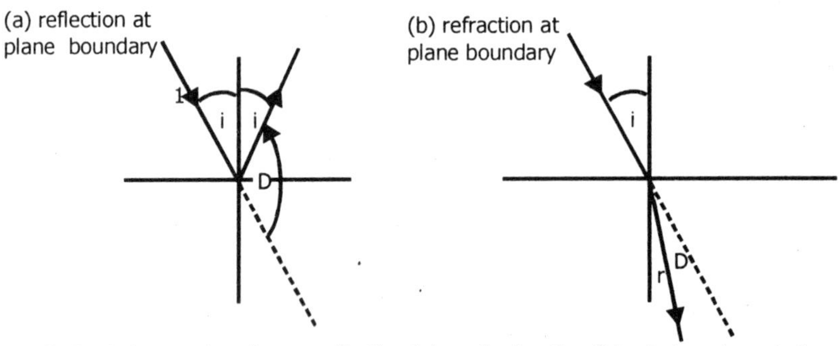

Fig. 5.3 Deviation produced upon reflection (a), and refraction (b), at plane boundaries.

When multiple reflections and refractions occur, the deviations are combined

(i) by addition if they are in the same sense (clockwise or counterclockwise), and
(ii) by subtracting the deviations if they are in opposite sense.

5.4 TOTAL INTERNAL REFLECTION

When a ray of light meets the plane boundary between two media, a fraction of the light intensity is reflected at the boundary whilst another fraction is refracted into the other medium. For a ray travelling from an optically less dense medium into a denser medium, the ray is refracted towards the normal, and the angle of refraction smaller than the angle of incidence. When light travels from a denser medium to a less dense medium, the refraction is away from the normal, and the angle of refraction in this case is greater than the angle of incidence. This, however, only takes place for certain angles of incidence.

Consider a ray of light incident in a denser medium and travelling into a less dense medium – medium 2. At small angles of incidence, there is a reflected ray, and a refracted ray. If the angle of incidence is gradually increased the refracted ray becomes weaker and weaker. Beyond a certain angle of incidence the refracted ray disappears, and we see a strong reflected ray in medium 1. The light is said to be **totally internally reflected**. This phenomenon is called **total internal reflection**. The angle beyond which total internal reflection occurs is known as the **critical angle**. **Fig. 5.4** shows the paths of rays of light from an object passing from a denser medium to a less dense medium.

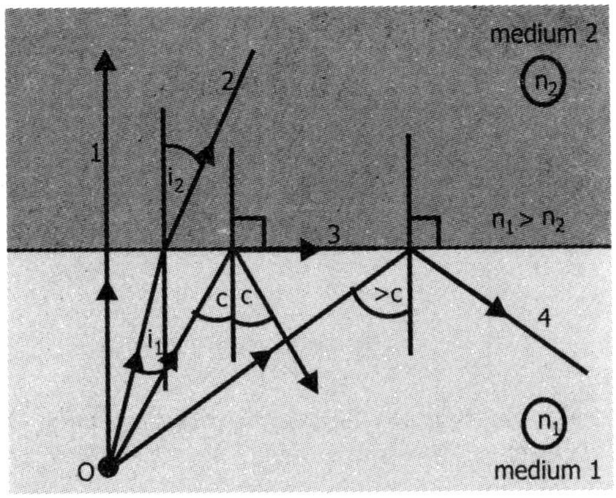

Fig. 5.4 Rays of light incident in a denser medium and refracted into a less dense medium. Total internal reflection takes place for angles of incidence greater than the critical angle.

Rays 1 and 2 produce refracted rays in medium 2. Ray 3 is incident at the critical angle c, producing a refracted ray with angle 90° to the normal. For ray 4, the angle of incidence is greater than the critical angle. Total internal reflection occurs, and there is no refracted ray. The ray is totally reflected internally at the boundary of medium 1. Applying Snell's law for ray 3, we have $n_1 \sin c = n_2 \sin 90^\circ$, giving

$$\sin c = \frac{n_2}{n_1}. \tag{5.6}$$

Example 5.2

Fig. 5.5 shows a cross section through a glass block in the form of a half-cylinder, mounted on a spectrometer table, such that a ray of light from a monochromatic source is always incident normally on the curved surface of the block as the table is rotated. It is found that when the face AOB is clean, the critical angle on the face AB is 39.6°. A small quantity of cooking oil is now smeared on the face AOB, and the critical angle found to be 69.4°. Calculate the absolute refractive indices of the glass and the cooking oil.

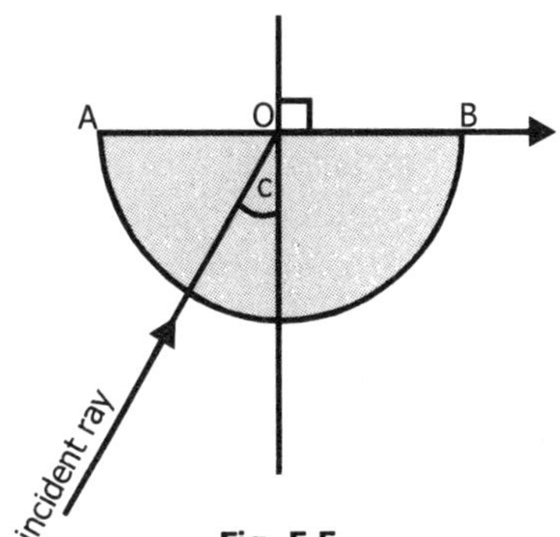

Fig. 5.5

Solution

Fig. 5.6 shows the passage of light incident on the face AOB at the critical angle in the two cases. From the data given, $c_1 = 39.6^{o}$, $c_2 = 69.4^{o}$.

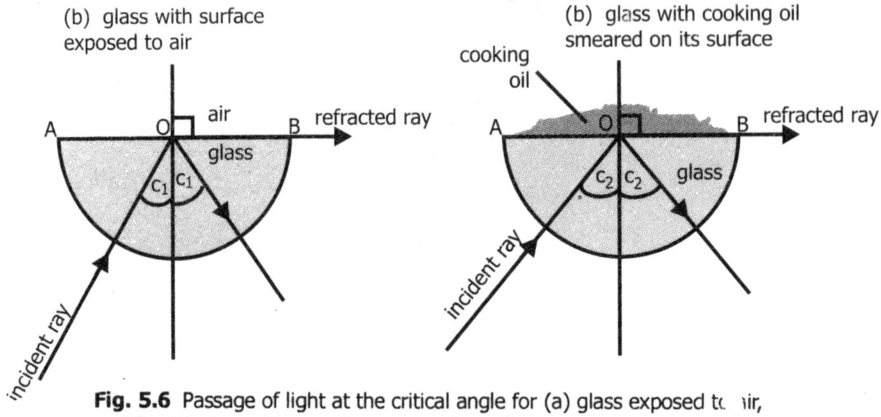

Fig. 5.6 Passage of light at the critical angle for (a) glass exposed to air, and for (b) glass with sample of oil on its surface.

When the glass is exposed to air, we write

$$\sin c_1 = \frac{1}{n_g}, \qquad (1)$$

where n_g is the absolute refractive index of the glass, and the refractive index of air has been taken as unity. If n_o is the refractive index of the oil, we have

$$\sin c_2 = \frac{n_o}{n_g}, \qquad (2)$$

Substituting the given data into equations (1) and (2), we get $n_g = 1.569$, and $n_o = 1.469$.

5.5 OPTICAL FIBRES

Example 5.2 above illustrates one laboratory use of total internal reflection – that of determining the refractive index of some transparent liquid of which only a small quantity is available. Well-known prism binoculars add to the growing list of uses. Modern communication needs have added yet another use of total internal reflection. This use is in optical fibres. Optical fibres are used in telephone communications, computer-networking applications, and in medical instruments called endoscopes, used by doctors and surgeons to examine internal organs of the body.

In telecommunications and computer applications, the advantages of a fibre-optic cable include the possibility of transmitting information at a high speed, and the fact that being an insulator, a fibre-optic cable is immune to electrical interferences. Since wave crests occur at a rate depending on the frequency, light of high frequency can be made to transmit data at a much faster rate than by any other means. This is the principle used in telecommunications employing a laser beam, or an infrared beam.

An optical fibre consists of a two types of glass, of different refractive indices. These are referred to as the **core**, and **cladding**. **Fig. 5.7** is a sectional view of a segment of fibre-optic cable. The diagram is highly magnified for purposes of clarity.

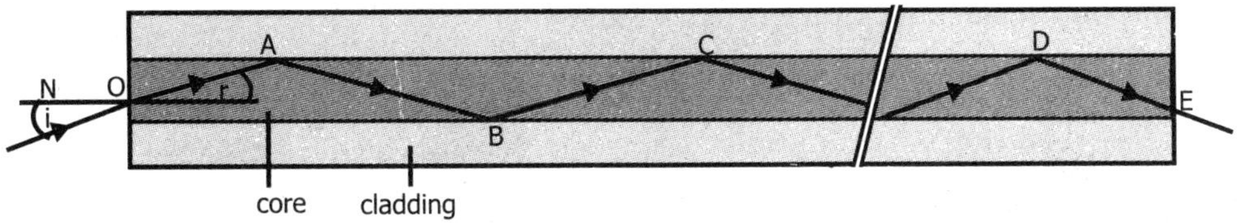

Fig. 5.7 Fibre-optic cable arrangement showing the core and cladding. A ray of light incident at O is totally internally reflected at the core-cladding boundaries before exiting the cable.

A typical fibre-optic cable would be about 50 μm in diameter, with a core about 10 μm in diameter. The core is made of glass of higher refractive index than the cladding. A ray of light incident at an angle i at the entrance to the core would be refracted toward the normal N. The ray travels toward A, at which it gets totally internally reflected into the core. Similarly, total internal reflections occur at the B, C, etc., before the light exits the fibre at E. Once the light enters the fibre, it appears "trapped" in the core, and travels the whole length of the fibre until it finally leaves.

Knowing the absolute refractive indices of the core and cladding, it is possible to determine the minimum angle of incidence i in the air-core boundary for light to be totally internally reflected at the core-cladding interfaces. For this minimum angle, the ray meets the cladding interface at the critical angle c, as shown in **Fig. 5.8**. We hence write

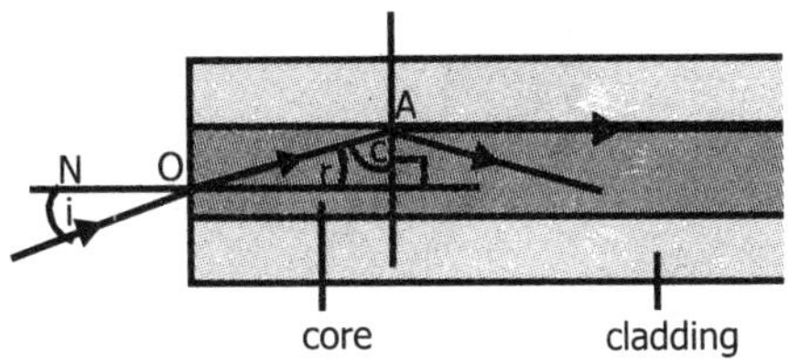

Fig. 5.8 Minimum angle of incidence in air at the core of a fibre-optic cable. The ray meets the cladding interface at the critical angle.

$$\sin i = n_1 \sin r, \text{ or } \sin r = \frac{\sin i}{n_1},$$

and, with c and r being complementary, we further have $\sin c = \frac{n_2}{n_1} = \cos r$, where n_1 and n_2 are the refractive indices of the core and cladding, respectively, and c is the critical angle at the core-cladding interface. From these relations, we get

$$\sin i = \sqrt{n_1^2 - n_2^2}\,. \qquad (5.7)$$

Example 5.3

A fibre-optic cable consists of a core made of glass of refractive index 1.62, surrounded by cladding material of refractive index 1.49. Calculate the minimum angle of incidence of a ray of light incident in air on the core material.

Solution

This example illustrates the use of equation (5.7) above. With reference to **Fig. 5.8**, and using the given data, we have $\sin i = \sqrt{n_1^2 - n_2^2} = \sqrt{1.62^2 - 1.49^2}$. This gives the minimum angle of incidence in air as 39.5°.

In fibre-optics application, the source of the radiation is a pulsed laser or infrared beam. In today's digital communication, each pulse becomes a coded message, and a bundle of optical fibres can carry within it over tens of millions of messages over long distances at the speed of light. At the other end of the line is an infrared or laser receiver which receives the coded light pulses and transforms them into electrical signals. This way information can be transmitted and received over long distances.

5.6 REFRACTION BY PRISMS

A prism is a solid figure bounded by three rectangular faces. For our present purposes in optics, we take the solid to be made of transparent material. A section perpendicular to both the top edge and the rectangular base of the prism is referred to as a **principal section** of the prism. Any principal section is a triangle, as shown in **Fig. 5.9**. The angle marked A is known as the **refracting angle** of the prism.

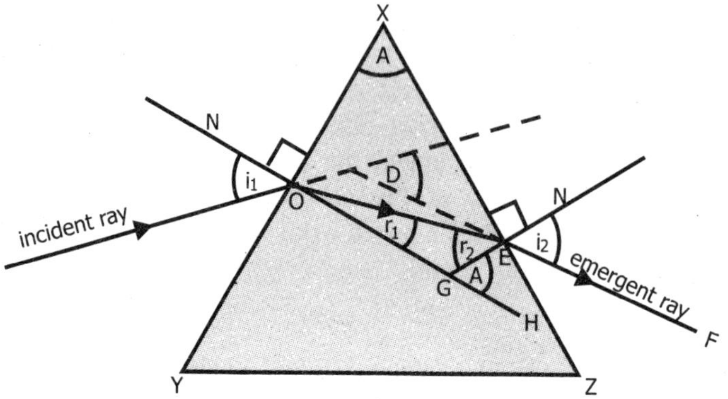

Fig. 5.9 Passage of light through a prism.

A ray of light is incident at O on the face XY. The angle of incidence i_1 is shown. The refracted ray travels along the path OE to the face XZ, at which it may emerge along EF. Refraction takes place at O and E, giving angles of refraction r_1, and r_2, respectively. The normal at O meets the normal at E at the point G. Thus by geometry, the angle EGH is equal to the refracting angle A.

The deviation produced at any face of the prism is the difference between the angle of incidence and the angle of refraction at that face. For the face XY, the deviation d_1 is given by

$$d_1 = i_1 - r_1.$$

Similarly, the deviation d_2 produced at the face XZ is given by,

$$d_2 = i_2 - r_2.$$

The overall deviation D caused by the prism is then the sum of the deviations. Thus,

$$D = d_1 + d_2 = (i_1 + i_2) - (r_1 + r_2) \tag{5.8}$$

Also,

$$A = r_1 + r_2. \tag{5.9}$$

We note that at some angle of incidence, the refracted ray will meet the face XZ at an angle r_2 greater than the critical angle, for which the light will be totally internally reflected in the face XZ. In this event no light emerges through the face XZ.

5.6.1 Symmetrical passage through the prism

A common experiment to determine the refraction properties of a glass prism involves varying the angle of incidence and measuring the corresponding angle of deviation. A typical plot of deviation against the angle of incidence is shown in **Fig. 5.10** (a).

The plot shows that as the angle of incidence is increased from small values, the angle of deviation decreases up to a point at which deviation is a minimum, beyond which the angle of deviation increases as the incidence angle is further increased. At minimum deviation, the angle of incidence is equal to the angle of emergence, and the two angles of refraction are also equal. This condition is often described by saying that the light passes symmetrically through the prism. This is shown in **Fig 5.10** (b).

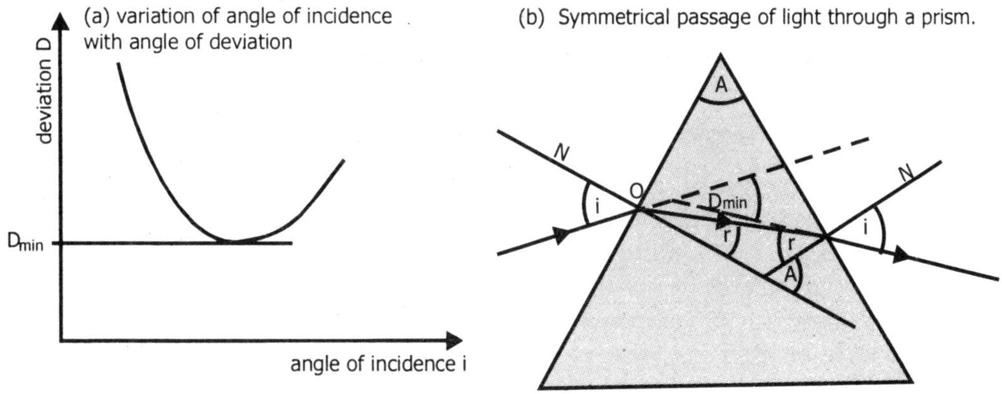

Fig. 5.10 Minimum deviation and symmetrical passage of light through a prism.

At minimum deviation D_{min}, we can set $i_1 = i_2 = i$, and $r_1 = r_2 = r$. Equations (5.6) and (5.7) are then transformed to read $D_{min} = 2(i - r)$, and $A = 2r$. From the latter two relations, $i = \frac{1}{2}(A + D_{min})$, and $r = \frac{1}{2}A$. If the prism is placed in air or vacuo, the absolute refractive index n of its material is given by

$$n = \frac{\sin \frac{1}{2}(A + D_{min})}{\sin \frac{1}{2}A}. \qquad (5.10)$$

If a prism with refractive index n_1 is surrounded by a medium with refractive index n_2, equation (5.10) is modified to read

$${}_1n_2 = \frac{n_1}{n_2} = \frac{\sin \frac{1}{2}(A + D_{min})}{\sin \frac{1}{2}A}. \qquad (5.11)$$

Example 5.4

A glass prism of refracting angle 60° and refractive index 1.72 is immersed in a liquid of refractive index 1.29. Calculate the angle of minimum deviation for a parallel beam of light passing through the prism.

Solution

This is solved by a direct application of equation (5.9). Representing the refractive index of the prism material by n_1, and that of the surrounding liquid by n_2, we write

${}_1n_2 = \frac{n_1}{n_2} = \frac{\sin \frac{1}{2}(A + D_{min})}{\sin \frac{1}{2}A}$. From the data given, $n_1 = 1.72$, $n_2 = 1.29$, and $A = 60°$.

With these values, we get $\frac{1.72}{1.29} = \frac{\sin \frac{1}{2}(60^\circ + D_{min})}{\sin 30^\circ}$. Thus, $D_{min} = 23.62^\circ$

5.6.2 Small angle prism

By definition, a small angle prism is one with a small refracting angle, as shown in **Fig. 5.11**. For light to emerge along the face XZ, the angles of incidence and refraction must be very small – i.e. the ray must be travelling very close to the normal at both the incident and emergent faces of the prism. Using the method of the previous section, we can write an expression for the deviation D produced from the prism as

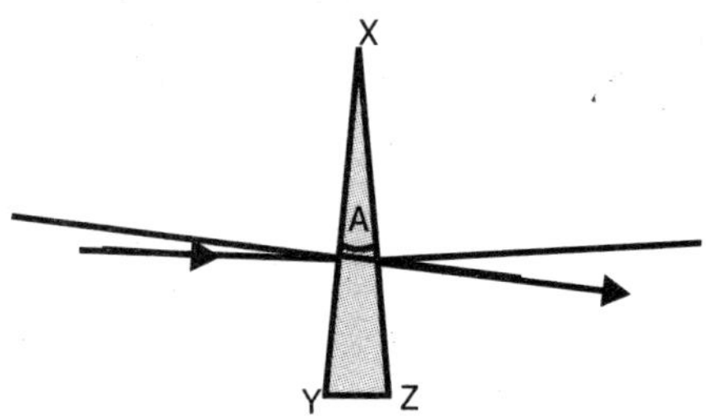

Fig. 5.11 Small angle prism

$$D = (i_1 + i_2) - (r_1 + r_2), \tag{5.12}$$

With small angles of incidence and refraction, we have the following approximate relations: $i_1 \approx nr_1$, $i_2 \approx nr_2$, and $r_1 \approx r_2 = r$. This gives $A = 2r$. Putting these into equation (5.10), we get

$$D = 2nr - 2r = 2r(n-1).$$

Substituting for A, we get the expression for the deviation produced as

$$D = (n-1)A. \tag{5.13}$$

The deviation produced is therefore independent of the angle of incidence.

5.6.3 Dispersion

The refractive index of a medium depends on the wavelength or colour of the light passing through the medium. If white light, for example, passes through a medium, the component colours suffer different deviations, the larger wavelength components being deviated more than the smaller wavelength components. The separation of the wavelength components of a beam of light consisting of more than one colour is called **dispersion**. The study of dispersion focuses on three colours: the red, with wavelength 656 nm, the blue colour with wavelength 587 nm, and the yellow colour, with wavelength 599 nm. Triangular prisms are very effective in producing dispersion. We define three parameters relating to dispersion through a prism as follows:

(i) ***the dispersion angle is the angle between the red and blue colours emerging from the prism,***

(ii) ***the dispersive power is the ratio of the dispersion angle to the mean deviation, the mean deviation being that occurring for yellow light,***

(iii) ***the constringence is the reciprocal of the dispersive power.***

Consider a beam of white light falling on one face of a small angle prism. If n_r, n_b, and n_y, are the refractive indices of the prism for red, blue, and yellow light, respectively, the corresponding deviations D_r, D_b, and D_y, are given by

$$D_r = (n_r - 1)A \quad \text{(for red colour)}, \tag{5.14}$$

$$D_b = (n_b - 1)A \quad \text{(for blue colour)}, \tag{5.15}$$

and

$$D_y = (n_y - 1)A \quad \text{(for yellow colour)}. \tag{5.16}$$

From equations (5.14), (5.15), and (5.16), the dispersion angle α can be written as

$$\alpha = D_b - D_r = (n_b - n_r)A \tag{5.17}$$

The mean deviation D_{mean} is that for yellow light:

$$D_{mean} = (n_y - 1)A . \tag{5.18}$$

The dispersive power ω is given by

$$\omega = \frac{\alpha}{D_{mean}} = \frac{(n_b - n_r)}{(n_y - 1)}, \tag{5.19}$$

and the constringence of the medium is given by ν, where

$$\nu = \frac{1}{\omega} = \frac{(n_y - 1)}{(n_b - n_r)}. \tag{5.20}$$

5.6.4 Achromatic and direct vision prisms

Small angle prisms are used in many optical devices – e.g. binoculars, viewfinders of cameras, and in spectrometers. Often in these applications white light, or a source with more than one colour component is used. It is desired that the image be produced without dispersion effects. This is done by using two small-angle prisms put together in such a way to eliminate or minimise dispersion. This arrangement produces what is called an **achromatic**, or **non-dispersive** prism. **Table 5.2** lists the refractive indices for colours in white light in crown and flint glass.

Table 5.2 Refractive indices of crown and flint glass for the colours in white light.

Approximate colour	Wavelength in vacuum (nm)	Refractive index in crown glass	Refractive index in flint glass
Red	660	1.520	1.662
Orange	610	1.522	1.665
Yellow	580	1.523	1.667
Green	550	1.526	1.674
Blue	470	1.531	1.684
Violet	410	1.538	1.698

An achromatic prism produces deviation but no dispersion. The arrangement is such that the dispersion produced by one prism is equal and opposite to that produced by the other prism. If we let A and B represent the refracting angles of the two prisms 1, and 2, in the combination, we write

$$(n_{b1} - n_{r1})A = (n_{b2} - n_{r2})B, \tag{5.19}$$

where the suffices 1 and 2 refer to the two prisms of the combination, n_b and n_r being the refractive indices of the prism materials for blue and red light, respectively. From (5.19), the ratio of the refracting angles of the two prisms is given by

$$\frac{A}{B} = \frac{(n_{b2} - n_{r2})}{(n_{b1} - n_{r1})}. \tag{5.20}$$

The deviation D produced for the blue and red colours is the same for both prisms.

$$D = (n_{b1} - 1)A - (n_{b2} - 1)B. \tag{5.21}$$

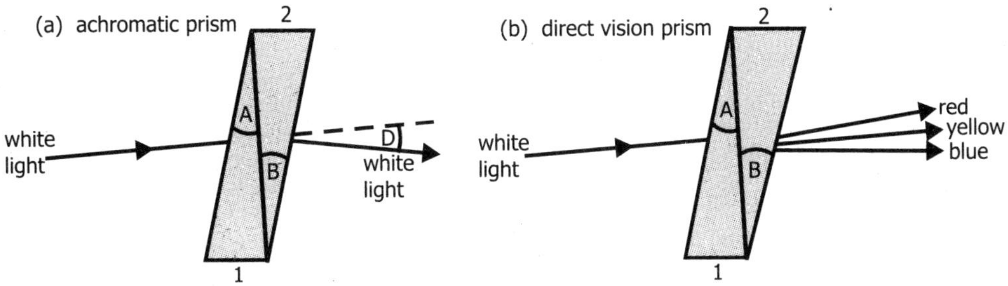

Fig. 5.12 Paths of rays through achromatic and direct vision prism combinations

In other spectroscopic devices, it is desired that the image be viewed directly without any deviation of the light. Again, two small-angle prisms are used to achieve this effect – an arrangement known as the **direct vision** prism. **Fig 5.12** (a) and (b) show these two arrangements. In both, the two prisms are oriented opposite to each other. In direct vision prisms, the deviations produced for yellow light by the two prisms are the equal and opposite. Using the same symbols as before, we have

$$(n_{y1} - 1)A = (n_{y2} - 1)B, \tag{5.22}$$

from which we can deduce the ratio of the refracting angles of the prisms. In direct-vision prisms, there is still some dispersion of the red and blue colours.

EXERCISES

1. If the speed of light in vacuum is 3.0 x 10^8 m/s, and the refractive index of water is 1.33, calculate the speed of light in water.

2. The speed of light is 50% larger in material A than it is in material B. Determine the ratio of the refractive indices of the two materials.

3. A layer of oil, refractive index 1.45, floats on the surface of an unknown liquid. A ray of light shines from the oil into the liquid. The angles of incidence and refraction are 65° and 53°, respectively. Calculate the refractive index of the unknown liquid.

4. Red light is incident on a block of glass. The refractive index for the red light is found to be 1.67, and the angle of refraction is 22.6°. Green light is also incident on the glass at the same angle of incidence. If the angle of refraction for the green light is 21.9°, calculate the ratio of the refractive indices of the material for the green and red light.

5. If the absolute refractive indices of glass and water are 1.52, and 1.33, respectively, find the relative refractive index between water and glass for
   (i) a ray incident in water,
   (ii) a ray incident in glass.

6. A ray of light strikes a parallel-sided glass plate of refractive index 1.50, at an angle of incidence of 45°.
   (a) Calculate the angle of refraction in the glass.
   (b) If the glass plate is now covered by a thin layer of transparent oil of refractive index 1.2, calculate the angles of refraction in the oil and in the glass.

7. A prism of refracting angle 45° has refractive index 1.50 for yellow light. Calculate
   (i) the angle of incidence of the ray which suffers minimum deviation,
   (ii) the smallest possible angle of incidence of a ray that can pass through the prism without being totally internally reflected.

8. For a glass sample, the refractive indices for red, blue, and yellow light are respectively 1.5210, 1.5320, and 1.5250. Calculate
   (i) the dispersive power and constringence of the material,
   (ii) the dispersion angle of a thin prism of refracting angle 7°, made of this glass.

9. Horizontal beams of red light of wavelength 660 nm (in vacuum) and violet light of wavelength 410 nm (in vacuo) are incident on a crown glass prism in the direction shown in **Fig. 5.13**. Determine the angle of refraction for each ray as it emerges from the prism. You may use data from **Table 5.2**.

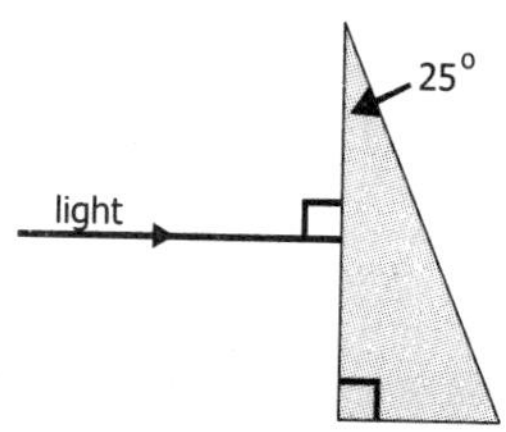

Fig. 5.13

10. Yellow light of approximate wavelength 589 nm strikes crown glass at an angle of incidence of 45°. Blue light strikes a piece of flint glass at the same angle of incidence. Calculate the angle of refraction of the blue light in the flint glass. You may use data from **Table 5.2**.

11. **Fig. 5.14** shows a ray of light incident in the direction shown, at the plane boundary between two types of glass of refractive indices n_1 and n_2.

(a) For an angle of incidence of 30°, calculate the angle of refraction in the second glass plate.

(b) Determine the critical angle for a ray of light travelling from the top plate to the bottom plate shown.

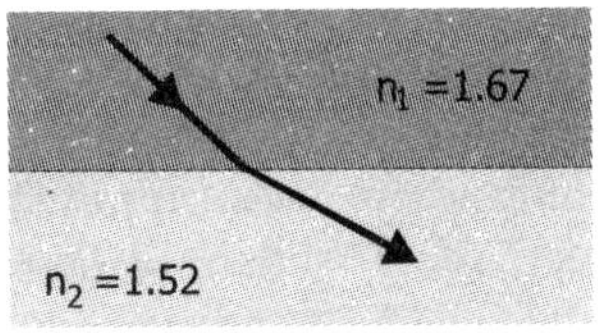

Fig. 5.14

12. **Fig. 5.15** shows a right-angled glass prism placed in air, at the shortest face of which light is incident normally. The ray of light travels through the glass, and is totally internally reflected at the critical angle at the boundary between the glass and a small quantity of liquid smeared on the longest face of the prism. Calculate the refractive index of the liquid, given that the refractive index of the glass is 1.52.

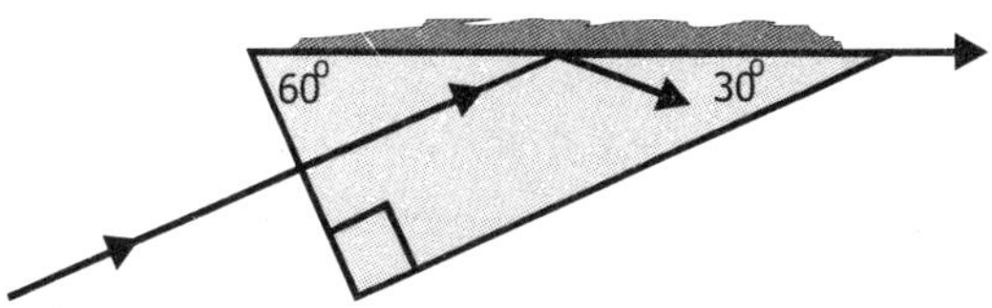

Fig. 5.15

13. An equiangular glass prism of refractive index 1.67 is immersed in a tank of water of refractive index 1.33. Calculate the angle of a ray of light incident in water on one face of the prism such that the light can just emerge from the opposite face.

14. The critical angle for total internal reflection at the interface of a certain liquid and air is 37°. A ray of light is incident at an angle of 30° at the interface between the liquid and air. Determine the angle of refraction of the ray when
(i) it is incident in air,
(ii) it is incident in the liquid.

15. A point source of light is placed at a point 10.0 cm below the surface of some light oil of refractive index 1.473. Calculate the diameter of the largest circle at the surface through which the light can emerge.

16. A ray of light passes symmetrically through an equiangular prism made of material of refractive index 1.60. Calculate
(i) the angle of incidence,
(ii) the angle of minimum deviation.

17. A ray of light is incident in air at an angle of 40° to the normal on one face of an equiangular glass prism of refractive index 1.60. Calculate the angle through which the ray will be deviated by the time it emerges from the opposite face of the prism.

18. The refractive indices of the core and cladding of an optical fibre are 1.60, and 1.40, respectively. Calculate
   (i) the critical angle at the core-cladding interface,
   (ii) the minimum angle of incidence in the air of a ray which enters the fibre and is then incident at the critical angle at the core-cladding interface.

19. A film of oil of refractive index 1.20 lies on the surface of water of refractive index 1.33. A ray of light is incident at an angle of 30° in the oil on the oil-water boundary. Calculate the angle of refraction in the water.

6

CHAPTER

Geometrical Optics 2
(Reflection and Refraction at Curved Surfaces)

6

CHAPTER

Geometrical Optics 2 (Reflection and Refraction at Curved Surfaces)

In the previous chapter we studied the effects of reflection and refraction at plane surfaces. We learned from these the phenomena of deviation and dispersion, and how these are applied in simple systems and optical gadgets. Many optical devices use curved surfaces to produce the reflection and refraction effects germane to the behaviour of these instruments. Our study of reflection and refraction will therefore be incomplete if limited to plane surfaces.

In this second chapter on geometrical optics, we study the reflection and refraction of light at curved surfaces. The reflection by curved mirrors give useful effects that can help us analyse the behaviour of optical systems based on these mirrors. There are two types of such mirrors: the converging (or concave) mirrors, and the diverging (or convex) mirrors. They are used either individually, or as combinations, to produce the desired results.

Refraction at curved surfaces is used to help us understand the behaviour of lenses. Putting together the desired combinations of lenses and mirrors can lead to an understanding of various optical instruments – the microscope, the camera, and the telescope – to name a few. These instruments are studied in the next chapter.

6.1 REFLECTION AT SPHERICAL MIRRORS

These are referred to as spherical mirrors because any such mirror can be considered to be part of a sphere. The behaviour of the mirror to light (and its use) depends on which surface has the silver coating. When the outer surface is silvered, the mirror is said to be **concave**, and we have a **concave mirror**. Coating the inner surface of the mirror produces a **convex mirror**. Spherical mirrors have the advantage over plane mirrors that, whereas plane mirrors always produce images the same size as the object, spherical mirrors can produce enlarged (i.e. magnified), as well as diminished images. This ability to produce images of a range of sizes has made spherical mirrors more versatile in their use. **Fig. 6.1** (a) and (b) show these two mirrors.

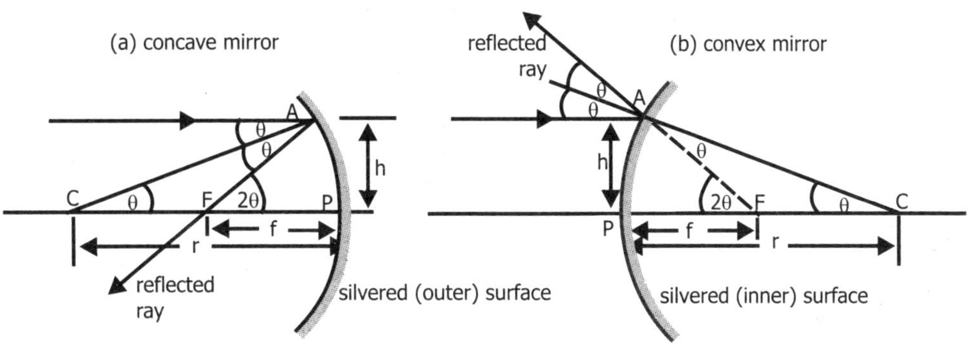

Fig. 6.1 Reflection by spherical mirrors: (a) concave mirror, and (b) convex mirror. Note the directions of the reflected rays from each surface.

The centre P of the spherical mirror is called the **pole**, whilst the **centre of curvature** C of the mirror is the centre of the sphere of which the mirror forms a part. The distance from the pole of the mirror to the centre of curvature is known as the **radius of curvature**. The line joining the pole and the centre of curvature is called the **principal axis**. Consider a ray of light incident at A parallel to the principal axis of a spherical mirror. Since a line from the centre of a sphere is normal to the sphere, we consider CA as the normal to the surface. The ray PA strikes the surface at an angle θ. By the law of reflection, this ray is reflected at an angle θ, as shown.

It is found that for a concave mirror, all rays parallel to the principal axis will upon reflection pass through a common point, whilst for a convex mirror, all rays parallel to the principal axis will appear to come from a common point when produced backwards. This common point is called the **focus** of the mirror, and the distance from the pole to the focus is known as the **focal length**.

In studying the behaviour of spherical mirrors, we consider rays that are parallel and close to the principal axis. The angles of incidence and reflection are taken as very small. For such small angles, we have

$$\tan 2\theta \approx 2\tan\theta = \frac{h}{f},$$

and

$$\tan\theta = \frac{h}{r}.$$

Thus,

$$2\frac{h}{r} = \frac{h}{f},$$

giving the relationship between the focal length and the radius of curvature as

$$r = 2f. \qquad (6.1)$$

Thus the radius of curvature of a spherical mirror is twice its focal length.

6.1.1 The mirror formula

Fig. 6.2 is a schematic diagram showing an object O placed on the principal axis of a concave mirror, forming an image I. The object is a distance u from the pole of the mirror, whilst the image is a distance v from the pole. We refer to these as the object and image distances, respectively.

From the geometry, $$\frac{h_O}{h_I} = \frac{u}{v} = \frac{u-r}{r-v}.$$

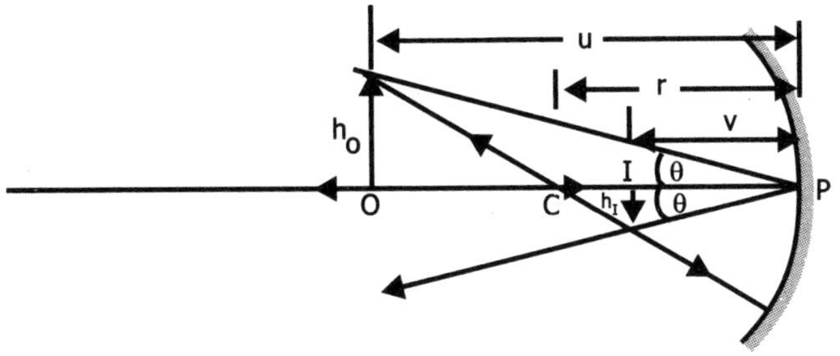

Fig. 6.2 Object and image distances in a concave mirror.

Bearing in mind that r = 2f, we transform the last equation to get $\frac{u}{v} = \frac{u-2f}{2f-v}$, or

$$\frac{1}{u}+\frac{1}{v}=\frac{1}{f}=\frac{2}{r}. \tag{6.2}$$

Equation (6.2) is the mirror formula, and relates the object and images with the focal length of the mirror. Whilst it had been derived assuming a concave mirror, the equation nonetheless holds for the convex mirror as well. In applying it, we adopt a **sign convention** for the object and image distances from the pole, and for the focal length of the mirror. With the reflecting surface taken as the origin, the following apply for the signs of the object and image distances, and of focal lengths:

1. ***Object distance is positive if the object is on the same side as the incoming light.***

2. ***Image distance is positive if the image is on the same side as the outgoing light.***

3. ***The radius of curvature is positive if the centre of curvature is on the same side as the outgoing light.***

The above sign convention is known as the **New Cartesian** convention. Another sign convention that had been in use is the **Real-Is-Positive** convention. This convention assigns a positive sign to all distances to real objects and images, and a negative sign to virtual object and image distances. Under this convention focal lengths and radii of curvature of concave mirrors are positive. This latter convention has been replaced by the Cartesian convention; we shall thus be using the Cartesian convention in our analyses.

6.1.2 Magnification

As mentioned earlier, the advantage of the spherical mirror over the plane mirror lies in the fact that the spherical mirror is capable of producing images of different sizes, whereas the plane mirror is limited to giving an image the same size as the object. We use the term magnification to describe the relative sizes of the image and the object.

Magnification is defined as the ratio of the size of the image to the size of the object.

If the magnification is greater than unity, we say that the image is magnified, and if the magnification is less than unity, the image is diminished. In symbols, the magnification m is given by

$$m = \left|\frac{h_I}{h_o}\right|, \tag{6.3}$$

where h_I and h_o are the heights of the image and object, respectively. As equation (6.3) expresses the ratio of linear dimensions, this magnification is commonly known as the **linear magnification**, to distinguish it from angular magnification, which we shall study in the next chapter under optical instruments. Images formed in spherical mirrors are either upright or inverted, as we can verify from **Fig. 6.2**. From the figure, we write

$$m = -\frac{v}{u}. \tag{6.4}$$

A negative sign indicates an inverted image, and a positive sign shows an upright image.

6.1.3 Images in concave mirrors

Based on certain simple rules, we can draw ray diagrams to locate the image in a spherical mirror for various distances of the object. These rules are as follows:

(i) Rays parallel to the principal axis will, after reflection, pass through (or appear to pass through) the focus,
(ii) A ray passing through the focus is reflected parallel to the principal axis,
(iii) A ray of light from the centre of curvature is reflected along the incident direction,
(iv) An image is formed when two or more rays intersect.

Using these rules, and a concave mirror as example, we draw ray diagrams for the formation of images as the object is moved from a point close to the pole of the mirror to a point very far away. **Fig. 6.3** gives these ray diagrams. The nature of the image produced (i.e. magnified, inverted, real, etc.) in each case is also given.

6.1.4 Image in convex mirror

The principles used for image formation in the concave mirror work equally well for the convex mirror. In this case parallel rays striking the mirror appear to diverge from the focus of the mirror. Also, the ray directed toward the centre of curvature of the mirror

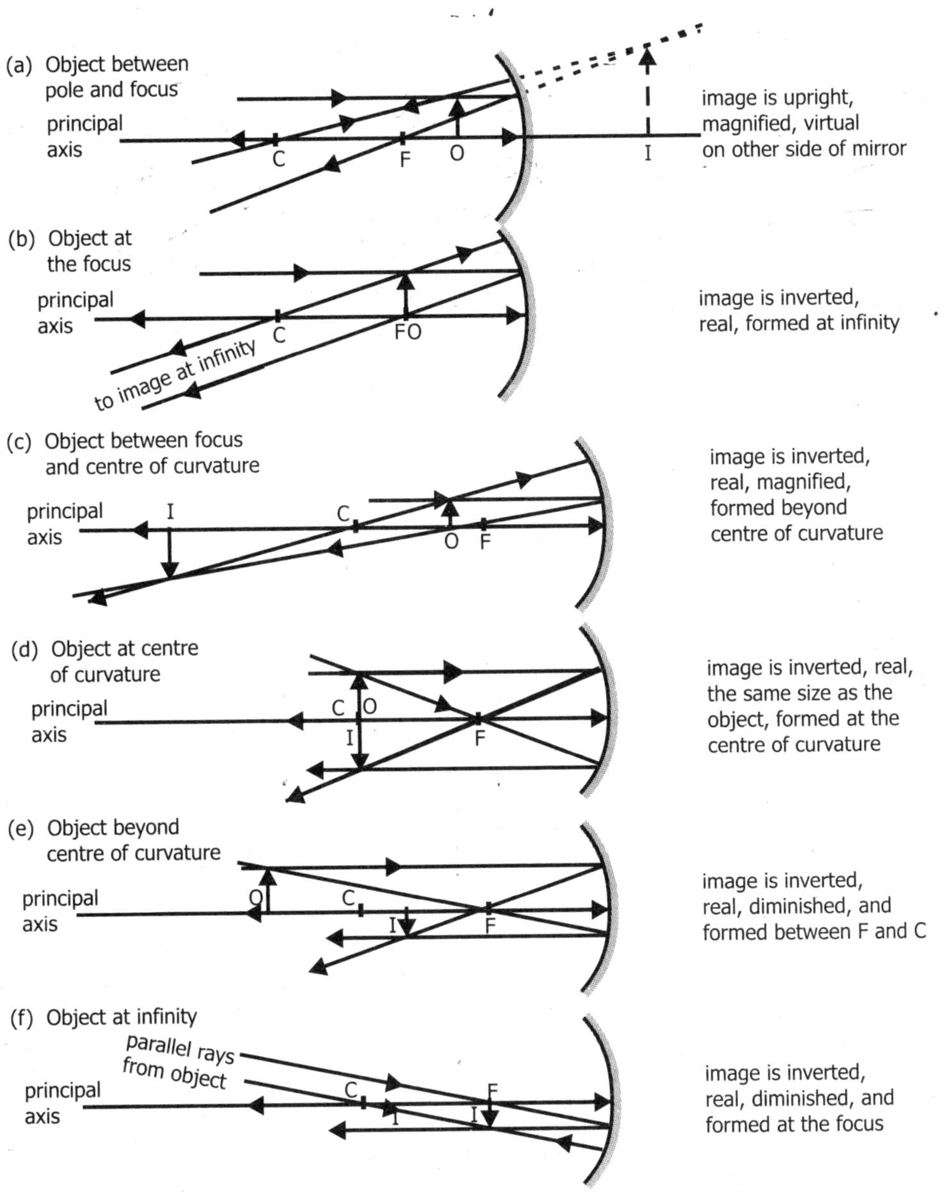

Fig. 6.3 Images formed in a concave mirror for various object distances.

strikes the mirror at an angle of 90°, and hence returns along the same path. There is only one case we need to consider. The nature of the image is independent of the position of the object. The ray diagram is shown in **Fig. 6.4**.

The image formed in a convex mirror is erect, diminished, virtual, and on the opposite side of the mirror as the object. While the equations remain the same as those for the concave mirror, the sign convention is most important in distinguishing the two. The following numerical examples serve to illustrate the applications of these equations.

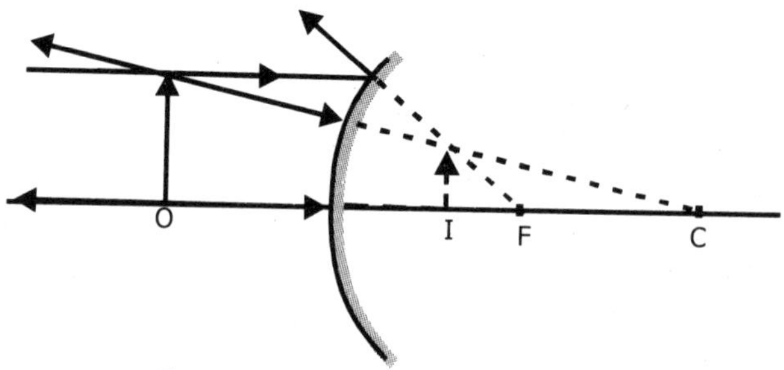

Fig. 6.4 Image formed in a convex mirror.

Example 6.1

An optical pin of height 2 cm is placed at a distance of 40 cm in front of a concave mirror of focal length 15 cm. Determine
(i) the position of the image from the pole of the mirror,
(ii) the height of the image produced.

Solution

Referring to **Fig. 6.3**, it is seen that the object in this question is placed at a point beyond the centre of curvature. We use the mirror equation to find the image distance, and hence the magnification. From the latter, we calculate the height of the image.

(i) From the equation $\frac{1}{u}+\frac{1}{v}=\frac{1}{f}$, we have the image distance v from

$$\frac{1}{v}=\frac{1}{f}-\frac{1}{u}=\frac{1}{15}-\frac{1}{40}=\frac{5}{120}\text{ cm}^{-1}, \text{ or } v = 24.0 \text{ cm}.$$

(ii) The magnification m is given by $m=-\frac{24.0}{40.0}=-0.60$. Thus the size of the image is 1.20 cm. The negative sign indicates that the image is inverted.

Example 6.2

If, with the same mirror as in **Example 6.1** above, the pin is now placed at 24 cm from the mirror, calculate the position of the image produced from the mirror.

Solution

Using u = 24.0 cm, and the focal length the same as before – i.e. 15.0 cm, we get the image distance v from $\frac{1}{v}=\frac{1}{f}-\frac{1}{u}=\frac{1}{15}-\frac{1}{24}=\frac{3}{120}\text{ cm}^{-1}$, giving v = 40.0 cm.

Thus an object placed 40 cm in front of the mirror produces an image 24 cm, and, conversely, an object at 24 cm from the mirror produces an image of 40 cm. This simple exercise illustrates **the principle of reversibility of light**, which stipulates that if an object placed at one (first) point forms an image at a second point, then an object placed at the second point would form an image at the first point. By this principle, it follows that object and image positions are interchangeable.

Example 6.3

An object is placed 75 cm in front of a convex mirror of radius of curvature 55 cm. Calculate

(i) the position of the image from the mirror,

(ii) the magnification produced by the mirror.

Solution

(i) From the data given, u = 75.0 cm, r = 2f = - 55.0 cm.

Using $\frac{1}{v} = \frac{1}{f} - \frac{1}{u}$, we have $\frac{1}{v} = \frac{1}{-27.5} - \frac{1}{75}$, giving v = - 20.12 cm.

(ii) The magnification produced is given by $m = -\frac{v}{u} = -\left(\frac{-20.12\text{ cm}}{75.0\text{ cm}}\right) = 0.27$ (or 27%).

Thus, the image is upright and diminished.

6.1.5 Experimental determination of focal lengths

The focal length of a mirror plays an important role in the application of the mirror. There are standard ways of determining this important quantity. One of these involves measurements of object and image distances from the mirror, and applying the mirror equation. Although one set of measurements is sufficient, for purposes of accuracy, it is normal practice to measure many corresponding values of object and image distances, and apply a graphical method. From the mirror equation, a plot of $1/v$ as ordinate against $1/u$ as abscissa, produces a straight line (**Fig. 6.5**), with intercept on both axes equal to $1/f$. From this graph the focal length of the mirror is calculated.

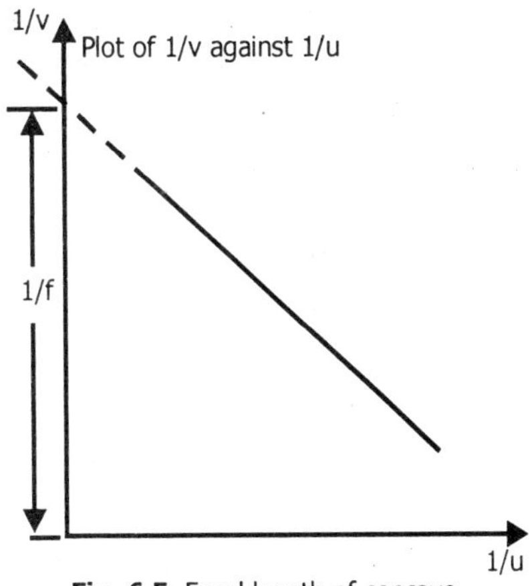

Fig. 6.5 Focal length of concave mirror via a plot of 1/v against 1/u.

A more direct method involves focusing a beam of parallel rays from a distant object (e.g. a tree) on a screen held in front of the mirror. The distance from the pole of the mirror to the screen is then the focal length. Due to problems associated with focusing, this method only gives approximate values. It is used only to give an estimate of the focal length, pending the use of a more accurate method.

Alternatively, one can determine the focal length of the mirror by measuring its radius of curvature. The radius can be determined by making direct measurements using a spherometer. More advanced methods exist, some making use of the linear magnification. The details of all these can be obtained from practical laboratory manuals.

Example 6.4

An object is situated 40 cm from a convex mirror. When a plane mirror is inserted between the object and the convex mirror at a distance of 32 cm from the object, the images in the two mirrors coincide. Find the radius of curvature of the convex mirror.

Solution

Fig. 6.6 is a schematic arrangement of the optical system referred to in this example. The plane mirror M is inserted between the object O and the convex mirror, as shown.

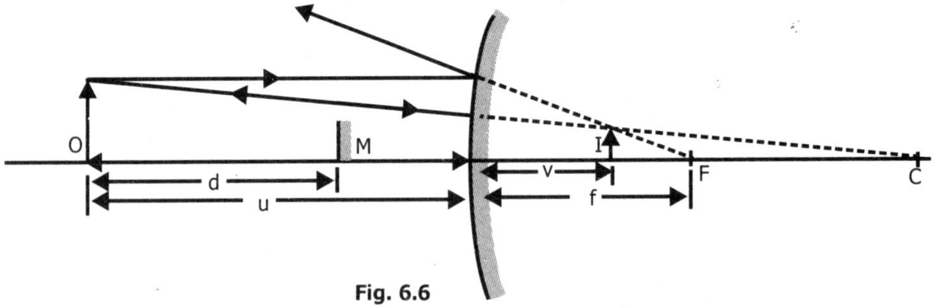

Fig. 6.6

We note that the image formed in a plane mirror is the same distance behind the mirror as the object is in front of the mirror. Thus the distance from the mirror to the common image is the same as d = 32 cm. From the geometry,

$$u + v = 2d, \text{ giving the image distance } v = 2d - u = 64 - 40 = 24 \text{ cm}.$$

Having got the image distance, we use this to determine the focal length of the mirror, and hence the radius of curvature. From the mirror equation relating object and image distances with the radius of curvature, we get $\frac{2}{r} = \frac{1}{u} + \frac{1}{-v} = \frac{1}{40} - \frac{1}{24} = -0.01666 \text{ cm}^{-1}$.

This gives the radius of curvature of the mirror as - 120 cm. Note that inn this example, we have used the convention that the image distance is negative if it is on the side opposite to the outgoing ray. This gives the correct (negative) sign to the radius of curvature.

6.2 REFRACTION AT CURVED BOUNDARIES

Curved spherical surfaces can also be used to produce images by refraction. Studies of refraction at these surfaces have led to the creation of many optical systems – lenses to correct vision, telescopes and microscopes used for observing distant, and minute near objects, and cameras for still and motion pictures. As a prelude to the understanding of these complex systems, we start with the refraction of light at a single spherical surface.

Fig. 6.7 shows the passage a ray of light from a point object on the axis of two media, with refractive indices n_1 and n_2, the media being separated by a thin spherical boundary. In this illustration, we assume the ray is incident in the less dense medium, and refracted into the denser medium.

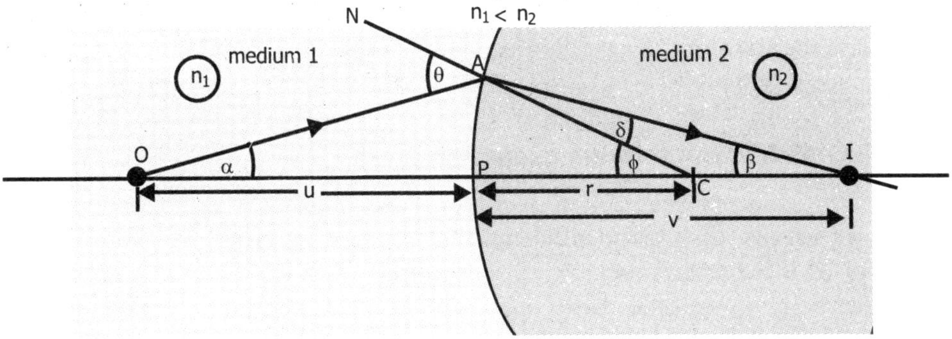

Fig. 6.7 Refraction at a single spherical boundary between two media.

A ray of light close to the axis from the object O meets the surface along the line OA that makes an angle α with the axis. The ray is refracted towards the normal CN in the direction AI, forming image at I. The angles made by the incident and refracted rays with the axis and the normal are shown in the figure. The object and image distances are u and v, respectively, and the radius of curvature of the surface is r.

From the figure, we have $\theta = \alpha + \phi$, and $\phi = \beta + \delta$. Using Snell's law, we get

$$n_1 \sin\theta = n_2 \sin\delta. \tag{6.5}$$

From the geometry, $\tan\alpha = \frac{h}{u}$, $\tan\phi = \frac{h}{r}$, and $\tan\beta = \frac{h}{v}$. Substituting into (6.5), we get

$$\frac{n_1}{u} + \frac{n_2}{v} = \frac{n_2 - n_1}{r}. \tag{6.6}$$

Equation (6.6) gives the object and image distances from a spherical refracting surface.

6.3 THIN LENSES

A lens consists of a refracting medium bounded by at least one spherical surface. The most common types of lenses consist of two spherical surfaces. Rays entering the lens on one side are brought to a focus, or appear to go out from a focus, on the other side of the lens. **Fig. 6.8** shows the common lenses in use. In (a) is shown the convex (or converging) lens. It is also known as biconvex lens, on account of the fact that both refracting surfaces are convex. In (b) is the diverging or concave lens, also referred to as biconcave lens. Less common are the plano-convex lens and plano-concave lens, shown

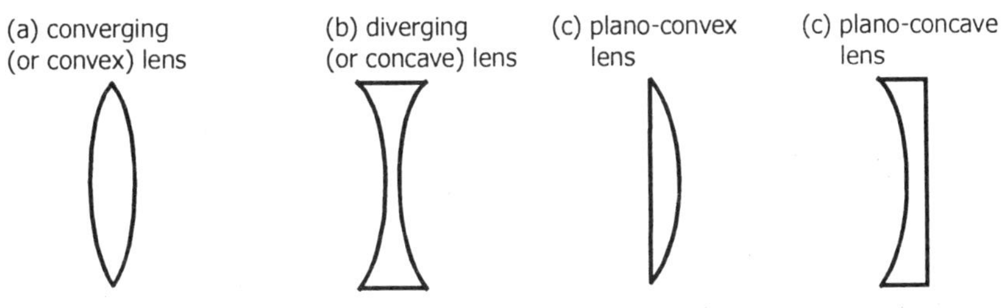

Fig. 6.8 Common lenses

in (c) and (d), respectively. In this section we study the image forming properties of lenses, the nature of images formed, and the effects of combining two or more lenses to produce images.

6.3.1 Refraction through a lens

We can understand the action of a lens if we consider it as consisting of many small-angled prisms, each prism refracting light toward its base. For converging lenses, the bases of these prisms are toward the centre, whilst for diverging lenses the refracting angles point towards the centre, and the bases are away from the centre.

Fig. 6.9 (a) shows the paths of parallel rays close to the principal axis meeting a converging lens, whilst (b) shows the paths of such rays passing through a diverging lens.

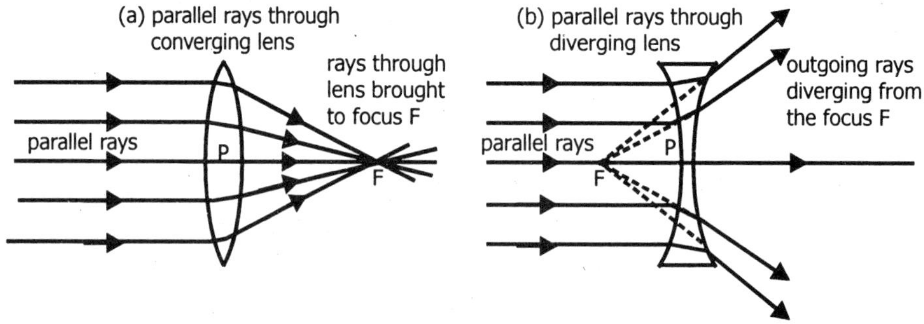

Fig. 6.9 Passage of parallel rays through a converging lens (a), and a diverging lens (b).

For the converging lens, the parallel rays are brought to a focus at F. In the diverging lens, the rays diverge on passing through the lens, and seem to be coming from a common focus F. In addition to the focus, we identify the pole, or centre of the lens P, and the principal axis joining the pole to the focus, as shown in the diagram. As there are two curved surfaces, there are two foci, one on each side of the lens. Both foci and centres of curvature take part in the formation of the image.

6.3.2 Images in a thin lens

In a manner similar to spherical mirrors, we can construct ray diagrams for images in both converging and diverging lens. The following rules hold for the construction of rays in lenses.

(i) Rays parallel to the principal axis will, after refraction, pass through (or appear to pass through) the focus,

(ii) A ray passing through the focus is refracted parallel to the principal axis,

(iii) A ray of light from the centre of curvature passes straight through the lens,

(iv) An image is formed when two or more rays intersect.

In the formation of the images in lenses, we use the distance 2F, and not the radius of curvature. This difference is due to the fact that the focusing power of a lens depends more on the refractive index of the material than the geometrical radius of the spherical surface. **Fig. 6.10** shows the ray diagrams.

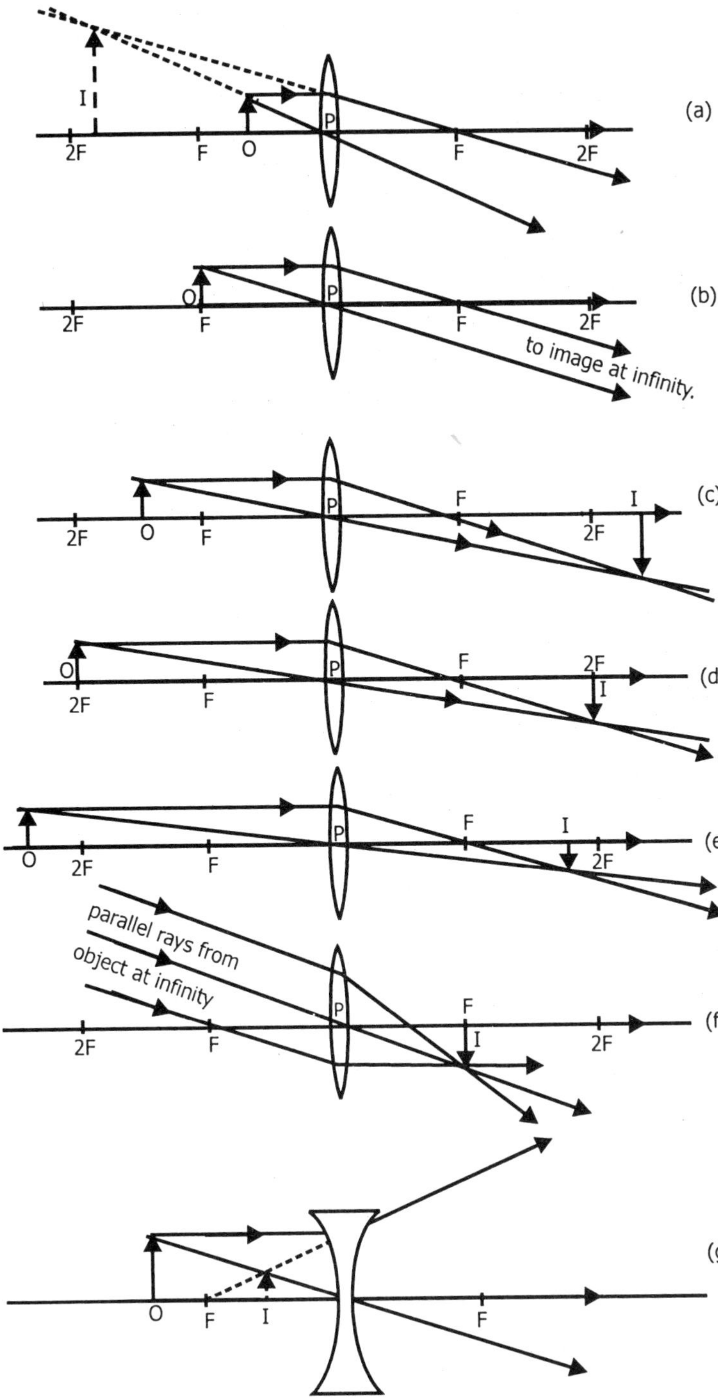

(a) object between pole and focus: Image is upright, magnified, and virtual.

(b) object at the focus: Image is inverted, formed at infinity.

(c) object between F and 2F: Image is real, inverted, magnified formed beyond 2F.

(d) object at 2F: Image is real, inverted, same size as object, formed at 2F.

(e) object beyond 2F: Image is real, inverted, diminished, between F and 2F.

(f) object at infinity: Image is real, inverted, diminished, formed at focus.

(g) In diverging lens: For any object distance, Image is virtual, upright, diminished, between P and F

Fig. 6.10 Images formed in a converging lens: (a) - (f), and diverging lens (g). For converging lens, the nature and size of the image depends on the position of the object. In the diverging lens the image is always virtual, diminished, and upright, irrespective of the position of the object.

6.3.3 Lens formulae

We study relationships between the distances of objects and images, the refractive index of the material of the lens, the focal lengths of lenses, the magnifications of images formed in lenses, and separations of objects and real images. These relationships are analysed in the sections that follow.

(a) Object and image distances

Fig. 6.11 shows a typical ray diagram for locating the image IB of an object OA in a thin convex lens. The distances of the object and image from the lens are u and v, respectively, whilst the focal length of the lens is f. From the diagram it follows that triangles PFD and OAF are similar.

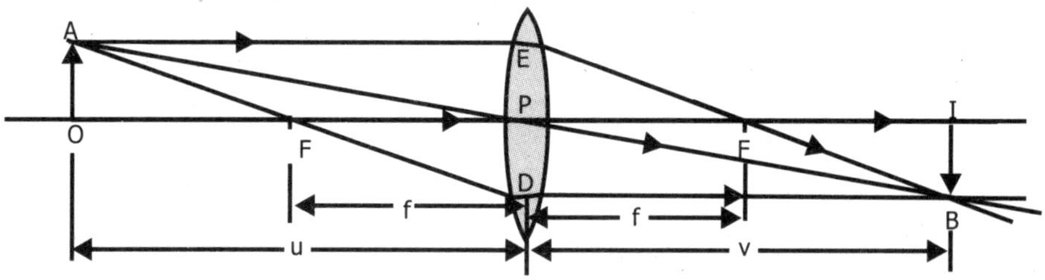

Fig. 6.11 Object and image locations in a thin convex lens

Thus, we write

$$\frac{PD}{OA} = \frac{FP}{OF} = \frac{f}{u-f}.$$

Also, triangles PFE and FIB are similar. Hence, $\frac{IB}{PE} = \frac{FI}{PF} = \frac{v-f}{f}$. Since PD = IB, and PE = OA, we have $\frac{f}{u-f} = \frac{v-f}{f}$, which on rearranging gives the lens formula connecting object and image distances with the focal length as

$$\frac{1}{f} = \frac{1}{u} + \frac{1}{v}. \qquad (6.7)$$

In applying equation (6.7), we use a sign convention similar to that employed for mirrors. The focal length of a diverging lens is given a negative sign.

Example 6.5

An object is 12 cm in front of a lens of focal length 8 cm. Find the position and nature of the image formed if the lens is

(a) a converging lens, and
(b) a diverging lens.

Solution

In both parts to this problem, we use the lens formula stated in equation (6.7).

(a) For the converging lens, we have the following data: $f = 8.0$ cm, $u = 12.0$ cm. From $\frac{1}{f} = \frac{1}{u} + \frac{1}{v}$, we have $\frac{1}{v} = \frac{1}{f} - \frac{1}{u}$, or $v = \frac{fu}{u-f} = \frac{8.0 \times 12.0}{12.0 - 8.0} = 24.0$ cm.
The image is hence real, and formed 24.0 cm behind the lens.

(b) For the diverging lens, $f = -8.0$ cm, $u = 12.0$ cm. From $-\frac{1}{f} = \frac{1}{u} + \frac{1}{v}$, we have $\frac{1}{v} = -\frac{1}{f} - \frac{1}{u}$, or $v = -\frac{fu}{u+f} = -\frac{8.0 \times 12.0}{12.0 + 8.0} = -4.8$ cm. With the image distance negative, this means a virtual image, formed 4.8 cm on the same side as the object.

(b) Magnification

As we saw for mirrors, magnification is the ratio of the height of the image to the height of the object. Using **Fig. 6.11**, we write the magnification m as

$$|m| = \frac{IB}{OA} = \frac{f}{u-f}.$$

Using the lens formula on image and object distances, we get

$$m = -\frac{v}{u}, \qquad (6.8)$$

the same expression obtained for the spherical mirror. A negative sign indicates an inverted image, while a positive sign means the image is upright.

Example 6.6

When an object is moved along the principal axis of a thin converging lens, the same magnification is obtained when the object is 8.0 cm from the lens, and again when the object is 16.0 cm from the lens. Determine the focal length of the lens.

Solution

Let m be the magnification produced. Any particular value of the magnification is obtained for one and only one value of the object distance. It follows therefore that if two positions are obtained, the magnifications at these positions are $+m$ and $-m$. One of the images formed will be real, while the other will be virtual. Further, the smaller object distance produces a virtual image, while the larger object distance produces a real image. This is due to the fact that a virtual image is produced when the object is close to the lens. We denote the smaller object distance by u_1, and the larger object distanced by u_2.
If we denote the corresponding image distances by v_1 and v_2, then

$$m = \frac{v_1}{u_1}, \text{ and } -m = \frac{v_2}{u_2}. \text{ This gives } v_1 = mu_1, \text{ and } v_2 = -mu_2.$$

From the lens equation, $\frac{1}{u_1} + \frac{1}{v_1} = \frac{1}{f} = \frac{1}{u_2} + \frac{1}{v_2}$. Thus $\frac{1}{u_1} + \frac{1}{mu_1} = \frac{1}{f} = \frac{1}{u_2} - \frac{1}{mu_2}$.

Rearranging these, we get $u_1 = \frac{m+1}{m} f$, and $u_2 = \frac{m-1}{m} f$. This gives $f = \frac{u_1 + u_2}{2}$.

Substituting the given numerical values, we get $f = \frac{8.0 \text{ cm} + 16.0 \text{ cm}}{2} = 12.0 \text{ cm}$ as the focal length of the lens.

(c) Minimum distance between object and real image

From our ray diagrams in **Fig. 6.10**, it is seen that an object placed at 2F gives a real, inverted image of unit magnification at the point 2F on the other side of the lens. This gives the minimum distance apart of two points such that an object placed at one of them produces an image at the other. Thus, for a lens of focal length f, the minimum distance between the two points is 4f.

(d) The Lens-makers formula

This formula connects the focal length of a lens to the radii of curvature of its surfaces, and the refractive index of the material of the lens. We consider refraction through the two spherical surfaces of the lens in succession. Thus the image formed by refraction through the first surface becomes the object for the second surface. **Fig. 6.12** shows the passage of rays through the system, magnified for purposes of clarity.

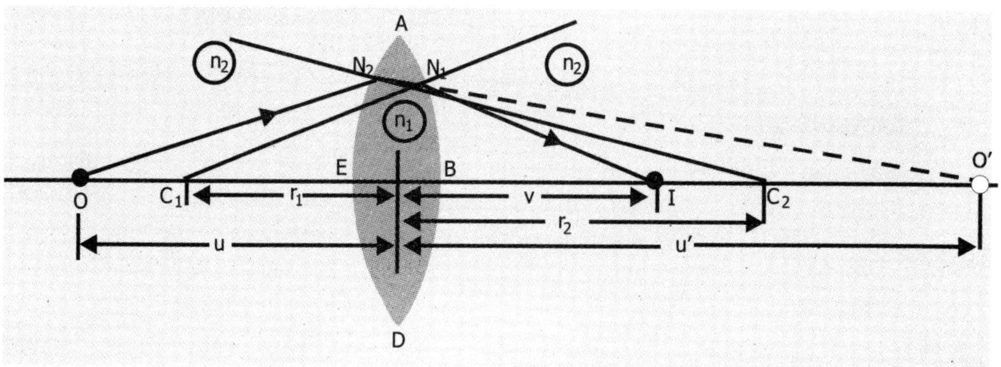

Fig. 6.12 Successive refractions by two spherical surfaces.

A lens with material of refractive index n_1 is in a medium of refractive index n_2 ($n_1 > n_2$). An object O placed a distance u from the spherical surface AED forms an image O′ at distance u′. This image acts as virtual object for the second spherical surface ABD, producing the final image I at a distance v from the centre of the two surfaces. From equation (6.6), we set up expressions for the refraction at the surfaces:

$$\frac{n_2}{u} + \frac{n_1}{u'} = \frac{n_1 - n_2}{r_1}, \qquad (6.9)$$

and

$$\frac{n_1}{-u'} + \frac{n_2}{v} = \frac{n_1 - n_2}{r_2}. \qquad (6.10)$$

Adding the last two equations, we get

$$\frac{n_2}{u} + \frac{n_2}{v} = (n_1 - n_2)\left(\frac{1}{r_1} + \frac{1}{r_2}\right). \qquad (6.11)$$

For reasons that will become clear shortly, equation (6.11) is usually written as

$$\frac{1}{u} + \frac{1}{v} = \left(\frac{n_1}{n_2} - 1\right)\left(\frac{1}{r_1} + \frac{1}{r_2}\right). \qquad (6.12)$$

In its most usual applications, a lens made of material with refractive index n is placed in air, for which $n_2 = 1$. For this system, we have

$$\frac{1}{u} + \frac{1}{v} = (n - 1)\left(\frac{1}{r_1} + \frac{1}{r_2}\right). \qquad (6.13)$$

Using the lens equation for object and image distances and the focal length, we re-write equation (6.13) to read

$$\frac{1}{u} + \frac{1}{v} = \frac{1}{f} = (n - 1)\left(\frac{1}{r_1} + \frac{1}{r_2}\right). \qquad (6.14)$$

Equation (6.14) is known as the **lens-makers formula**. Its usefulness lies in the fact that the lens manufacturer can make lenses of precise focal lengths by suitable choice of materials, and machining the surfaces to precise radii of curvature. For most cases, these radii are equal. However, the equation works just as well if the radii are different.

Example 6.7

The absolute values of the radii of curvature of the surfaces of a convex lens are 20.0 cm, and 15.0 cm. If the lens is made of glass of refractive index 1.52, find its focal length.

Solution

From the data given, and equation (6.14) the focal length of the lens is given by

$$\frac{1}{f} = (n - 1)\left(\frac{1}{r_1} + \frac{1}{r_2}\right) = (1.52 - 1)\left(\frac{1}{20} + \frac{1}{15}\right) = \frac{0.52 \times 35}{300}, \text{ or } f = \frac{300}{0.52 \times 35}\text{ cm} = 16.48 \text{ cm}.$$

(e) Thin lenses in contact

In many optical devices, it is desired to use a combination of lenses in order to give a system with the desired qualities. Two thin lenses placed in contact produce an effect equivalent to a single lens with a combined focal length related to the focal lengths of the two lenses. **Fig. 6.13** can be used to derive this relationship.

The diagram shows two thin converging lenses L_1 and L_2, with focal lengths f_1 and f_2, respectively, in contact along their common axis. An object O placed in front of L_1 forms an image I′, a distance v′ from the lens combination. This image acts as a virtual object for the lens L_2, and a final image I at a distance v from the lenses.

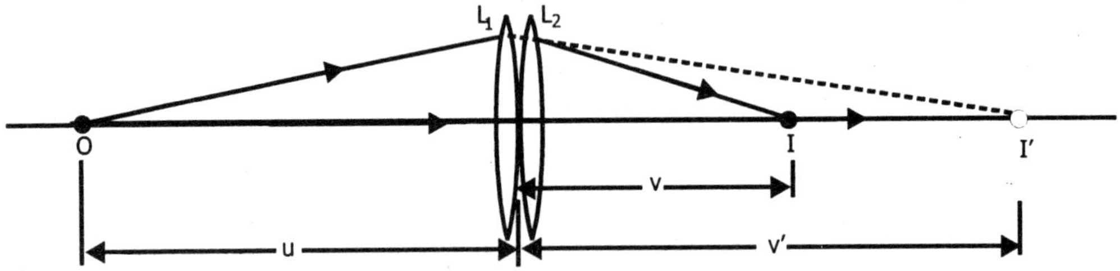

Fig. 6.13 Image produced by two thin lenses in contact.

We set up equations for the refractions through the lenses, as follows:

$$\frac{1}{f_1} = \frac{1}{u} + \frac{1}{v'} \tag{6.15}$$

for lens L_1. For the lens L_2, we have

$$\frac{1}{f_2} = \frac{1}{-v'} + \frac{1}{v} \tag{6.16}$$

Adding equations (6.15) and (6.16) above, and rearranging, we get

$$\frac{1}{f_1} + \frac{1}{f_2} = \frac{1}{u} + \frac{1}{v}. \tag{6.17}$$

Comparing (6.17) with the general lens formula, we can conclude that the two lenses taken together is equivalent to a single lens with focal length f, given by

$$\frac{1}{f} = \frac{1}{f_1} + \frac{1}{f_2}. \tag{6.18}$$

All the above equations are equally applicable to concave lenses. In applying them to concave lenses, we must remember to use the correct sign convention for the focal length. The following worked examples demonstrate how this equation can be used.

Example 6.8

A lens system consists of two converging lenses of focal lengths 10.0 cm and 12.0 cm in contact along a common axis. An object is placed 15 cm in front of the first lens. Determine the location and magnification of the final image produced.

Solution

If we let f_1 and f_2 represent the focal lengths of the first and second lenses, respectively, then f_1=10.0 cm, and f_2=12.0 cm. From (6.18), their combined focal length f is given by $\frac{1}{f} = \frac{1}{f_1} + \frac{1}{f_2} = \frac{1}{10\text{ cm}} + \frac{1}{12\text{ cm}}$, or $f = \frac{f_1 f_2}{f_1 + f_2} = \frac{120}{22}\text{ cm} = 5.45$ cm. An object 15.0 cm in front of this combination, produces an image a distance v given by $\frac{1}{v} = \frac{1}{f} - \frac{1}{u} = \frac{1}{5.45} - \frac{1}{15.0}$, or v = 8.62 cm. The magnification $m = -\frac{v}{u} = -\frac{8.62}{15.0} = -0.57$. The image is hence real, inverted, and diminished.

Example 6.9

A thin converging lens of radius of curvature 20.0 cm is placed on a plane mirror in a laboratory experiment for the location of images. The image of an optical pin mounted above the combination coincides with the object when the pin is 10.0 cm above the glass plate. When a small quantity of strong transparent liquid is smeared between the lens and the mirror, the image of the pin coincides with the object with the pin 15.0 cm above the mirror. Determine the refractive index of the liquid from the above data.

Solution

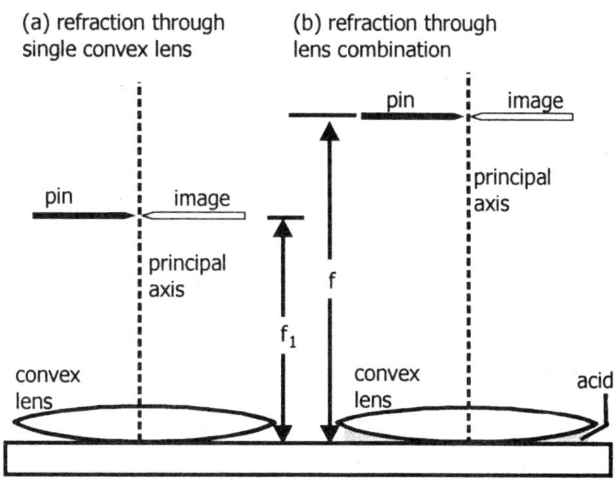

Fig. 6.14

This example illustrates a practical application of the lens equations. **Fig. 6.14** (a) and (b) give the schematics of the experimental arrangement. In this figure, (a) represents the refraction through the convex lens alone; (b) shows the refraction through the two thin lenses in contact. One of these is the convex lens. The other lens is a plano-concave lens formed by the liquid between the convex lens and the glass plate. We shall call this latter lens the liquid lens. In both cases, for the image of the pin to coincide with the pin, rays passing through the lens(es) must strike the glass plate at right angles to the plate. The rays would hence appear to be coming from an object at infinity, thereby forming an inverted image at the focus. Thus, the distance above the plate in each case is the focal length of the lens system.

Representing these focal lengths by f_1 and f, as shown, we have $\frac{1}{f} = \frac{1}{f_1} - \frac{1}{f_2}$, where f_2 is the focal length of the liquid lens. Notice that we have used the negative sign, since this is a diverging lens. Thus, we have $\frac{1}{f_2} = \frac{1}{f_1} - \frac{1}{f} = \frac{1}{10} - \frac{1}{15} = 0.0333\,\text{cm}^{-1}$, giving $f_2 = 30.0\,\text{cm}$. Using the Lens makers' formula, we determine the refractive index n of the liquid as follows: $\frac{1}{f_2} = (n-1)\left(\frac{1}{r} + \frac{1}{\infty}\right) = (n-1)\frac{1}{r}$, where r is the given radius of curvature. Substituting the given values and rearranging, we get $n = 1.667$

6.3.4 Experiments to find focal lengths

From our equations developed in the preceding sections, it is clear that the focal length of a lens is about the most important property of the lens. Experimental methods have been developed for determining the focal length of a lens. These methods are described in detail in practical manuals. We survey a few of these techniques.

(a) Distant object method

This is the most direct method, as it does not involve any calculations. The method is based on the fact that parallel rays from a distant object are brought to a focus after passing through the lens. A distant object is focused on a screen held behind the lens. The distance from the lens to the screen is then the focal length of the lens. While it is quick, the method is not considered very accurate, as the lens may not in fact be receiving parallel rays. The method is useful mainly as an initial test in more accurate experiments to give a rough guide of the value of the focal length. As concave lenses do not form real images on a screen, the method cannot be used to find the focal length of a concave lens.

(b) Measurement of object and image distances

Using an illuminated object, the image is cast on a screen placed behind the lens. Corresponding object and image distances are measured. A plot of suitable variables is drawn from which the focal length is obtained. Examples of such corresponding variables are:

(i) A plot of 1/v as ordinate against 1/u as abscissa. The extrapolated intercept on the axes gives the reciprocal of the focal length,

(ii) A plot of uv as ordinate against (u+v) as abscissa. Here, the slope of the graph gives the focal length of the lens.

Fig. 6.15 shows typical plots.

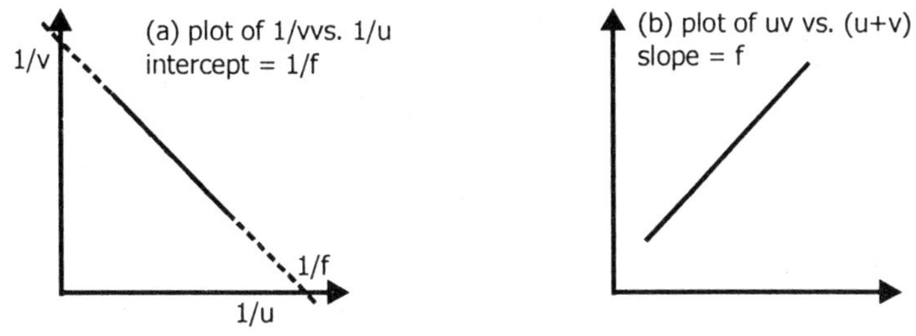

Fig. 6.15 Determination of the focal length of a lens by measuring object and image distances.

(c) The displacement method

This method normally preceded by the distant-object method to find an estimate of the focal length of the lens. Using this estimate, an illuminated object and a screen are placed such that their distance apart is at least four times this rough focal length value. A converging lens placed between the two is moved slowly back and forth until a clear image is cast on the screen. The position of the lens is noted. The lens is then moved in the opposite direction until a sharp image is again formed on the screen. The distance moved by the lens – i.e. the displacement of the lens – is measured. The theory of this method can be understood with reference to **Fig. 6.16**.

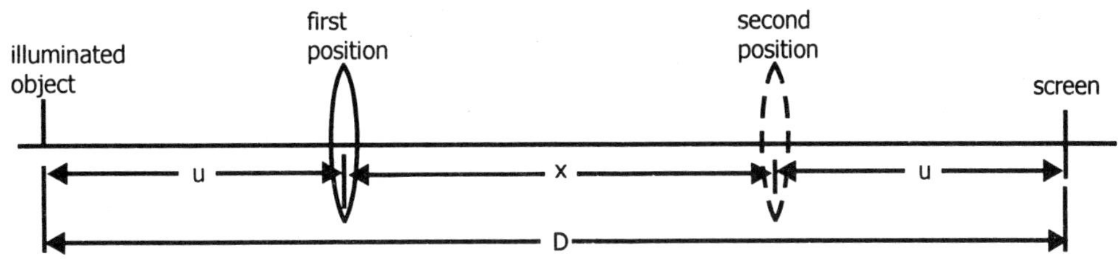

Fig. 6.16 Displacement method of finding focal length of a converging lens.

The diagram shows the lens in two positions between the object and the screen, giving real images. The distance between the two positions is x, whilst the distance between the object and the screen is D. For the first position, the object distance is u, and the image distance is x+u. Thus, we have $x+2u = D$, from which $u = \frac{1}{2}(D - x)$. The image distance v is given by $v = \frac{1}{2}(D + x)$. Substituting these in the lens equation, we get

$$f = \frac{D^2 - x^2}{4D}. \qquad (6.19)$$

6.3.5 The power of a lens

In the optical industry, a useful term used in connection with lenses is the **power** of a lens. This term describes the ability of the lens to focus images. The greater the power the higher the focusing ability.

The power of a lens is the reciprocal of its focal length.

Thus the smaller the focal length f the lens the more powerful it is. The power P is expressed in units of **dioptres**. Thus

$$P\text{ (in dioptres)} = \frac{1}{f\text{ (in metres)}}, \tag{6.20}$$

where f is the focal length of the lens, expressed in metres. Hence the shorter the focal length of a lens the greater the focusing ability. The power of a converging lens is positive, while that of a diverging lens is negative. For two thin converging lenses in contact, the total power of the combination is equal to the sum of the powers of the separate lenses.

Example 6.10

A lens system consists of a converging lens of focal length 25 cm and a diverging lens of focal length 125 cm. Calculate

(i) the power of the combination,

(ii) the focal length of a single converging lens that would be equivalent to this combination.

Solution

(i) Representing the powers of the converging and diverging lenses, respectively by P_1 and P_2, we have $P_1 = \frac{1}{0.25} = 4$ dioptres, and $P_2 = -\frac{1}{1.25} = -0.8$ dioptres. The power P of the combination is hence given by $P = 4.0 + (-0.8) = 3.2$ dioptres.

(ii) From equation (6.20), the focal length f of the single converging lens is given by $f = \frac{1}{P} = \frac{1}{3.20}\text{ m} = 31$ cm.

EXERCISES

1. Two spherical mirrors of radii of curvature −20 cm and +24 cm are placed coaxially 41.33 cm apart. An object is placed 20 cm from the convex mirror. Find the nature and position of the final image formed after reflection in the convex mirror, and then in the concave mirror.

2. When an object is moved along the axis of a spherical mirror, a magnification of 4.0 is obtained when the object is 9.0 cm from the pole of the mirror. Calculate the radius of curvature of the mirror.

3. Two mirrors are placed coaxially 32 cm apart with their reflecting surfaces facing each other. The image of a small object placed midway between the mirrors coincides with the object. If one mirror is concave, and its radius of curvature is 24 cm, determine
   (i) what kind the other mirror is,
   (ii) the magnification produced by the double reflection.

4. A converging lens with a focal length of 10.0 cm forms a real image of height 1.0 cm, at a distance of 15.0 cm from the lens. Determine the position and size of the object.

5. A family shows colour slides of dimensions 35 mm by 35 mm by projecting the image on to a screen 10.0 m from the projector lens. The image is 1.0 m by 1.0 m in size. Calculate
   (i) the distance of the slide from the lens,
   (ii) the focal length of the lens

6. An object is placed 12 cm in front of a diverging lens of focal length 10 cm. Find the position, nature, and magnification of the image formed.

7. Show that if an object is placed on the axis of a lens at a distance x_1 from the first focal point, and forms an image at a distance x_2 from the second focal point, then $x_1 x_2 = f^2$, where f is the focal length of the lens.

8. A photographer uses a camera with a lens of focal length 5.0 cm to photograph a man of height 2.0 m. If the height of the photo on the film is 3.0 cm, calculate
   (i) the distance of the man from the camera lens,
   (ii) the distance from the camera lens to the film.

9. The distance between an object and its image in a converging lens is 50.0 cm. If an inverted image four times the size of the object is formed, calculate the focal length of the lens.

10. An object and a screen are 48.0 cm apart. A converging lens of focal length 9.0 cm is inserted between them. For what positions will clear images be formed on the screen? Compare the magnifications produced in the two cases.

11. With the aid of a suitable ray diagram, show that two thin lenses of focal lengths f_1 and f_2 in contact are together equivalent to a single lens of focal length f, given by $f = \dfrac{f_1 f_2}{f_1 + f_2}$.

12. A telephoto lens consists of a converging lens of focal length 30.0 cm and a diverging lens of focal length −10.0 cm, the two lenses being 27.5 cm apart. An object to be photographed is placed 100.0 cm from the converging lens. Calculate the distance of the photographic plate from the converging lens.

13. A narrow beam of parallel rays in air enters a solid glass sphere of radius 2.0 cm along the radial direction. If the refractive index of the glass is 1.52, determine the point at which the rays will be brought to a focus.

14. Two thin identical converging lenses of focal length 20.0 cm are aligned along a common axis. The second lens is 30.0 cm from the first. A small object is placed 40.0 cm to the left of the first lens. Determine the position of the final image from the first lens.

15. A real image of magnification 0.5 is to be obtained in a concave spherical mirror of radius of curvature 40.0 cm. Determine the image and object distances.

16. An object and a screen are placed at a fixed distance L apart. A converging lens of focal length f forms an image on the screen for two positions separated by a distance x. Show that the ratio of the image sizes for the two positions is given by $\left(\frac{L-x}{L+x}\right)^2$.

17. An optical pin 2 cm tall is placed 20 cm from a screen. A converging lens of focal length 3 cm is to be used to project an image of the pin on the screen. Determine
(i) the distances from the pin at which the lens must be placed,
(ii) the height of the image produced for each position of the lens.

18. The focal length of a lens placed in air for any particular lens in air is given by the lens-makers' formula: $\frac{1}{f} = (n-1)\left(\frac{1}{r_1} + \frac{1}{r_2}\right)$, where r_1 and r_2 are the radii of curvature of the spherical surfaces of the lens, which is made of material of refractive index n for the particular colour. Crown glass has refractive indices for blue and red light of 1.523 and 1.517, respectively. Calculate the distance between the blue and red focal points for a convex lens of crown glass whose radii of curvature are both 10.0 cm.

7

CHAPTER

Optical Instruments

7

CHAPTER

Optical Instruments

The many phenomena in geometrical optics are put to practical use in optical instruments. These instruments enable the eye to see images of objects with such clarity that could not be achieved by the ordinary or unaided eye. With the use of these instruments it had been possible to see images of the planets and stars that are very far away, and the images of objects too small to be seen by the naked eye.

With the advancement of technology, devices to help us discern minute details of objects are being continually developed. Nowadays, many of these devices use a combination of optical principles and principles from other branches of science. For instance, many microscopes and telescopes incorporate electronic attachments that help improve their overall functionality.

Whatever other segments that are incorporated in an optical instrument, the most important single component is the optical part – i.e. the part that deals with the propagation of light through the instrument, and the formation of the image. All optical instruments are made up of lenses and mirrors, combined in such a manner as to produce clear images that would otherwise not be seen by the unaided eye.

In this chapter we study the most common optical instruments. Starting from single lens instruments – the camera and the slide projector, we study systems made up of two or more lenses – the compound microscope, the astronomical telescope, and the terrestrial telescope. Defects of lenses are also covered. Our study concludes with a description of perhaps the most efficient natural optical system we sometimes take for granted – the human eye. We describe the salient features of the eye, and proceed to its wavelength response, the defects of vision, and how some of these defects are corrected.

7.1 VISUAL ANGLE AND ANGULAR MAGNIFICATION

The main purpose of an optical instrument is to aid the eye in viewing the object with a clarity that is not possible with the naked eye. An object may appear clearer perhaps because it is bigger, or because it is closer to the observer. In the study of optical instruments the term **visual angle** is used to describe the clarity with which an image is viewed. The visual angle is defined as the angle subtended by the object or image at the eye. **Fig. 7.1** illustrates the concept of visual angle.

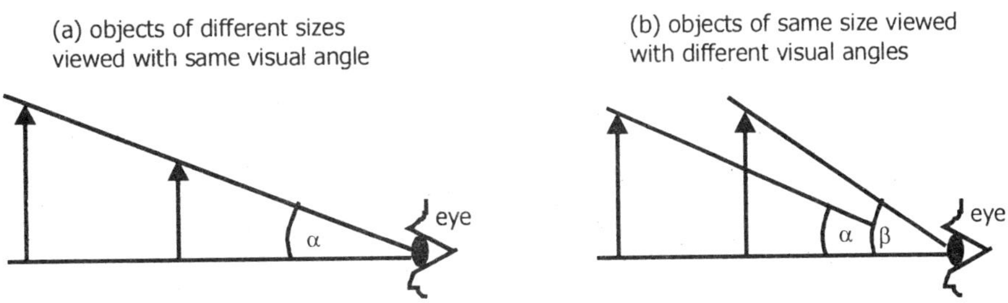

Fig. 7.1 Visual angle and angular magnification.

In (a), two objects of different sizes are shown subtending the same visual angle. The eye sees these two with the same clarity, even though one of them is larger than the other. In (b), two objects of the same size subtend different visual angles. The object that subtends the larger visual angle is seen more clearly than the one subtending the smaller visual angle.

The purpose of an optical instrument is to increase the visual angle, such that near objects that are very small (e.g. bacteria) can be seen, and far objects that cannot be seen because of their distances away (e.g. planets, stars) can be clearly seen. An important quantity used for describing the performance of an optical instrument is the **angular magnification**, defined as

the ratio of the visual angle subtended by the image at the eye with the instrument in use to the visual angle subtended by the object at the unaided eye.

In symbols, if the visual angle with the instrument in use is α', and the visual angle without the instrument is α, the angular magnification M is given by

$$M = \frac{\alpha'}{\alpha}. \qquad (7.1)$$

In our study of optical instruments, we shall be finding expressions for the angular magnification of each instrument.

A notion we often come across in the study of optical instruments is the **least distance of distinct vision**. Often represented by D,

the least distance of distinct vision is the distance from the unaided eye that an object is seen with maximum clarity.

Objects within or farther away will not be seen as clearly. For the normal human eye, this distance has been determined to be about 25.0 cm (or 0.25 m). In the sections that follow, we study each class of optical instrument in some detail. We start with the single lens instruments, and proceed to the double and multiple lens systems.

7.2 THE CAMERA

The camera consists of a single lens system enclosed in a light-tight chamber with a photosensitive film at the other side of the chamber. **Fig. 7.2** shows the image formation in the camera. The image formed on the film is real and inverted. In order to provide the real images for a wide range of object distances, it is arranged that the lens can be moved backward and forward, so a clear image is focused on the film.

The film is made up of very fine grains of alkyl halide (e.g. silver iodide) crystals uniformly distributed on plastic backing. The silver iodide has the chemical property of changing (or blackening) on interaction with the light photons. Development of the film

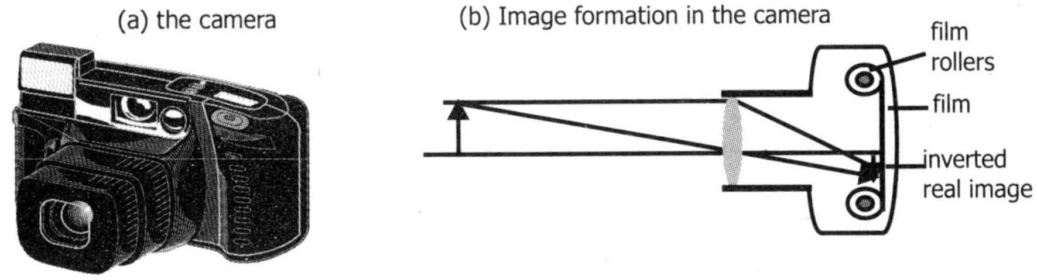

Fig. 7.2 Features of the camera, and image formation

following exposure would reveal differences between the areas of the film that are exposed and areas not exposed.

The quality of the image formed on the film depends on the amount of light passing through the lens. Too much light leads to an overexposed or "burnt" image. Too little light produces a blurred faded image. The light entering the lens is controlled by two factors: the **shutter speed**, and the **lens aperture**. The shutter speed is varied in steps from 1.0 s to 1.0 ms. The intensity of the light falling on the film depends on the area of the lens that is exposed. This is done by means of a nearly circular hole of adjustable diameter d in the **diaphragm**. A parameter known as the **f-number** is used to express the extent of this opening. The f-number is defined by the equation

$$\text{f}-\text{number} = \frac{\text{focal length of lens}}{\text{diameter of aperture}} = \frac{f}{d}. \qquad (7.2)$$

The f-number is indicated by f-stops. In some cameras, these stops are 2, 2.8, 4, 5.6, 8, 11, 16, 32, and so on. The larger the f-number, the smaller the aperture. The light intensity falling on the film is proportional to the area of the aperture. Thus doubling the diameter of the aperture increases the light intensity four times. The actual exposure is proportional to both the aperture area and the time of exposure. Thus the f-number is set to correspond to the shutter speed.

Example 7.1

The lens for a 35-mm camera has a focal length of 50 mm, and a range of f-stops from f/2 to f/16. Calculate

(i) the corresponding range of aperture diameters,

(ii) the light intensity ratio corresponding to this aperture range.

Solution

(i) From equation (7.2) we have $d = \frac{\text{focal length of lens}}{\text{f}-\text{number}} = \frac{f}{\text{f}-\text{number}}$.

The maximum diameter of the aperture is hence given by

$$d_{max} = \frac{\text{focal length}}{\text{smallest f-number}} = \frac{50\text{ mm}}{2} = 25\text{ mm}.$$

Similarly, the smallest aperture is

$$d_{min} = \frac{\text{focal length}}{\text{largest f - number}} = \frac{50 \text{ mm}}{16} = 3.1 \text{ mm}.$$

(ii) Since the intensity is proportional to the area, and hence proportional to the square of the diameter of the aperture, the intensity ratio corresponding to this aperture range is $\left(\frac{d_{max}}{d_{min}}\right)^2 = \left(\frac{25}{3.1}\right)^2 = 64.0$

7.3 THE SLIDE PROJECTOR

The slide project is used to project the image on a film slide on to a viewing screen. We can understand the principles of the projector by considering it very much like the reverse of the camera. In the projector, light from a source (e.g. carbon arc) shines through the film, and casts an enlarged inverted image of the film on the projection screen. Additional lenses, known as **condenser lenses**, are placed between the lamp and the film. Their role is to ensure that most of the light from the source goes through the film slide. The condenser lenses must be large enough to cover the entire area of the film. A concave mirror behind the lamp also helps to direct the light towards the film. **Fig. 7.3** shows the formation of the image on the projector screen.

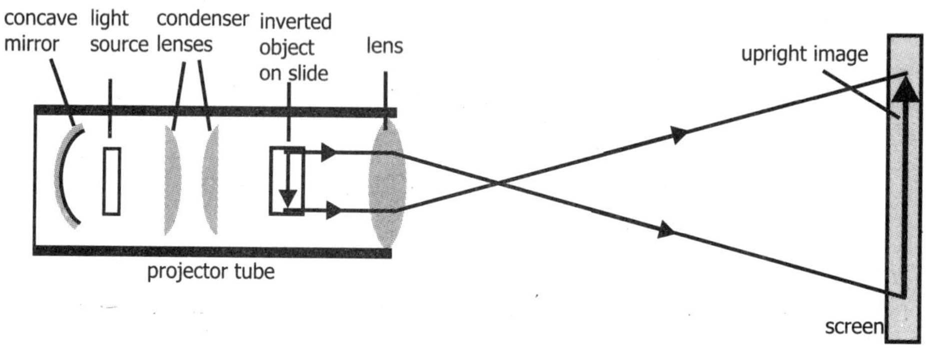

Fig. 7.3 Image formation in the slide projector

As the image formed is inverted, the film is normally inserted with the object inverted. The position and size of the image projected on the screen are determined by the distance of the screen from the projector lens, and by the focal length of the lens.

Example 7.2

A 35 mm colour slide with a picture area 24 x 24 mm is inserted in a slide projector. An image 1.2 m x 1.2 m is to be projected on to a screen 5.0 m from the lens. Calculate the focal length of the projector lens.

Solution

The lateral magnification produced is of magnitude 50 (i.e 1.2 m/24 mm). The image distance v = 5.0 m. The object distance u is hence 0.10 m. Using the lens formula, we can calculate the focal length f of the lens as $\frac{1}{f} = \frac{1}{u} + \frac{1}{v} = \frac{1}{0.1} + \frac{1}{5}\ m^{-1}$ or f = 9.8 cm.

In the camera and the projector, the image is formed on a screen. These images are hence real. In other classes of optical instruments, the final image is viewed through the system. In many of these cases the images are virtual. Microscopes and telescopes fall into this category. In their normal use, microscopes form the image at the least distance of distinct vision. The instrument set in this mode is described as being in **normal adjustment**. Any other adjustment is abnormal. The telescope, on the other hand, is used for viewing distant objects. In normal adjustment of this instrument, the final image is formed at infinity. Sometimes the lenses might be arranged to produce an image at some other point. The instrument is then said to be in abnormal adjustment.

7.4 MICROSCOPES

Microscopes are used to view near objects, often too small to be seen clearly by the naked eye. In normal adjustment, the final image is formed at the least distance of distinct vision. There are two types of microscopes. The simple microscope, also known as the magnifying glass, consists of a single lens. The other type is the compound microscope, which uses a combination of lenses aligned along a common axis to give a final highly magnified image of the object. This latter microscope is used for viewing very small objects – e.g. blood cells, and bacteria.

7.4.1 The simple microscope

The simple microscope uses one converging lens, and arranged such that a virtual magnified image is produced, as shown in **Fig. 7.4.** In this, both the object and image subtend the same visual angle. The magnification is achieved when we consider that without the glass, the object is to be viewed at the least distance of distinct vision, and not at the position O.

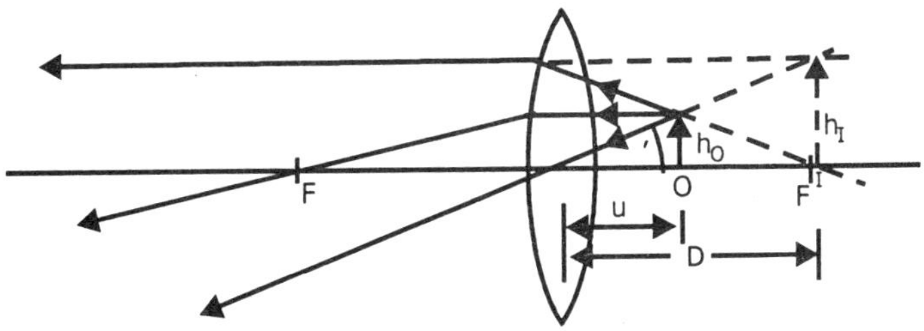

Fig. 7.4 Image formation in the simple microscope

The angular magnification is given by the expression

$$M = \frac{\text{angle subtended by the image}}{\text{angle subtended by the object at the least distance of distinct vision}}.$$

If α is the visual angle subtended by the object at the least distance of distinct vision, then

$$M = \frac{\alpha'}{\alpha} = \frac{\tan\alpha'}{\tan\alpha} = \frac{h_I/D}{h_O/D} = \frac{h_I}{h_O} = \frac{D}{u}, \tag{7.3}$$

since the angles are small. Expressing the object distance in terms of the image distance and the focal length f of the lens, we get the angular magnification of the microscope as

$$M = \frac{D}{f} - 1. \tag{7.4}$$

From (7.4) it follows that the instrument gives a magnified image when the focal length of the lens is less than the least distance of distinct vision. Magnifying glasses have small focal lengths. In an alternative arrangement, the image can be formed at infinity. For this case, the object to be viewed is placed at the focus of the lens.

Example 7.3

A convex lens of focal length 5.0 cm is used as a magnifying glass. Find the angular magnification when an object is viewed using this lens and the image is formed

(i) at the least distance of distinct vision, 25 cm,

(ii) at infinity.

Solution

(i) From (7.2), the angular magnification is given by $M = \frac{D}{u}$, where D is the least distance of distinct vision, and u is the object distance. Using the lens equation, $\frac{1}{u} = \frac{1}{f} - \frac{1}{v} = \frac{1}{5} - \frac{1}{-25} = \frac{6}{25}\,\text{cm}^{-1}$, giving $M = \frac{D}{u} = 25\,\text{cm} \times \frac{6}{25}\,\text{cm}^{-1} = 6.0$

(ii) When the final image is at infinity, the object distance becomes $\frac{1}{u} = \frac{1}{f} - \frac{1}{v} = \frac{1}{5} - \frac{1}{-\infty} = 0.2\,\text{cm}^{-1}$, giving $M = \frac{D}{u} = 25 \times \frac{1}{5} = 5.0$

From the above calculations, it is seen that the magnification produced with the image at infinity is less than that when the image is formed at the least distance of distinct vision.

7.4.2 The compound microscope

The compound microscope consists of two thin converging lenses, L_o and L_e, arranged coaxially, as shown in **Fig. 7.5**. The lens closer to the object to be viewed is called the object lens, or objective, whilst the other lens is called the eye lens, or eyepiece.

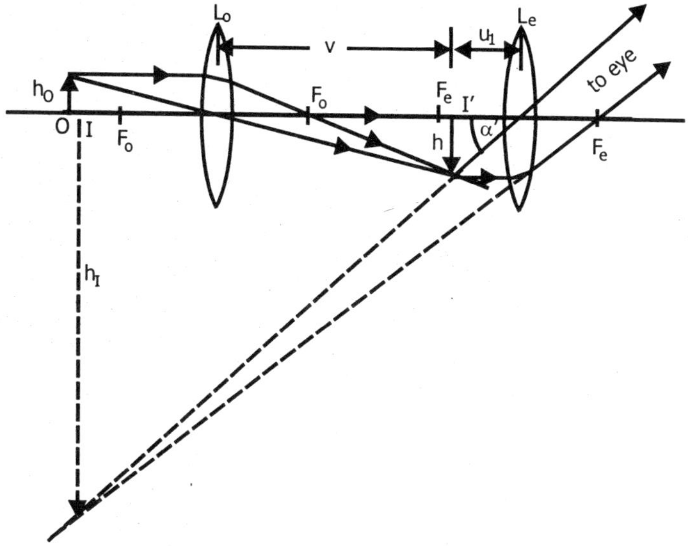

Fig. 7.5 Image formation in the compound microscope

The distance of the objective from the object O is a little greater than the focal length of the objective. A magnified inverted image I′ is formed at a distance beyond $2f_o$, where f_o is the focal length of the objective. This image acts as a virtual object for the eyepiece. The eyepiece is placed such that is serves as a magnifying glass for this virtual object, producing the magnified final image I. In normal adjustment, the final image is inverted, and formed at the least distance of distinct vision. This inversion, however, is not considered a disadvantage of the instrument, as objects usually viewed (e.g. bacteria, blood cells) need not be upright.

The angular magnification can be determined in terms of the magnifications produced by the two lenses. The magnification produced in the final image is the product of the magnifications in the two lenses. If u is the object distance from the objective, and u_1 is the distance from the image I′ to the eye lens, then

$$M = -\frac{v}{u} \times \frac{D}{u_1}, \tag{7.5}$$

where v is the distance of I′ from the objective. The length L of the microscope is the distance between the objective and the eyepiece. Thus,

$$L = v + u_1. \tag{7.6}$$

Like the simple microscope, it is also possible to form the final image at infinity. This is not the normal adjustment for the instrument; in this condition, the image I′ is at the focal point of the eye lens.

Example 7.4

A compound microscope consists of an objective of focal length 0.50 cm and an eyepiece of focal length 1.50 cm. An object is placed 0.55 cm from the objective, and the final image is formed at infinity. Calculate

(i) the angular magnification produced by the instrument,

(ii) the angular magnification that would be produced if the instrument was in normal adjustment, and find the length of the microscope in this adjustment.

Solution

(i) For the object lens, we have $\frac{1}{v} = \frac{1}{f_o} - \frac{1}{u} = \frac{1}{0.50\text{ cm}} - \frac{1}{0.55\text{ cm}} = 0.18\text{ cm}^{-1}$.

Thus, v = 5.5 cm. Since the final image is at infinity, the image formed by the objective lens is hence at the focus of the eye lens. The angular magnification is hence given by $M = \frac{v}{u} \times \frac{D}{u_1} = \frac{5.5\text{ cm}}{0.55\text{ cm}} \times \frac{25\text{ cm}}{1.50\text{ cm}} = 166.67$.

(ii) In normal adjustment, $\frac{1}{u_1} = \frac{1}{f_e} - \frac{1}{-D} = \frac{1}{1.50} + \frac{1}{25} = \frac{25 - 1.50}{25 \times 1.50}$. The magnification is hence given by $M = \frac{v}{u} \times \frac{D}{u_1} = \frac{5.5\text{ cm}}{0.55\text{ cm}} \times 25 \times \frac{25 \times 1.50}{23.5} = 398.90$

Again, in normal adjustment, the magnification is higher that that for which the final image is formed at infinity. From (7.5), the length of the microscope is given by $L = v + u_1 = 5.5\text{ cm} + 1.59\text{ cm} = 7.09\text{ cm}$.

7.5 TELESCOPES

Whilst the compound microscope is used for viewing near objects, the telescope is used for viewing objects that are far away. We study here two versions of the telescope. One type is used for viewing the planets and other similar bodies. Its major advantage is to aid astronomers in their studies of the planets and the galaxies. As a result, these telescopes are referred to as **astronomical** telescopes. The other class of telescopes is used to view distant objects on the earth. These are known as **terrestrial** telescopes.

7.5.1 Astronomical telescope

In normal adjustment, the astronomical telescope forms the final image is formed at infinity. The telescope consists of an objective L_o, and an eyepiece L_e, arranged coaxially as in **Fig. 7.6**.

Parallel rays from the object (e.g. a distant star) are brought to a focus by the objective lens to form an image I′ of height h at the focus. This image becomes the virtual object for the eye lens. The eye lens is placed such that its focus coincides with the position of the image I′. The final image is thus formed at infinity.

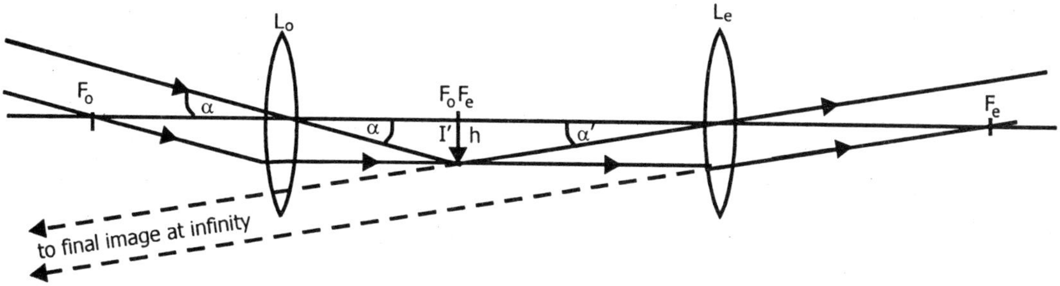

Fig. 7.6 Image formation in the astronomical telescope in normal adjustment

In the usual notation, the angular magnification M produced by the astronomical telescope is given by

$$M = \frac{\alpha'}{\alpha}, \tag{7.7}$$

where α is the visual angle subtended by the object at the objective lens, and α′ is the visual angle subtended by the image at the eye lens. From the above ray diagram for the instrument, we get $\alpha \approx \tan\alpha = \frac{h}{f_o}$, and $\alpha' \approx \tan\alpha' = -\frac{h}{f_e}$, since the angles α and α′ are small. The angular magnification therefore becomes

$$M = -\frac{f_o}{f_e}. \tag{7.8}$$

Thus the angular magnification is the ratio of the focal length of the objective to the focal length of the eyepiece; the greater this ratio the higher the angular magnification. It is important to bear in mind that the telescope does not really magnify the object. The object is already huge – a star, or some planet. What the instrument does is to increase the visual angle at which the object is viewed. Increase in visual angle implies greater clarity at which the object can be viewed. This brings out yet again the distinction between linear magnification and angular magnification.

Example 7.5

An astronomical telescope in normal adjustment is constructed using an objective of focal length 1.0 m, and an eyepiece of focal length 0.50 cm. Determine

(i) the angular magnification of the telescope,

(ii) the approximate length of the telescope.

Solution

The angular magnification can be found by a direct application of equation (7.8). The approximate length of the telescope tube is the sum of the focal lengths of the objective and the eyepiece.

(i) The angular magnification M is given by $M = -\frac{f_o}{f_e} = -\frac{1.0\text{ m}}{0.005\text{ m}} = -200$, the negative sign indicating an inverted image.

(ii) The approximate length of the telescope tube is $L = f_o + f_e = 1.005$ m.

7.5.2 Terrestrial telescope

These telescopes are used for viewing distant objects on earth – e.g. buildings, geological formations, mountains and other similar structures. In the compound microscope, and the astronomical telescope, the final images are inverted. This does not appear to be a serious drawback for these instruments, since the objects usually viewed do not appear to have upright position. In contrast, an upright image is required when viewing distant terrestrial objects.

To get an upright image, the terrestrial telescope makes use of an extra lens that re-inverts the image formed in the objective lens. The image formed in the objective is at a distance 2f from the re-inverting lens, where f is the focal length of the re-inverting lens. An upright image the same size is formed at distance 2f behind the lens. This latter image is then viewed in the eyepiece to produce a final upright image. **Fig. 7.7** shows the ray diagram of the terrestrial telescope. In this telescope, the final image is formed at the least distance of distinct vision.

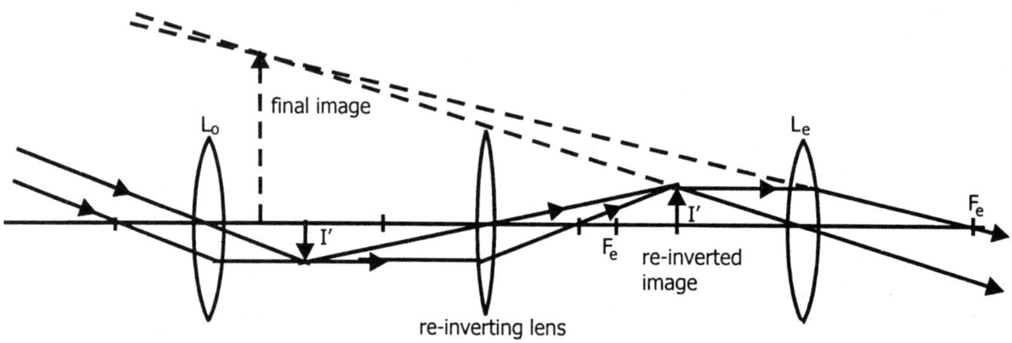

Fig. 7.7 Image formation in the terrestrial telescope in normal adjustment

We note that since the re-inverting lens produces an image the same size as its object, the inclusion of this lens does not affect the magnification of the final image. In some versions of the terrestrial telescope, the re-inversion is achieved with the use of right-angled totally reflecting prisms.

7.6 LENS ABERRATIONS

In optical systems such as lenses and curved mirrors, the deviation of light rays through lenses sometimes causes the images of objects to be blurred. In an ideal system, every point on the object will focus to a point of zero size on the image. In practice, however, each image point occupies a small volume of finite size, and unsymmetrical shape, causing some blurring of the whole image. This image blurring is described as **aberration**. Aberration can arise in many ways, the major ones of which are described below. A plane mirror is an exception to this aberration, as the images formed in plane mirrors are free from aberrations. A lens, however, is an imperfect image producer, becoming ideal only for rays passing through its centre and parallel to the optical axis. Consequently, the equations developed for object-image distances in a lens having spherical surfaces are only approximate, and hold true only for those rays very close to the principal axis.

With a source of monochromatic light, there are five aberrations for lenses. These are **spherical aberration**, **coma**, **astigmatism**, and **curvature of field** and **distortion**. When the source of light is not monochromatic, another aberration, called **chromatic aberration**, occurs for lenses, but not for mirrors. Below is a brief description of each of these aberrations; more detailed account of each aberration can be obtained in more specialized books on optics.

(a) Spherical aberration

In spherical aberration, rays of light from a point on the optical axis of a lens having spherical surfaces do not all meet at the same image point. Rays passing through the lens close to its centre are focused farther away than rays passing through a circular zone near the rim. For every cone of rays from an axial point on the object meeting the lens, there is a cone of rays that converges to form an image point, the cone being different in length according to the diameter of the circular zone. Hence a circular object appears elongated, very much like an ellipse.

(b) Coma

Coma is so called because a point image is blurred into a comet shape. The aberration is produced when rays from an object off the axis of a lens are brought together by different zones of the lens. In coma, the images of an off-axis object point are circular in shape, of varying size, but displaced with respect to each other. The usual way to reduce coma is to use a diaphragm that eliminates the outer cones of rays.

(c) Astigmatism

Unlike spherical aberration and coma, astigmatism results from the failure of a single zone of a lens to focus the image of an off-axis point at a single point. The two planes which are at right angles to each other passing through the optical axis are often called the meridian plane, and the sagittal plane. The meridian plane is the one containing the off-axis object point. Rays not in the meridian plane are focused farther away from the lens than those lying in the plane. The rays do not meet in a point focus, but rather meet as lines perpendicular to each other. Between these two positions the images appear elliptical in shape, rather than circular.

(d) Curvature of field

Curvature of field refers to the location of image points with respect to one another. In curvature of field, the image of a plane object perpendicular to the principal axis will lie on a paraboloidal surface. In photography, flat image fields are preferred in order to match the film plane and projection when the enlarging paper and the projection lie on a flat surface during printing of the film.

(e) Distortion

Distortion refers to the deformation of the image. There are two kinds of distortion, either of which may be present in a lens. These are barrel distortion, in which magnification decreases with distance from the principal axis of the lens, and pincushion distortion, in which magnification increases with distance from the axis of the lens.

(f) Chromatic aberration

The last aberration is chromatic aberration. This aberration results in the failure of a lens To focus all colours in the same plane. Because the refractive index is least at the red end of the electromagnetic spectrum, the focal length of a lens in air will be greater for red and green than for blue and violet. Because of this difference in the degrees of focussing, magnification is affected by chromatic aberration.

7.7 THE EYE AND VISION

We conclude this study with a closer analysis of the oldest, and most efficient optical instrument of all – the eye. The eye is the most fascinating of all optical instruments because of its perfection both as a sensory organ and as an optical system. The eye operates in a wide range of situations. It can detect light as small as one quantum, but will register this fact only if there are enough quanta to build up a pattern. It operates continuously over large intensity ranges, and has protection mechanisms designed to ensure the system as a whole is not overloaded. Images can be viewed over 180° angles with sufficient clarity in both the horizontal and vertical directions. Any part of the vision spectrum can be chosen and concentrated upon to derive the relevant information on that small area, while at the same time suppressing information from stimuli generated from the surroundings. The lens has an automatic adjustment mechanism, so that the focal length can be varied depending on the object distance. Most times, all the above fascinating yet complicated mechanisms are at work – and we do take it all for granted.

7.7.1 Parts of the eye

Fig. 7.8 shows the internal structure of the human or mammalian eye. The eye works like a camera, consisting of a lens L, which forms an inverted image of an object placed in front of it. The image is formed on the part known as the retina R. In the normal eye the eyeball is a sphere of radius about 2.3 cm, with a transparent bulge at the front, known as the cornea C, through which light enters the eye.

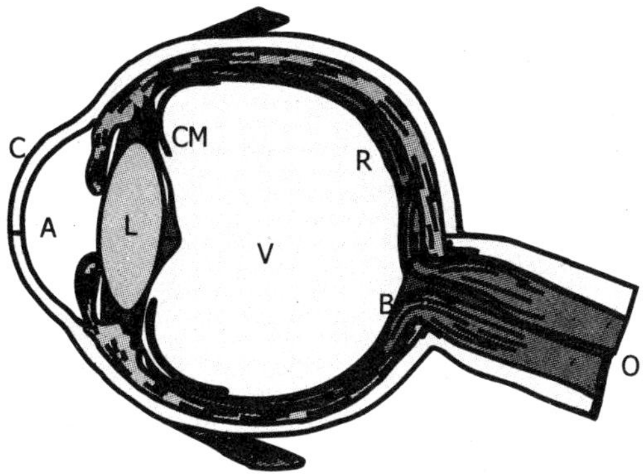

Fig. 7.8 The human eye

Most of the refractive power of the lens is thought to be due to the curved surface of the cornea. The coloured portion of the eye is the iris, having a central opening, the pupil. This controls the amount of light entering the eye at any one time. The lens is cellular in nature, and, with the aid of the ciliary muscles CM, has a focal length that can be adjusted. This property of focal length variation is referred to as **accommodation**, and enables the eye to view near as well as far objects.

The lens sits between two sets of liquids: the aqueous humour A in front, and the vitreous humour V behind the lens. The optic nerve O transmits signals from the retina to the brain. The region where the eye is not sensitive to light is known as the **blind spot** B.

7.7.2 Response of the eye

The light reaching the eye from the sun and other sources often spans a range of wavelengths both within and beyond the visible region of the electromagnetic spectrum. Studies show that the retina does respond to wavelengths far into the ultraviolet region; however, we do not see beyond the visible region.

In the short wavelength region (i.e. below about 300 nm), the cornea absorbs most of the light. In many instances, this is the cause of damage to the cornea. For this reason, it is suggested that dark glasses be worn when ultraviolet (UV) light is to be used, or when one is in intense sunlight. These glasses absorb UV radiation. Nowadays, specific glasses are manufactured which can cut off almost all the UV radiation.

In the long wavelength region, the water molecules in the cornea and aqueous humour absorb most of the energy in the light with wavelengths above 1200 nm. Also, the eye pigments do not respond to wavelengths above 800 nm. In the visible region (i.e. from 300 nm to 700 nm), the eye does not respond equally to all colours. The response varies with the intensity of the light, and also with the previous history of the individual.

7.7.3 Defects of vision

Whilst the normal eye is about the most efficient optical instrument, the eyesight of many people suffer from one defect or another, depending on the age and other circumstances of the individual. The major defects of the eye are **myopia, hypermetropia, presbyopia**, and **astigmatism**.

(a) Myopia

This defect is also known as **short-sightedness**. With this defect, the individual is able to clearly see near objects, but cannot see far objects with clarity. For people with this defect, the eyeball becomes a bit too long, and parallel rays are brought to a focus in front of the retina, instead of on the retina. Near objects can, however be focused properly to fall on the retina. This defect is corrected by wearing a diverging lens in front of the eye. The lens diverges the rays further before falling on the lens of the eye. Ray diagrams showing the defect and its correction are given in **Fig. 7.9**.

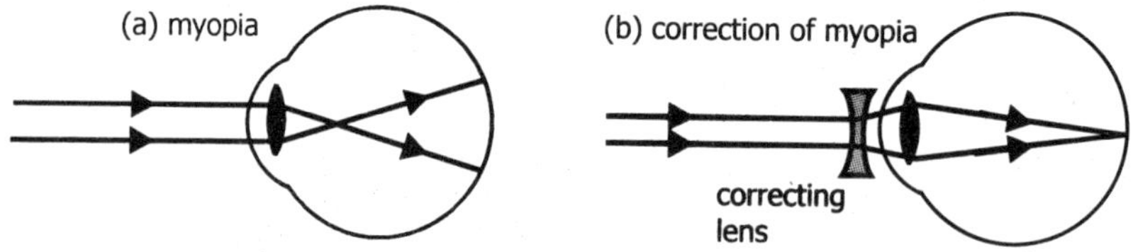

Fig. 7.9 Ray diagram showing (a) myopia, and (b) its correction using a concave lens.

(b) Hypermetropia

Also called **long-sightedness**, this defect exhibits itself in the individual not being able to focus near objects on the retina, but the eye works well for far objects. The defect is corrected by using a convex spectacle lens. **Fig. 7.10** shows ray diagrams for the defect and its correction.

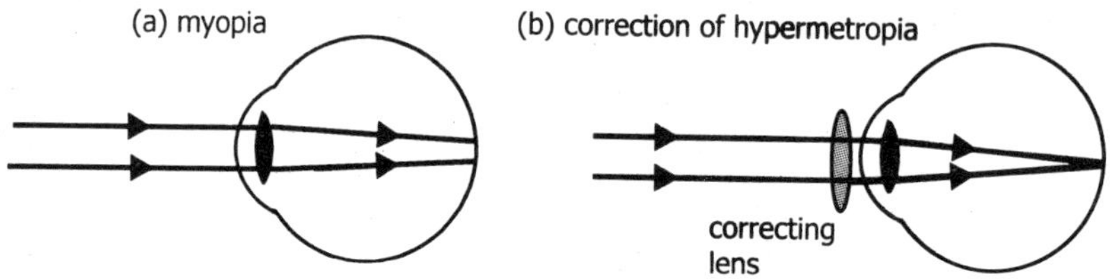

Fig. 7.10 Ray diagram showing (a) hypermetropia, and (b) its correction using a convex lens.

(c) Presbyopia

This defect is often called old sight. As people get older, the ciliary muscles weaken and the eye lens loses some of its elasticity. The eye therefore loses some of its power of accommodation, as the individual grows older. To correct this defect, converging lenses are used, as in the case of hypermetropia. The vision of the individual for distant objects, however, remains good; it is the vision for near objects that are impaired. Thus the correcting glasses are used only in those circumstances when the individual needs to view near objects – e.g. in reading.

(d) Astigmatism

In astigmatism, point objects do not form point images on the retina. This defect results from the defect on the cornea. The cornea of a person suffering from astigmatism has unequal curvature in different directions. If the curvature is larger in the horizontal direction than in the vertical direction, rays are brought to a focus quicker in the horizontal direction than in the vertical direction. The image of a point thus appears elongated in the vertical plane. Astigmatism is corrected by a cylindrical lens. The extra curvature of the lens in one direction compensates for the lack of curvature in the eye in that direction.

Often, a person's eyesight suffers from more than one defect – e.g. astigmatism and myopia. A lens that corrects for both defects is rather complex, and results from careful observations of the retina under varying light conditions. An instrument designed to observe the retina is the **ophthalmoscope**.

7.8 THE OPHTHALMOSCOPE

The ophthalmoscope is an instrument used to view the retina of the eye. From this viewing, defects of the retina, if present, can be discerned, and appropriate measures taken to correct the defect. The ophthalmoscope is an optical instrument in its own right, even though it is employed only to view one object – the retina.

In its basic form, the ophthalmoscope consists of a plane mirror M attached to a handle, the latter containing a small light bulb B, and a lens L, whose position can be adjusted. Light from B is converged by the lens L to fall just before the retina inside the eye being studied. A diverging beam of light that originated from B illuminates the retina. **Fig. 7.11** is a ray diagram illustrating the operation of the ophthalmoscope.

With the system, the ophthalmologist is effectively using the optical system of the eye as a simple magnifying glass in order to obtain a magnified image of the retina. If the observer has defective eyesight, a collection of converging lenses C is available to correct this defect. These correcting lenses are arranged such that they can easily be swung into position for viewing.

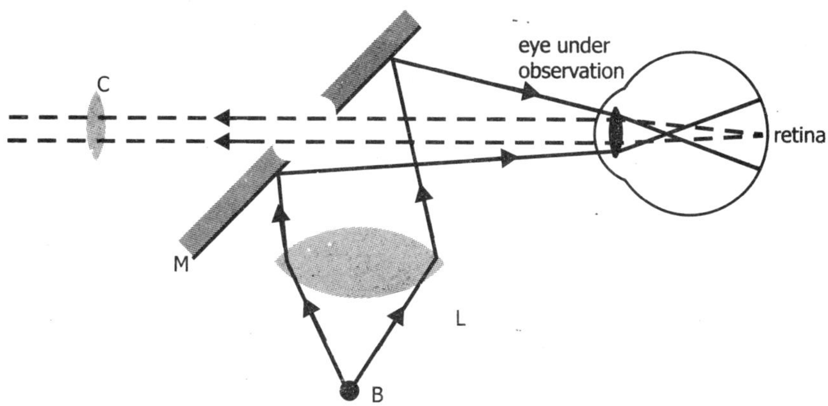

Fig. 7.11 Optical diagram of the ophthalmoscope

EXERCISES

1. A page of print is viewed through a magnifying glass of focal length 5 cm, the final image bring formed at the least distinct vision, 25 cm. Calculate
   (i) the lateral magnification produced,
   (ii) the angular magnification produced.

2. An insect on a leaf is examined through a magnifying glass of focal length 10 cm, the glass being held in such a position that the final image is formed at a distance of 30 cm from the lens. Determine the angular magnification produced.

3. A compound microscope has an eyepiece of focal length 2 cm, at a distance of 16 cm from the objective lens of focal length 1 cm.
   (a) Where must an object be placed in order that the final image is formed at infinity?
   (b) Calculate the magnification of the instrument in this condition.

4. The objective and eyepiece of a compound microscope have focal lengths of 3 mm and 1.5 cm, respectively. The tube length of the instrument is 14.2 cm, and the final image of the object is viewed at a distance of 25 cm from the eyepiece. Determine the position of the object.

5. An astronomical telescope which has an objective lens of focal length 50 cm and an eyepiece of focal length 1 cm is used to view the moon when in normal adjustment. Calculate
   (i) the angular magnification produced by the instrument,
   (ii) the visual angle subtended by the image at the eyepiece, given that the moon has diameter 3.50×10^6 m, and the distance from the earth to the moon is 3.84×10^8 m.

6. A compound microscope consists of two lenses of focal lengths 0.50 cm and 3.0 cm, at a distance of 20.0 cm apart. The microscope is used by an observer whose least distance of distinct vision is 27.0 cm. Calculate the magnifying power of the instrument if the final image is formed at infinity.

7. An astronomical telescope has an objective lens of focal length 1.0 m, and an eyepiece of focal length 2.0 cm. Determine the magnifying power of the instrument
   (i) in normal adjustment,
   (ii) when the image is viewed at the least distance of distinct vision of 25 cm.

8. Describe **normal adjustment** for an astronomical telescope. Show that the angular magnification for such a telescope is given by the ratio of the focal lengths of the lenses.

9. Two lenses having powers of +5.0 and +33.3 dioptres are used to construct a telescope. Calculate
   (i) the angular magnification of the telescope in normal adjustment,

(ii) the distance by which the separation of the lenses must be altered in order to produce an image at the least distance of distinct vision, 25.0 cm.

10. The lenses of an astronomical telescope adjusted for normal use are 21.0 cm apart, and the angular magnification is 6.0. Find the focal lengths of the lenses.

11. An astronomical telescope consisting of an objective of focal length 75.0 cm and an eyepiece of focal length 3.0 cm is focused on a distant object. Calculate the magnification produced when

(i) the final image is formed at infinity,

(ii) the final image is formed at the least distance of distinct vision, 25.0 cm.

12. The dimensions of the picture on a slide film are 24 mm x 36 mm. The projector forms an image on a screen 6.0 m from the projector lens. If the focal length of the projector lens is 12.0 cm, calculate

(i) the distance from the slide to the projector lens,

(ii) the dimensions of the image formed on the screen.

13. The moon subtends an angle of 0.009 rad at the unaided eye. An astronomical telescope in normal adjustment using an objective of focal length 1.80 m and an eyepiece of focal length 42.0 cm is used to observe the moon. Calculate the angle subtended at the eye by the final image.

14. A magnifying glass is used to view a small object in normal adjustment. The lens produces an image with magnification 3.4, at a distance of 25.0 cm from the lens. Calculate the focal length of the lens.

15. On a roll of movie film, each picture has a width of 70.0 mm. The projector lens has a focal length of 305 mm. If the screen is 60 m away from the projector lens, determine the width of the image projected on the screen.

16. To focus a camera on objects at different distances, the converging lens is moved toward or away from the film, so a sharp image will always fall on the film. A camera with a lens of focal length 20 cm is to be focused on an object located at distances ranging from 3.5 m to 50.0 m from the lens. Find the range of movement of the camera lens.

Part 2

Electricity, Electromagnetism, and Electronics

CONTENTS

8

CHAPTER

Electrostatics

8
CHAPTER

Electrostatics

We begin our study of electricity and electromagnetism with electric charges at rest, a subject known as **electrostatics**. Although most of the common applications of electricity involve charges in motion, a lot can be learnt about electric charges in motion from a study of their behaviour at rest.

The study of electrostatics has given us an insight into many applications. Many molecular bonds, for instance result from electrostatic forces between charged positive and negative centres in the atoms that make up the molecule. High-energy "atom-smashing" experiments involve collisions of positive and negative charges travelling at high speeds. The results of these experiments have yielded much of the information we know today on the nature of molecules and molecular forces. In addition, we have many everyday applications of electrostatic charges – electrostatic precipitation in chemical industries, the office photocopier, and the use of polythene food wrappers, to name a few.

8.1 METHODS OF CHARGING

All matter are made up of atoms, each atom having positive and negative electrical charges. The positive charges are called **protons**, and the negative charges are called **electrons**. A third class of particles, the **neutrons**, is also present in the atom. The electrons have the same magnitude of charge as the proton, but the protons are about 2000 times heavier than the electron. The neutron carries no charge, and has nearly the same mass as the proton. The neutrons and protons occupy the central **nucleus**, and electrons orbit the nucleus. **Fig. 8.1** is a pictorial representation of the atom.

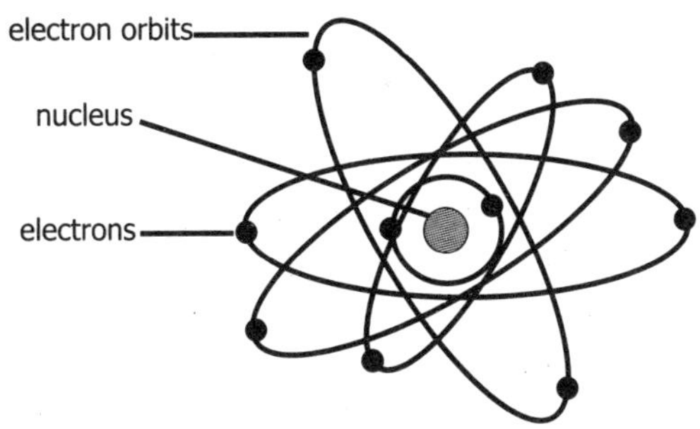

Fig. 8.1 Pictorial representation of the atom. This is a highly simplified representation of the atom. The electron orbits are shown here as closed loops, whilst in actual fact, the orbits are a lot more complicated.

A neutral atom has equal numbers of protons and electrons. The sodium atom, for example, has 12 neutrons and 11 protons in the nucleus, surrounded by 11 electrons in orbits. A material is said to be charged if there are more of one kind of charge than the other. A negatively charged body has excess electrons over protons, whilst a positively charged body has excess positive charges over electrons. The protons are tightly bound in the nucleus, making them very difficult to remove. Charging a body therefore involves the removal, addition, and rearrangement of the orbital electrons. In the sections that follow, we study ways in which this addition, removal or rearrangement is achieved in practice. A body becomes positively charged if it loses electrons, and negatively charged if electrons are added to it. Removal of electrons usually involves the outermost or valence electrons, as these are the least bound to the nucleus. . Bodies are charged by **friction**, by **contact** with a charged body, and by **induction**.

8.1.1 Charging by friction

Charging by friction is the oldest form of charging – it dates back to the very discovery of electrostatics. It was found that when an amber rod is rubbed with fur, the rod became negatively charged. Also, a glass rod rubbed with silk cloth became positively charged. The mechanical energy used in the rubbing goes to remove electrons from one of the bodies and transfer these electrons to the other. In the end, the two bodies acquire opposite signs of electricity; one gets positively charged, whilst the other becomes negatively charged. When two bodies are charged by friction, they acquire the same magnitude of charge. The charges acquired are opposite in sign. Furthermore, the bodies retain these excess charges even when they are separated from each other.

8.1.2 Charging by contact

In charging by contact, an uncharged body is brought into contact with a charged body. The uncharged body acquires the same sign of charge as the charged body. The total charge is distributed between the two bodies.

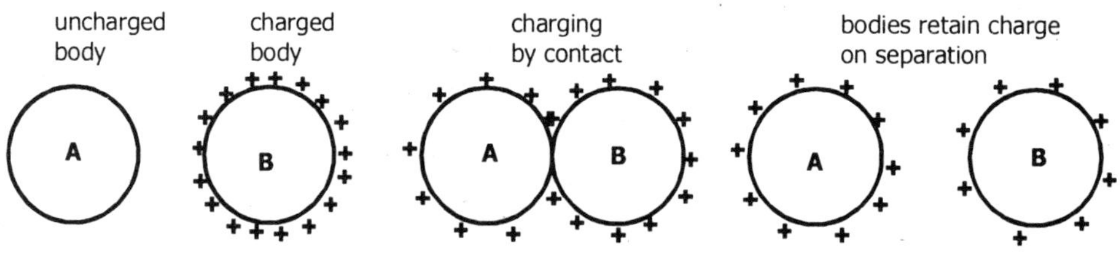

Fig. 8.2 Charging a body by contact with a charged body. Charge shared by between the two bodies, remain on the bodies after separation.

Fig. 8.2 depicts the distribution of charge. An uncharged sphere is brought into contact with a positively charged sphere B. The charge on B is distributed on the two spheres. On separation, A retains the charge acquired through the contact. Notice that the two spheres do not necessarily share the charge equally.

8.1.3 Charging by induction

In charging by induction, an uncharged body is brought close to, but not touching, a charged body. Charging by induction is an example of the rearrangement of charges between two bodies. The charge **induced** on the previously uncharged body is of the same magnitude as the **inducing** charge of the charged body. However, the sign of the induced charge is opposite to that of the inducing charge. Furthermore, the induced charges last only while the inducing charge is present. The induced charges disappear when the inducing charge is taken away.

We can understand the origin and behaviour of induced charges by considering a simple experiment, depicted by **Fig. 8.3**. The figure shows two metallic spheres, A and B, with A carrying a positive charge, with B originally uncharged. If sphere B is brought near A, the free electrons in B will be attracted to the surface nearest A. This will leave one side of B charged negatively, while the other side will have an equal positive charge. If we take B away, the positive and negative charges on its surface neutralise each other.

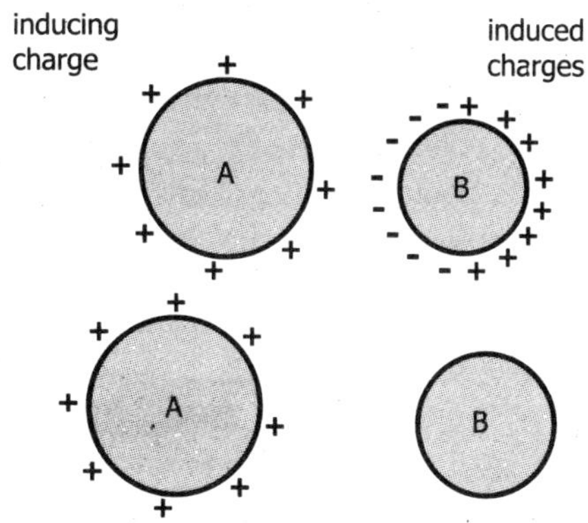

Fig. 8.3 Charging by induction

8.2 CONDUCTORS AND INSULATORS

The behaviour of bodies towards static electricity puts them into one of two categories: conductors, on the one hand, and insulators on the other. Insulators do have the ability to retain the charge placed on them even when touched with the fingers. Conductors, on the other hand lose their charge very easily when touched. Examples of conductors are metals, whilst organic compounds and ceramics are examples of insulators.

Experiments show that charges placed on a metallic conductor reside on the outside of the conductor. The surface distribution of charge (i.e. charge per unit area) varies from point to point on the surface, depending on the curvature at the point. The greater the curvature the greater the density of charge. A material with uniform curvature (e.g. sphere) would have its charge evenly distributed on the surface. **Fig. 8.4** illustrates this phenomenon. In (a), the charge is evenly distributed on the surface of the metal sphere; a pear-shaped metal body on the other hand shows a higher density of charge at those points of greater curvature than points of smaller curvature.

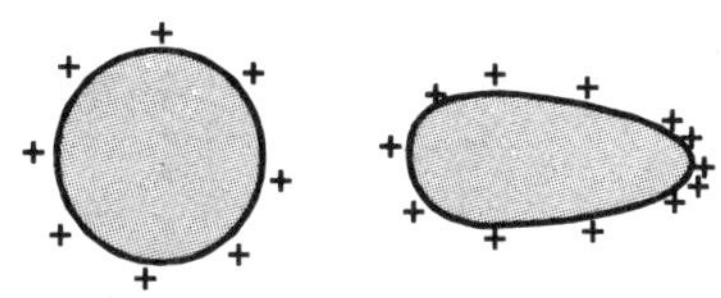
Fig. 8.4 Charge distributions on the surface of a conductor. Density of charge at any point depends on the curvature at that point.

8.3 THE GOLD LEAF ELECTROSCOPE

One of the earliest instruments designed to study the behaviour of static charges is the

gold-leaf electroscope. The electroscope consists of a pair of gold leaves suspended from an electrical conductor that leads to the outside of an insulating chamber. **Fig. 8.4** shows the features of this instrument. An electric charge brought near to, or on the top conducting plate causes the leaves to stand apart from each other at an angle. The like electric charge either induced in, or transferred to each leaf causes them to repel each other.

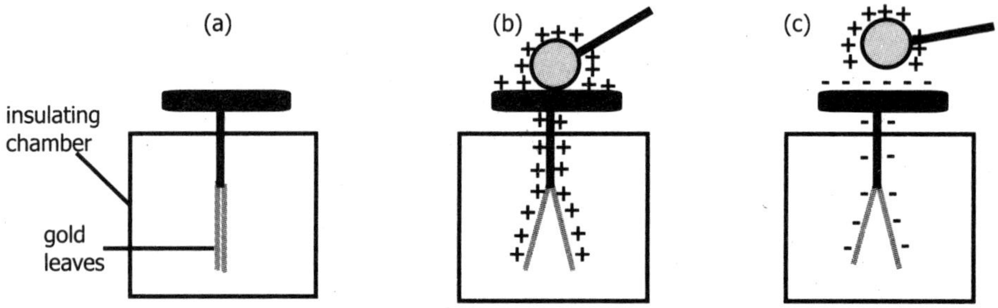

Fig. 8.4 The gold-leaf electroscope; (a) the uncharged electroscope, (b) charging electroscope by contact, (c) charging by induction.

A charged electroscope can be used to determine the sign of the charge on an object by bringing the object in close proximity to the electroscope. Induced charges on the electroscope would cause the leaves to either diverge further from each other, or move toward each other. Some of the charge on the leaves is neutralised by the induced charges if the body has the same sign of charge as that originally on the electroscope.

8.4 COULOMB'S LAW

Early experiments on static charges had established simple rules concerning the force between charged bodies. These were

(i) Like charges repel each other, whilst unlike charges attract each other,

(ii) Charged bodies attract uncharged bodies.

These rules could not, however, allow a quantitative description of the force acting on charged objects. Experiments to investigate the nature of the force between charges were performed by Charles Augustin de Coulomb (1736 – 1806), and the law resulting from these experiments became known as **Coulomb's law**.

Using an apparatus similar to the torsion balance, Coulomb measured the force between two charged bodies as functions of the quantity of charge placed on them, and also as a function of their separation. The results of these investigations indicated that the force between two charges has the following properties:

1. varies directly with the magnitude of the charges,
2. varies inversely with the square of the distance between the charges,
3. acts along the line joining the charges,
4. is independent of the presence of other charges within the vicinity.

These findings are summarised into Coulomb's law, which states:

The force between two point charges is directly proportional to the product of the charges, inversely proportional to the square of the distance between the charges, and acts along the line joining the charges.

Mathematically, we state Coulomb's law for two point charges Q_1 and Q_2, a distance r apart, as

$$\mathbf{F} = k\frac{Q_1Q_2}{r^2}\hat{\mathbf{r}}, \tag{8.1}$$

where $\mathbf{F}$ is the force between the charges, $\hat{\mathbf{r}}$ is a unit vector along the line joining the charges, and k is a constant. A repulsive force is given a positive sign, while an attractive force is assigned a negative sign.

The S.I. unit of electrostatic charge is the Coulomb. The Coulomb itself is a very large unit. Hence sub-multiples (e.g. nano-Coulomb, and micro-Coulomb, etc.) are in greater use. The charges on the atomic particles – the proton and the electron are as follows: proton charge $= +1.6021 \times 10^{-19}$ C; electron charge $= -1.6021 \times 10^{-19}$ C. Based on the S.I. system, for two charges placed in vacuum, or free space as it is usually known, the constant k in equation (8.1) is given by $k = \frac{1}{4\pi\varepsilon_o}$, where ε_o is a constant, known as the **permitivity of free space**. With this, Coulomb's law becomes

$$\mathbf{F} = \frac{1}{4\pi\varepsilon_0}\cdot\frac{Q_1Q_2}{r^2}\hat{\mathbf{r}}. \tag{8.2}$$

Whilst we always recognise the vector nature of the force between the charges, in most problems requiring the use of Coulomb's law, the direction of the force is already known, depending on the signs of the charges on the two bodies. In such cases, the use of (8.2) is simplified as we need concentrate only on the magnitude of the charges. The value of ε_o is 8.85×10^{-12} C²/N/m². Approximate values for easy calculation are $k = \frac{1}{4\pi\varepsilon_o} = 9 \times 10^9$ Nm²/C², and $\varepsilon_0 = \frac{10^{-9}}{36\pi}$ C²/N/m².

Example 8.1

Compare the magnitudes of the electrostatic and gravitational forces between the two protons in the helium nucleus, given that they are a distance 1.0×10^{-15} m apart. Mass of the proton $= 1.667 \times 10^{-27}$ kg, charge $= 1.602 \times 10^{-19}$ C. Use any of the figures above for the permitivity of free space. Universal gravitational constant $= 6.67 \times 10^{-11}$ Nm²/kg².

Solution

Using Coulomb's law, we calculate the electrostatic force between the two protons as

$$F_e = \frac{1}{4\pi\varepsilon_o}\cdot\frac{Q_1Q_2}{r^2} = 9\times10^9 \times \frac{(1.602\times10^{-19})^2}{(1.0\times10^{-15})^2} = 9\times10^9 \times (1.602\times10^{-4})^2 = 230.976\text{ N.}$$

The gravitational force between two bodies is given by an expression very much similar to that for the electrostatic force. In this case, this force F_g is given by

$$F_g = G\frac{m_1m_2}{r^2} = 6.67\times10^{-11} \times \frac{(1.667\times10^{-27})^2}{(1.0\times10^{-15})^2} = 6.67\times10^{-11} \times (1.667\times10^{-12})^2\text{ N.}$$

This reduces to 1.854×10^{-34} N. Comparing these two figures we see why the gravitational force is ignored when we carry out calculations of the force between charged bodies.

8.5 DISCRETE CHARGES

The force between two charges is independent of the presence of other charges. This notion can lead us to find the net force acting on a single charge as a result of forces exerted by other charges in the same vicinity. This is done by adding the forces to produce a resultant. The principles used for vector addition and subtraction in our study of mechanics apply very well in these cases. The only difference here is that the forces are electrostatic, rather than mechanical. The resultant force acting on a charge Q by a set of discrete charges Q_i, distributed in space is hence given by

$$\mathbf{F} = \frac{Q}{4\pi\varepsilon_o}\cdot\sum_{i=1}^{n}\frac{Q_i}{r_i^2}\hat{\mathbf{r}}_i, \qquad \text{(vector sum)} \qquad (8.3)$$

where r_i are the corresponding distances of the charges from Q.

The illustration of **Fig. 8.5** shows a distribution of discrete charges, labelled Q_1, Q_2, and Q_3, in space. The resultant force on a charge Q in the vicinity is the vector sum of the forces on Q due to the presence of the other charges. Thus, the magnitude of the force F on the charge Q is as follows:

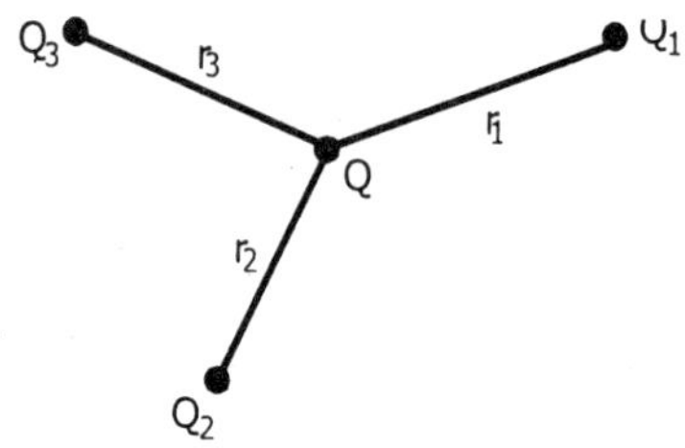

Fig. 8.5 Coulomb's law applied to a distribution of discrete charges in space

$$F = \frac{Q}{4\pi\varepsilon_o}\sum_{i=1}^{n}\frac{Q_i}{r_i^2} = \frac{Q}{4\pi\varepsilon_o}\left[\frac{Q_1}{r_1^2}+\frac{Q_2}{r_2^2}+\frac{Q_3}{r_3^2}\right]. \qquad (8.4)$$

The direction of the resultant is found by using the rules of vector algebra. Very often, the nature of the problem does offer certain simplifications for easy calculation. The following example illustrates the application of this summation principle.

Example 8.2

Fig. 8.6 shows two point charges, of +8.0 nC, and −8.0 nC, placed at points X and Y, 0.5 m apart in vacuo. Determine the magnitude and direction of the force that would act on a point charge Q of −2.0 nC

(i) placed at A

(ii) placed at B

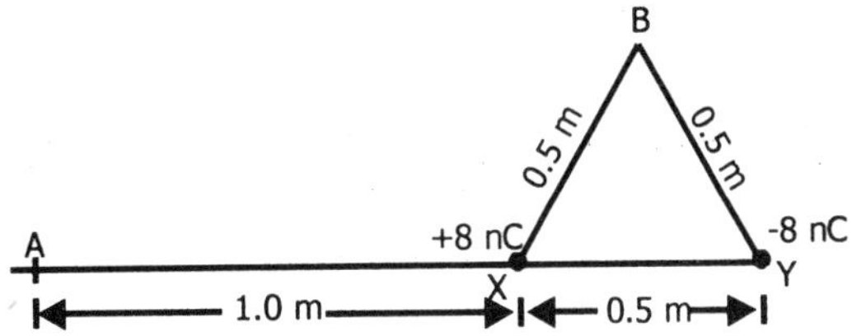

Fig. 8.6

Solution

When the −2.0 nC charge is placed at either of positions A and B, the net force acting on it is the resultant of the Coulomb forces F_X and F_Y due to X and Y, respectively. **Fig. 8.7** shows these forces for the two given positions of Q. In either position the force due to Y is repulsive, whilst that due to X is attractive. This results from the nature of the charges. We set up Coulomb's law equations to find the values of the forces, and then combine the forces together using vector addition methods to get the resultant force acting on Q.

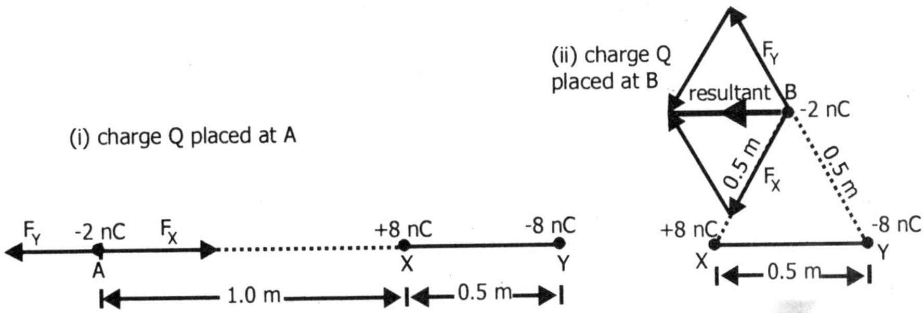

Fig. 8.7 Interaction for diagram for (i) Q placed at A, and (ii) Q placed at B. Force vectors not drawn to scale. F_Y is repulsive, whilst F_X is attractive.

(i) The resultant force at A has magnitude given by $F = F_X - F_Y$, where F_X and F_Y are in the directions shown. $F_X = \frac{1}{4\pi\varepsilon_0} \times 2 \times 10^{-9}\,C \times \frac{8 \times 10^{-9}\,C}{1^2} N = 1.44 \times 10^{-7} N$,

$$F_Y = \frac{1}{4\pi\varepsilon_0} \times 2 \times 10^{-9}\,C \times \frac{8 \times 10^{-9}\,C}{1.5^2} = 6.4 \times 10^{-8} N.$$

Thus, the net force at A =8.0 x 10^{-8} N, in the direction of the force F_X, as shown.

(ii) When the charge Q is placed at the position B, the forces due to the charges at X and Y are not along the same straight line, as we had in (i). We get the resultant force by applying the law of vector addition. We resolve both F_X and F_Y into horizontal and vertical components. Since the distance from each charge to B is the same, the vertical components are equal in magnitude and opposite in direction. They thus cancel each other out. Hence the resultant force at B is the sum of the horizontal components of F_X and F_Y, in the direction shown.

The resultant force F is given by

$$F = 2F_X \cos 60^o = 2 \times \frac{1}{4\pi\varepsilon_o} \times 2 \times 10^{-9}\,C \times \frac{8 \times 10^{-9}\,C}{0.5^2} \times 0.5000\,N = 5.76 \times 10^{-7}\,N,$$

in the direction shown.

8.6 ELECTRIC FIELD

8.6.1 Concept of a field

In the study of physics, we often come across the concept of a field.

A field in physics may be described as a continuous distribution of some observable quantity in space and time.

The observable quantity, which must be measurable, may be any of a variety of phenomena, including properties like colour in a liquid, density of dust in the atmosphere, temperature distribution in a body, the flow pattern in a stream, or the magnetic field surrounding the earth. Whatever is described as the field must be measurable using some technique. If the observed quantity varies in space but does not vary with time, the field is said to be **static**. Otherwise it is a **time-varying** field.

The quantity measured can either be material or non-material. Examples of material fields include dust storms, and fluid pressure at a point. Examples of non material fields are gravitational field, magnetic field, and electrostatic field. The corresponding forces in these fields are the gravitational force, the magnetic force, and the electrostatic force.

If the field considered is such that the work done (or energy expended) between two points in the field is independent of the path taken between the points, the field is said to be **conservative**. Gravitational, magnetic, and electrostatic fields are examples. If, on the other hand, the work done (or energy expended) between two points depends on the path, the field is said to be **dissipative**. A common feature with conservative force fields is that the force varies with the inverse-square of the distance from the origin of the field. In the study of electromagnetism, we deal with electric and magnetic fields.

8.6.2 Electric field intensity

The force of attraction or repulsion between charges is explained by considering that each charge creates around itself an electric field which acts on charges placed anywhere in the field. At any point, the strength of the field is described by the term "**electric field intensity**", also called **electric field strength**, or more often simply the **electric field**.

The electric field intensity at a point in an electric field is defined as the force acting per unit positive test charge placed at the point.

A test charge is considered to be a charge so small that its introduction in the field does not disrupt the field distribution at the point. Electric field intensity is a vector quantity having the same direction as the force acting on the test charge at the point. If $\mathbf{F}$ is the force on the test charge q, the electric field intensity $\mathbf{E}$ at the point is given by

$$\mathbf{E} = \frac{\mathbf{F}}{q}. \tag{8.5}$$

Since positive charges repel one another, the electric field around an isolated positive point charge is oriented radially outward. A more correct form of equation (8.5), which takes into account the condition that the charge must be as small as possible, gives

$$\mathbf{E} = \lim_{q \to 0} \frac{\mathbf{F}}{q}. \tag{8.6}$$

Whilst (8.6) is mathematically more accurate, equation (8.5) is the one used in most numerical calculations. The force acting on a charge placed in an electric field is

$$\mathbf{F} = q\mathbf{E}. \tag{8.7}$$

The unit of electric field intensity is N/C (i.e. Newton/Coulomb).

8.6.3 Electric field of a point charge

Using Coulomb's law we can derive an expression for the electric field due to a point charge placed some distance from a fixed charge Q. The force $\mathbf{F}$ experienced by a test charge q placed at a distance r from the charge Q is given by

$$\mathbf{F} = \frac{1}{4\pi\varepsilon_0} \frac{Qq}{r^2} \hat{\mathbf{r}}.$$

The electric field intensity $\mathbf{E}$, being defined as the force per unit charge, is given by

$$\mathbf{E} = \frac{\mathbf{F}}{q} = \frac{1}{4\pi\varepsilon_0} \frac{Q}{r^2} \hat{\mathbf{r}}. \tag{8.8}$$

From (8.8), it is seen that for a given point in the field, the bigger the charge, the larger the magnitude of the electric field. Similarly, for a given charge, the electric field is larger for points closer to the charge than for points farther away.

For a set of discrete point charges, the resultant electric field at a point is given by the vector sum of the fields created by the charges at the point. We thus have the resultant field **E** given by

$$\mathbf{E} = \frac{1}{4\pi\varepsilon_0}\sum_{i=1}^{n}\frac{Q}{r^2}\hat{\mathbf{r}}. \qquad \text{(vector sum)} \qquad (8.9)$$

8.6.4 Electric field lines and electric flux

Electric field lines are imaginary lines drawn to depict the direction of the electric field at the point. In this representation, the field lines are shown starting on a positive charge and ending on a negative charge. Electric field lines indicate the path a small positive test charge would take if placed in the field. The tangent to an electric field line at a point indicates the direction of the electric field at that point. Where the field lines are close together the electric field is stronger than where the field lines are far apart.

A related quantity, the **electric flux**, defines the total quantity of electric field lines emanating from a given point charge. For a point charge Q placed in vacuum,

$$\text{Electric flux} = \frac{Q}{\varepsilon_0}, \qquad (8.10)$$

and the electric field E at a point distant r from the charge has magnitude given by

$$E = \frac{\text{Electric flux}}{4\pi r^2}. \qquad (8.11)$$

Fig. 8.8 shows electric field lines around a positive point charge, and between a positive and a negative point charges.

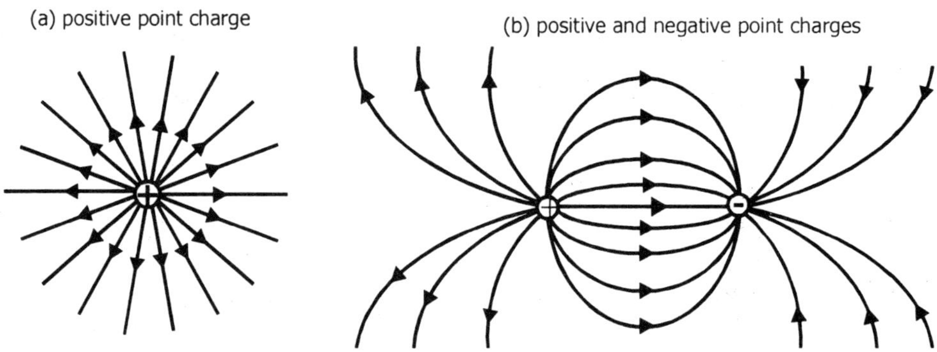

Fig. 8.8 Electric field lines (a) around a point positive charge, and (b) between a pair of positive and negative point charges.

Example 8.3

A proton starting from rest moves in the direction of a uniform electric field in which it attains a speed of 10^6 m/s in 40 cm. Calculate the magnitude of the electric field. Charge on proton = $+1.602\times10^{-19}$ C, mass of proton = $+1.667\times10^{-27}$ kg.

Solution

In solving this problem, we equate the work done by the electric field to the change in kinetic energy of the proton. If E is the electric field intensity and e the charge on the proton, the force on the electron is eE, and the gain in kinetic energy in moving a distance x, we have $eEx = \frac{1}{2}mv^2$, where eEx represents the work done by the field in moving the charge through the distance x. Thus we have $E = \frac{1}{2}\frac{mv^2}{ex}$. Substituting the numerical values in the problem gives $E = \frac{1}{2}\times\frac{1.667\times10^{-27}\,kg\times10^{12}\,m^2/s^2}{1.602\times10^{-19}\,C\times0.40\,m} = 1.30\times10^4$ N/C.

Example 8.4

A, B, and C are the vertices of a triangle, with the sides having the following measurements: AB = 13.0 cm, BC = 12.0 cm, and CA = 5.0 cm. Point charges of 10 nC and 20 nC are placed at A and B, respectively. Determine the magnitude and direction of the resultant electric field at C.

Solution

Fig. 8.9 is a sketch of the arrangement of the charges. From the figure, it is seen that the triangle is a right-angled triangle. The resultant field E at C is the vector sum of the fields E_A and E_B, due to the charges at A and B, respectively.

Using (8.8), the magnitudes of E_A and E_B are as follows:

$$E_A = \frac{1}{4\pi\varepsilon_0}\cdot\frac{10\times10^{-9}}{0.05^2} = 3.6\times10^4\ \text{N/C},$$

$$E_B = \frac{1}{4\pi\varepsilon_0}\cdot\frac{20\times10^{-9}}{0.12^2} = 1.25\times10^4\ \text{N/C}.$$

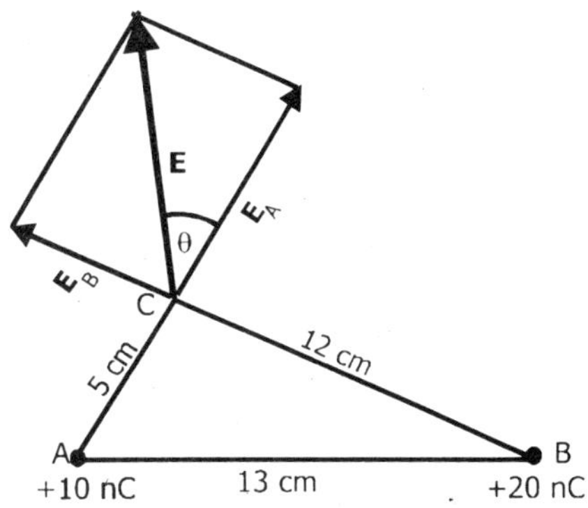

Fig. 8.9 Example of field due to a distribution of discrete point charges.

Thus, $E = \sqrt{E_A^2 + E_B^2} = \sqrt{(3.6\times10^4)^2 + (1.25\times10^4)^2} = 3.81\times10^4$ N/C. The direction of this resultant field is given by θ, where $\theta = \tan^{-1}\left(\frac{E_B}{E_A}\right) = \tan^{-1}\left(\frac{1.25\times10^4}{3.6\times10^4}\right) = 19.15^o$.

8.7 DIELECTRICS

When a mechanical force is applied to a system that is not free to move, the applied force sets up a mechanical strain within the material. The strain depends on the force applied, and the nature of the material. Similarly, if an electric field is applied to a material that has no free charges, an electric strain is produced. The magnitude of this strain depends on the electric field strength and the nature of the material.

Materials that do not conduct electricity are called **insulators**. For purposes of electrostatics, they are known more as **dielectrics**. Examples of such materials are ceramics, paper, glass, and certain pastes. These materials contain mainly covalent bonds between the atoms. In these bonds, each electron is part of an electron-pair bond, and therefore not free to move. By contrast, the valence electrons in metals are free to move, and contribute to the high conduction of both electricity and heat.

When a dielectric is placed in an electric field, there is a relative shift of the positive and negative charge in opposite directions within the dielectric. This phenomenon is known as **electric polarisation**. Polarisation occurs when an electric field distorts the negative cloud of electrons around the positive atomic nuclei, in a direction opposite to the field. This slight separation of the charges makes one side of the atom slightly positive, and the opposite side slightly negative.

The displacement of charges due to the electric field is described by a quantity called the **electric displacement**. The magnitude of the displacement vector is the displaced charge per unit area of the surface. Thus we have

$$D = \frac{Q}{A}. \qquad (8.12)$$

The electric displacement depends on the strength of the electric field, and the material of the dielectric. This relationship is given by the equation

$$\mathbf{D} = \varepsilon \mathbf{E}, \qquad (8.13)$$

where **D** is the displacement corresponding to the electric field intensity **E**. Displacement is a vector quantity, having the same direction as the electric field intensity vector. The constant of proportionality ε is known as the **absolute permitivity**, or simply the **permitivity** of the dielectric medium.

More usually, the dielectric property of the medium is described in terms of its **relative permitivity**, or **dielectric constant** ε_r, being the ratio of the permitivity of the medium to the permitivity of free space ε_0. Sometimes k is used instead of ε_r. Thus,

$$k = \varepsilon_r = \frac{\varepsilon}{\varepsilon_0}. \qquad (8.14)$$

A more practical definition of the dielectric constant is covered in the study of capacitors, done later in the chapter. The dielectric constant is a property of the insulator, and varies from one dielectric to another.

The effect of a dielectric on the force between charges, and on the electric field at a point, is to reduce both quantities by the factor equal to the dielectric constant. Thus two point charges Q_1 and Q_2 placed a distance r apart in a medium with dielectric constant ε_r, will exert a force **F** on each other, given by

$$\mathbf{F} = \frac{1}{4\pi\varepsilon_r\varepsilon_o} \cdot \frac{Q_1 Q_2}{r^2} \hat{\mathbf{r}}. \tag{8.15}$$

The electric field intensity **E** at a point distant r from a point charge Q in a dielectric is

$$\mathbf{E} = \frac{1}{4\pi\varepsilon_r\varepsilon_o} \cdot \frac{Q}{r^2} \hat{\mathbf{r}}. \tag{8.16}$$

Table 8.1 gives the dielectric constants of some electric materials.

Table 8.1 Dielectric constants of some materials

Material	Dielectric constant
Air (1 atm)	1.00059
Vacuum	1.00000
Water at 20 °C	80
Glass	5 – 10
Mica	3 – 6
Polyethylene at 20 °C	2.25 – 2.3
Rubber	2.5 – 3.0
Paper	3.0 – 6.0
Porcelain	6.0 – 6.8

8.8 ELECTRIC POTENTIAL

The electrical energy at a point due to an electric field is described in terms of the **potential** at that point. We can understand the electric potential at a point by first relating to the electric potential difference between two points in an electric field.

The electric potential difference between two points in an electric field is the electrostatic work done in taking a unit positive charge from one point to the other.

From the above definition, we get the electric potential at a point as the work done in conveying a unit positive test charge from an arbitrary reference zero potential to the point. We thus define the electric potential at a point as

the electrostatic work done in taking a unit positive test charge from a point of zero potential to the point of interest.

It is clear from the above that the unit of potential is the ratio of the unit for work to that for charge. Hence potential is expressed in Joule/Coulomb. This unit is also referred to as the Volt, represented by V. Thus if the potential of a point is 4 volts, it will take 40 Joules of work from an external source to move 10 coulombs of charge from a point at zero potential to the point. Theoretically, the point of zero potential is so far away from the charge that it is considered to be outside the field. Such a point is said to be an infinite distance away. Using this, we redefine the electric potential at a point as

the work done in taking a unit positive charge from infinity to the point.

From the practical standpoint, the earth is taken to be at zero potential. This is based on the fact that the earth is so large a body that it is not possible to raise its potential above zero by adding any amount of charge to it. Electric potential is a scalar quantity; it can be positive or negative. A positive potential indicates work done by the electric field, whilst a negative potential implies that some external agency does work against the direction of the electric field. Thus if a unit negative charge is taken from infinity to a point, the potential is positive; a similar case for a positive charge would yield negative potential, implying work has to be done against the field direction.

Since the potential at a point is work done per unit positive charge, the electrostatic work done, or energy U required, to bring a charge q from infinity to a point where the potential is V is given by

$$U = qV. \qquad (8.17)$$

8.9 POTENTIAL OF DISCRETE CHARGES

We can utilise the above definition for electric potential to derive an expression for the potential at a point distance r from a point charge, and a discrete distribution of point charges. Consider a point charge Q placed in space. The force acting on a unit positive test charge at this point is the electric field E at the point.

Consider a point charge Q, creating an electric field in a region of free space. At any distance from Q, the force exerted on a charge q has magnitude given by the expression

$$F = \frac{1}{4\pi\varepsilon_0} \cdot \frac{Qq}{r^2}.$$

The elemental work dW done in moving q through a distance dr is given by

$$dW = \frac{1}{4\pi\varepsilon_0} \cdot \frac{Qq}{r^2} dr$$

From our definition of potential difference as the work done per unit charge, it follows that the potential difference dV over this distance is given by

$$dV = \frac{dW}{q} = \frac{1}{4\pi\varepsilon_o} \cdot \frac{Q}{r^2} dr.$$

Thus, the potential at the point distance r from Q as

$$V = \frac{Q}{4\pi\varepsilon_o} \cdot \int_{\infty}^{r} \frac{dr}{r^2} = \frac{Q}{4\pi\varepsilon_o r}. \tag{8.18}$$

Comparing (8.17) to (8.8), we get a relationship between the electric potential at a point and the electric field intensity at that point, as

$$E = -\frac{dV}{dr}, \tag{8.19}$$

indicating that the electric field intensity is the gradient of the potential at a point. Thus we can calculate one of these quantities at a given point when the other is known. If a constant potential V exists in a region d of space (e.g. between two parallel conducting plates), the electric field intensity in the region would be constant, with magnitude

$$E = \frac{V}{d}. \tag{8.20}$$

For a distribution of discrete point charges, the potential at a point due to all the charges is the sum of the potentials at the point due to each charge in turn. Hence,

$$V = \frac{1}{4\pi\varepsilon_o} \sum_{i=1}^{n} \frac{Q_i}{r_i}. \tag{8.21}$$

From our study of dielectrics, it follows that if the charge or charges are placed in a medium other than vacuum, ε_o in equations (8.18) and (8.20) should be replaced by $\varepsilon_r\varepsilon_o$.

Example 8.5

In a particle accelerator, a charged particle of mass 4.69×10^{-27} kg and charge 3.2×10^{-19} C is accelerated in a field where the potential difference is 2.5 MV. Calculate

(i) the energy of the particle when it reaches its final speed,

(ii) the final speed of the charged particle.

Solution

(i) In this example, the potential is given. We find the electrostatic potential energy of the particle by using equation (8.16). Thus the energy U acquired by the particle is $U = qV = 3.2\times10^{-19}C\times2.5\times10^{6}V = 8.01\times10^{-13}J.$

(ii) The final speed of the particle is obtained by equating its kinetic energy to the electrostatic energy obtained in (i). Thus, we write

$$\tfrac{1}{2}mv^2 = 8.01\times10^{-13}J, \text{ from which } v = \sqrt{\frac{2\times8.01\times10^{-13}J}{4.69\times10^{-27}kg}} = 1.8\times10^{7}m/s.$$

Example 8.6

ABCD is a square of side 0.10 m. Point charges of 4nC, -6nC, and 2nC are placed at the corners A, B and D, respectively, in vacuum. Find the electric potential at C due to these charges. Take $\frac{1}{4\pi\varepsilon_o} = 9\times10^{9}Nm^2/C^2$.

Solution

Fig. 8.10 shows the arrangement of the charges. The potential at C is the sum of the potentials due to the charges at A, B, and D.
Potential V_{CA} at C due to charge at A is given by

$$V_{CA} = \frac{1}{4\pi\varepsilon_o}\cdot\frac{4\times10^{-9}C}{0.10\sqrt{2}\,m} = \frac{4\times10^{-9}\,C\times9\times10^{9}Nm^2/C^2}{0.1414\,m} = 254.6\text{ V}.$$

Similarly, potential V_{CB} at C due to charge at B is given by

$$V_{CB} = -\frac{1}{4\pi\varepsilon_o}\cdot\frac{6\times10^{-9}C}{0.10\,m} = \frac{6\times10^{-9}\,C\times9\times10^{9}Nm^2/c^2}{0.10\,m} = -540.0\text{ V}.$$

Potential V_{CD} at C due to charge at D is given by

$$V_{CD} = \frac{1}{4\pi\varepsilon_o}\cdot\frac{2\times10^{-9}C}{0.10\,m} = \frac{2\times10^{-9}\,C\times9\times10^{9}Nm^2/c^2}{0.10\,m}$$
$$= 180.0\text{ V}.$$

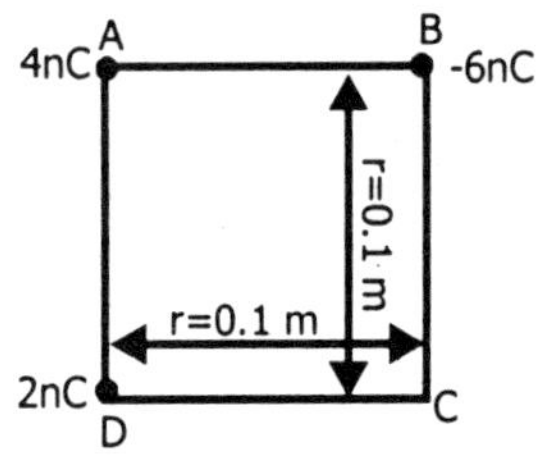

Fig. 8.10

Thus the potential at C = $V_{CA} + V_{CB} + V_{CD}$ = -105.4 V. The negative sign indicates that an overall work is done **on** the field in taking a unit positive charge from infinity to C.

8.10 ELECTRIC DIPOLE

An electric dipole consists of two charges equal in magnitude, opposite in sign, and separated by a distance. The study of the electric dipole is important for an understanding of the electrical behaviour of certain compounds. When a dipole is placed in an electric field, there is a torque on the dipole, tending to rotate it such that its axis becomes aligned with the electric field. The magnitude of the torque depends on

(i) the strength of the external field,
(ii) the magnitude of the charges,
(iii) the separation of the charges, and
(iv) the angle between the field direction and the axis of the dipole. For a given field strength, magnitude and separation of the charges, maximum torque occurs when the axis of the dipole is at right angles to the field direction.

If each charge has magnitude Q, and the distance from the centre of the negative charge to that of the positive charge is d, we define the **electric dipole moment** p by

$$p = Qd \tag{8.22}$$

The magnitude of the electric dipole moment represents the maximum torque exerted upon the dipole per unit value of the surrounding electric field. Electric dipole moment is a vector, directed along the line from the negative charge to the positive charge. The S.I. unit of electric dipole moment is Cm (Coulomb-metre). In the following section, we derive expressions for the field intensity and potential around a dipole.

Finding mathematical expressions for the electric field due to a dipole is normally done for very simple cases: for points along the axis of the dipole, and for points along the perpendicular bisector of the line joining the charges.

8.10.1 Field and potential along the axis

Fig. 8.11 illustrates this case.

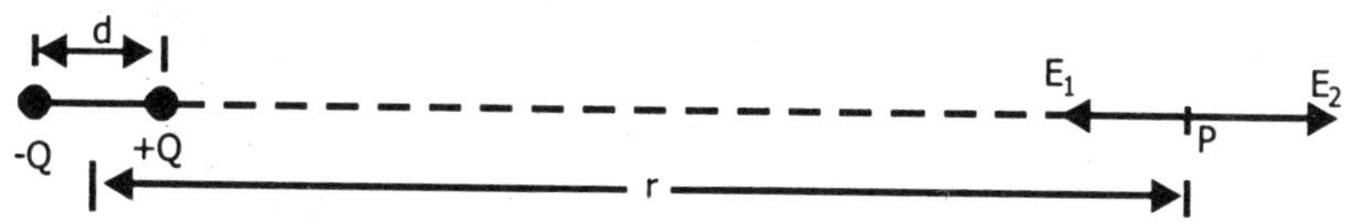

Fig. 8.11 Electric field and potential due to a dipole for points along the axis.

The resultant field at P has magnitude given by $E = E_2 - E_1$, where

$$E_2 = \frac{1}{4\pi\varepsilon_o}\frac{Q}{\left(r - \frac{d}{2}\right)^2}, \text{ and } E_1 = \frac{1}{4\pi\varepsilon_o}\frac{Q}{\left(r + \frac{d}{2}\right)^2}. \text{ Thus,}$$

$$E = \frac{Q}{4\pi\varepsilon_o}\left[\frac{1}{\left(r - \frac{d}{2}\right)^2} - \frac{1}{\left(r + \frac{d}{2}\right)^2}\right].$$

For points far away from either charge (i.e. points for which r>>d), we get

$$E = \frac{2Qd}{4\pi\varepsilon_o r^3} = \frac{2p}{4\pi\varepsilon_o r^3}, \tag{8.23}$$

where p (= Qd) is the dipole moment. The potential V at P is given by $V = V_2 + V_1$, where $V_2 = \frac{1}{4\pi\varepsilon_o}\frac{Q}{\left(r-\frac{d}{2}\right)}$, and $V_1 = -\frac{1}{4\pi\varepsilon_o}\frac{Q}{\left(r+\frac{d}{2}\right)}$.

Thus,

$$V = \frac{Qd}{4\pi\varepsilon_o r^2} = \frac{p}{4\pi\varepsilon_o r^2}. \qquad (8.24)$$

8.10.2 Field and potential at points along the perpendicular bisector

This case is illustrated by **Fig. 8.12**. As we saw in the previous case, the electric field at P is made up of contributions from the charges –Q and +Q, as shown. These are represented by $\mathbf{E}_1$ and $\mathbf{E}_2$, respectively. These two contributions are equal in magnitude, and act in the directions shown. If the dipole is in the vertical plane, the horizontal components of the two fields are equal and opposite. They hence cancel each other out. The vertical components, on the other hand, are in the same direction, and add up to each other. A dipole in the horizontal plane will have the vertical components cancelling out each other, while the vertical components add up.

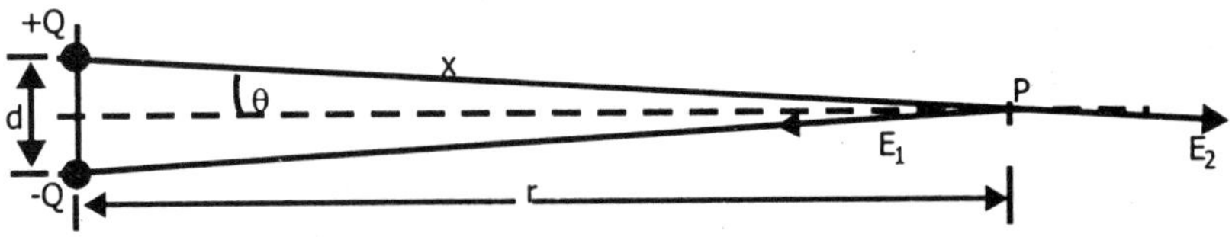

Fig. 8.12 Field and potential due to a dipole for points on the perpendicular bisector of the line joining the charges.

The magnitude of the resultant field E at P is given by

$$E = 2E'\sin\theta, \text{ where } E' = \frac{1}{4\pi\varepsilon_o}\frac{Q}{x^2} = \frac{1}{4\pi\varepsilon_o}\frac{Q}{\left(r^2+\frac{d^2}{4}\right)}, \text{ and } \sin\theta = \frac{d/2}{\sqrt{r^2+\frac{d^2}{4}}}.$$

Thus, $E = \frac{2Q}{4\pi\varepsilon_o}\frac{d/2}{\left(r^2+\frac{d^2}{4}\right)^{3/2}}$.

For points such that $r>>d$, the field has magnitude

$$E = \frac{2Q}{4\pi\varepsilon_o}\frac{d/2}{r^3} = \frac{p}{4\pi\varepsilon_o r^3}, \qquad (8.25)$$

where p is the dipole moment. From the above configuration, the direction of this field is vertically downwards.

Since the two charges are equal in magnitude, they will contribute equal potentials at P, one negative and the other positive. The net potential on the perpendicular bisector of a dipole is hence zero.

8.11 CAPACITANCE AND CAPACITORS

The capacitance of an electrical system represents the ability of the system to store electrical charge. A large capacitance implies a high ability to store charge. A system designed to store charge or electrical energy is called a **capacitor**.

The capacitance of a capacitor is the charge the capacitor can store per unit voltage.

A large capacitance means the capacitor can store a large amount of charge with little increase in its potential. From the above definition, we write the mathematical expression for the capacitance C of a capacitor as

$$C = \frac{Q}{V}, \tag{8.26}$$

where Q is the charge stored, and V is the potential across the capacitor. The unit of C is Coulomb/Volt, known as the **Farad**, in honour of Michael Faraday, one of the pioneers in the study of electricity. Because the Coulomb is a very large unit for practical purposes, the Farad too is very large. Submultiples of the Farad are normally used to express capacitance. The common units are the microfarad (μF), the nanofarad (nF), and the picofarad (pF), where $1\mu F = 10^{-6} F$, $1nF = 10^{-9} F$, and $1pF = 10^{-12} F$.

The ability to store charge by a capacitor depends on two factors: the dimensions of the capacitor, and the material between the plates. This material is always an insulating material, or dielectric. The simplest capacitor is constructed by having two metal plates separated by a small distance. This gives the **parallel-plate** capacitor. Other capacitor arrangements also exist. In the sections that follow, we survey some types of capacitors in use, and derive mathematical expressions for the capacitance of each.

8.11.1 The parallel plate capacitor

As mentioned above, the parallel plate capacitor consists of two metal plates separated by a gap, normally filled with dielectric material. **Fig. 8.13** shows the schematic arrangement of the parallel-plate capacitor. Examples of dielectric materials used are paper, mica, and some resins. The plates have a common area of overlap A, and are a small distance d apart.

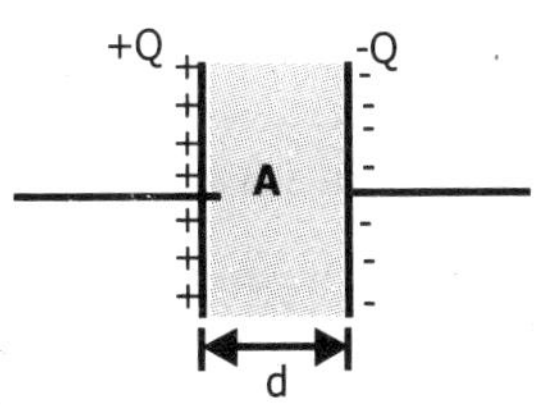

Fig. 8.13 The parallel-plate capacitor

A potential difference V between the plates produces a charge +Q on one plate of the capacitor. This charge in turn induces an equal and opposite charge on the other plate. The electric displacement D produced in the medium is

$$D = \varepsilon E = \varepsilon \frac{V}{d} = \frac{Q}{A},$$

where ε is the permitivity of the dielectric and E is the electric field intensity. Thus the capacitance C of the capacitor is given by

$$C = \frac{Q}{V} = \frac{A\varepsilon}{d}. \tag{8.27}$$

When expressed in terms of the dielectric constant ε_r, this expression becomes

$$C = \frac{A\varepsilon_r \varepsilon_o}{d}. \tag{8.28}$$

If we have vacuum between the plates, the capacitance C_o is

$$C_o = \frac{A\varepsilon_o}{d}. \tag{8.29}$$

From equations (8.27) and (8.28), we adopt a more practical definition of the dielectric constant of a material as

the ratio of the capacitance of a capacitor with the dielectric between the plates to the capacitance of the capacitor with vacuum between the plates.

It is seen from the above that the capacitance of a capacitor depends on the dimensions of the capacitor (i.e. the area, and the separation of the plates), and the material of the dielectric between the plates. The electrical symbol for a capacitor is based on the parallel-plate - **Fig. 8.14**. The dielectric material is not represented in the symbol.

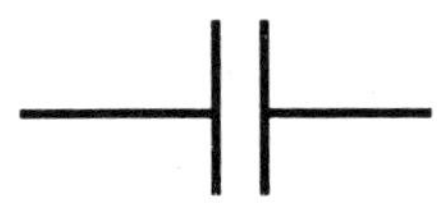

Fig. 8.14 Capacitor symbol

Example 8.6

A fine layer of gold is sprayed on each side of a sheet of paper of area 10.0 cm^2, and thickness 0.10 mm in order to make a parallel-plate capacitor. If the dielectric constant of the paper is 4.5, find the capacitance of the capacitor.

Solution

From the given data, area of overlap $A = 0.001\ m^2$, $\varepsilon_r = 4.5$, thickness $d = 0.0001$ m. Using equation (8.26), the capacitance is given by

$$C = \frac{A\varepsilon_r\varepsilon_o}{d} = \frac{0.001\,m^2 \times 8.85\times10^{-12}F/m\times4.5}{0.0001m}F = 0.398\,nF = 398\,pF$$

8.11.2 Concentric spheres

A system consisting of two concentric spheres is another common example of a capacitor. **Fig. 8.15** is a schematic representation of such a capacitor. A charge +Q on one sphere induces charge −Q on the other sphere.

The potential difference V between the spheres is given by

$$V = \frac{Q}{4\pi\varepsilon}\left(\frac{1}{a}-\frac{1}{b}\right) = \frac{Q}{4\pi\varepsilon}\left(\frac{b-a}{ab}\right),$$

where a and b are the radii of the spheres, and ε is the permitivity of the medium between them. The capacitance of the system is hence given by the expression

$$C = \frac{Q}{V} = 4\pi\varepsilon_r\varepsilon_o\left(\frac{ab}{b-a}\right). \qquad (8.30)$$

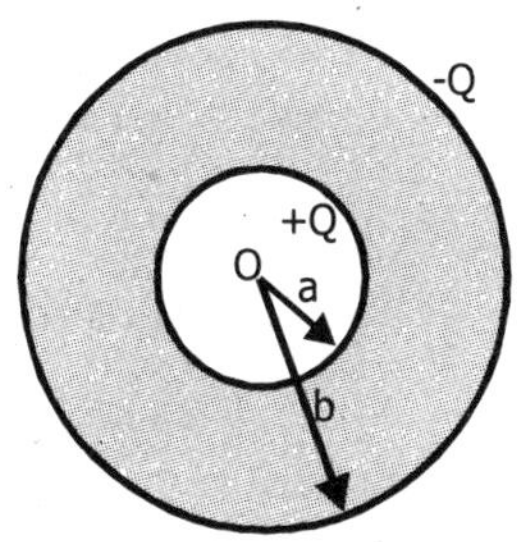

Fig. 8.15 Capacitance of concentric spheres

8.11.3 Isolated sphere

An isolated sphere too has the ability to store charge, and hence has capacitance. In order to find the capacitance of an isolated sphere, we consider the sphere as part of a system of concentric spheres, the outer one of which has infinite radius. Thus we use a modified form of equation (8.30) and consider that $b>>a$, or $b \to \infty$. The capacitance of an isolated sphere is

$$C = 4\pi\varepsilon_o a. \qquad (8.31)$$

Example 8.7

By considering the earth as an isolated sphere situated in vacuo, estimate the amount of charge that can be deposited on it in order to increase its potential by 1 volt. Radius of the earth = 6.40×10^6m. Use the approximation $\varepsilon_o = \frac{10^{-9}}{36\pi}F/m$.

Solution

We use the equation for the capacitance of an isolated sphere to determine the capacitance of the earth, and calculate the charge that can be deposited in order to achieve the required potential rise. From (8.31), the capacitance C is given by

$$C = 4\pi\varepsilon_o R_E = 4\pi \times \frac{10^{-9}}{36\pi} F/m \times 6.40 \times 10^6 m = 7.11 \times 10^{-4} F.$$

From the equation $C = \frac{Q}{V}$, the charge required for a potential difference of 1 volt becomes $Q = CV = 7.11 \times 10^{-4} C$. This is a very large charge indeed. Hence a large charge is needed to produce a noticeable change in the potential of the earth. For this reason, the earth is taken to be at zero potential at all times. An electrical device is said to be "earthed" if there is a connection between the device and earth. This serves very useful purposes in practical circuits.

In addition to the above capacitors, there are other arrangements of materials producing other types of capacitors. These include cylindrical arrangements, parallel wires, and line charges. For detailed analyses of these, the reader is advised to consult more advanced physics texts.

8.11.4 Energy stored in a capacitor

Since a capacitor stores up electrical charge, it can also be considered as a store of electrical energy. Consider a capacitor charging up steadily from zero to a charge Q. The potential across the capacitor would increase steadily from zero to a final value V. The average potential across the plates is then $\frac{1}{2}(0 + V) = \frac{1}{2}V$. For a charge Q, the energy stored in the capacitor is given by U, where

$$U = Q \times \frac{1}{2}V = \frac{1}{2}QV. \qquad (8.32)$$

Expressing equation (8.32) in terms of the capacitance C of the capacitor, we have

$$U = \frac{1}{2}QV = \frac{1}{2}CV^2 = \frac{1}{2}\frac{Q^2}{C}. \qquad (8.33)$$

Example 8.8

The plates of a parallel-plate capacitor with vacuum between them are 5.00 mm apart, and each plate caries a charge of magnitude 50 nC. If the electric field between the plates has magnitude $2.50 \times 10^6 V/m$, determine

(i) the potential difference between the plates,
(ii) the capacitance of the capacitor,
(iii) the area of overlap of the plates,
(iv) the energy stored in the capacitor.

Solution

Data: Separation of the plates $d = 0.005$ m, Electric field $E = 2.50 \times 10^6$ V/m, Charge stored in the capacitor $Q = 5.0 \times 10^{-8}$ C.

(i) Bearing in mind that electric field intensity is the potential gradient, we write $E = \frac{V}{d}$, or $V = Ed$, where V is the potential difference between the plates. Thus, $V = 2.50 \times 10^6 \text{ V/m} \times 0.005 \text{ m} = 1.25 \times 10^4 \text{ V}$.

(ii) Having found the potential difference between the plates, we can calculate the capacitance C from the ratio of the charge stored to the potential. Thus, we have

$$C = \frac{Q}{V} = \frac{5.0 \times 10^{-8} \text{ C}}{12500 \text{ V}} = 0.0004 \times 10^{-8} \text{ F} = 4.0 \text{ pF}.$$

(iii) The capacitance is given by $C = \frac{A\varepsilon_o}{d}$, where A is the area of overlap of the plates.

Thus, $$A = \frac{Cd}{\varepsilon_o} = \frac{4.0 \times 10^{-12} \text{ F} \times 0.005 \text{ m}}{8.85 \times 10^{-12}} = 0.00225 \text{ m}^2 = 22.5 \text{ cm}^2.$$

(iv) The energy U stored in the capacitor is given by

$$U = \frac{1}{2}QV = \frac{5.0 \times 10^{-8} \text{ C} \times 12500 \text{ V}}{2} = 3.125 \times 10^{-4} \text{ J}.$$

8.12 COMBINATIONS OF CAPACITORS

Often the capacitance required in a given connection cannot be obtained from a single capacitor. It then becomes necessary to use a combination of capacitors to produce the desired result. The two combinations used are the parallel, and the series connections.

8.12.1 Capacitors in parallel

The parallel combination has all the plates on the same side connected to the same electrode of the power supply. The potential difference across each capacitor is the same. For capacitors of different capacitances, the charges stored in the capacitors would be different. The total charge from the supply source is hence divided up among the capacitors, each capacitor having a charge depending on the value of its capacitance. The connection is depicted in **Fig. 8.16**.

Consider capacitances C_1, C_2, C_3,..., C_n, all connected in parallel. For the potential difference V, the charges in the capacitors are Q_1, Q_2, Q_3, ..., and Q_n. The total charge Q in the combination is

$$Q = Q_1 + Q_2 + Q_3 + ... + Q_n, \qquad (8.34)$$

where

$$Q_1 = C_1V,\ Q_2 = C_2V,\ Q_3 = C_3V,\ ....,\ \text{and}\ Q_n = C_nV.$$

We hence have

$$CV = C_1V + C_2V + C_3V + ... + C_nV,$$

where C is the effective capacitance of the combination. This gives

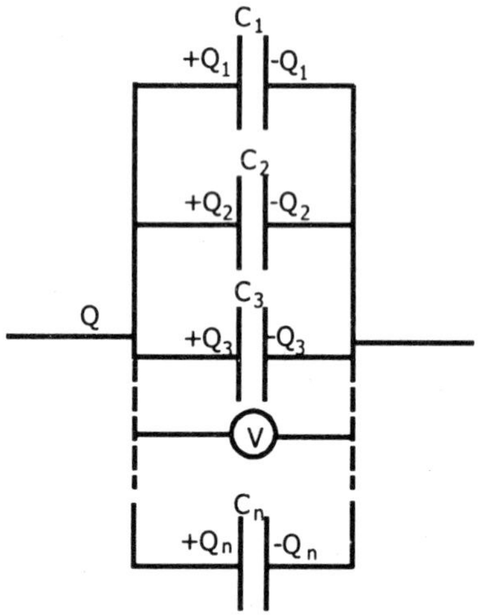

Fig. 8.16 Capacitors in parallel

$$C = C_1 + C_2 + C_3 + ... + C_n = \sum_{i=1}^{n} C_i \qquad (8.35)$$

Thus the effective capacitance of the combination is the sum of the capacitances of the individual capacitors.

8.12.2 Capacitors in series

Fig. 8.17 shows a system of capacitors connected in series. A potential difference V is established across the combination. The charge across each capacitor is the same. However, the potential differences across the capacitors are different, and add up to the total potential difference V.

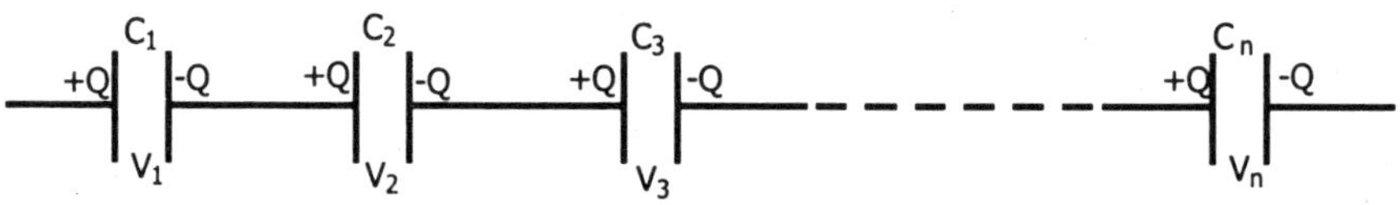

Fig. 8.17 Capacitors in series

Thus, we have

$$V = V_1 + V_2 + V_3 + ... + V_n. \qquad (8.36)$$

If the combined capacitance is C, we have

$$V = \frac{Q}{C},\ V_1 = \frac{Q}{C_1},\ V_2 = \frac{Q}{C_2},\ V_3 = \frac{Q}{C_3},\ ...\ V_n = \frac{Q}{C_n}.$$

Substituting these relations into (8.35), we get

$$\frac{Q}{C}=\frac{Q}{C_1}+\frac{Q}{C_2}+\frac{Q}{C_3}+....+\frac{Q}{C_n},$$

giving

$$\frac{1}{C}=\frac{1}{C_1}+\frac{1}{C_2}+\frac{1}{C_3}+....+\frac{1}{C_n}=\sum_{i=1}^{n}\frac{1}{C_i}, \tag{8.37}$$

8.12.3 Variable capacitors

A variable capacitor is one in which the capacitance of the capacitor can be varied over a range of values. Taking the parallel-plate capacitor as an example, it is easy to see that by having the option of varying the area of overlap continuously, the capacitance of the capacitor can be varied. A wider range of variation can be achieved if we have a set of capacitors in parallel. One common example is shown in **Fig. 8.18**, with air used as the dielectric. The area of overlap of the plates is varied by pulling the plates inward or outward. The sets of plates actually make up a combination of capacitors in parallel. In most variable capacitors one set of plates remain fixed, while the other can be rotated about an axis, producing the same effect as that depicted in the figure.

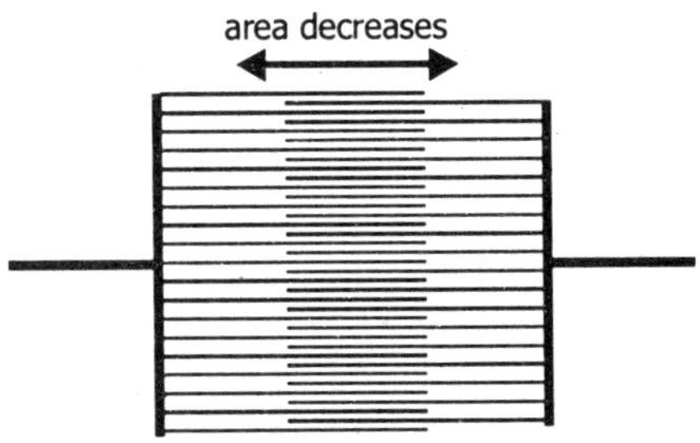

Fig. 8.18 Variable capacitor. The capacitanc changes as the two sets of plates are move toward or away from each other.

The major use of variable capacitors is in tuning circuits used in radio and television sets. In these circuits, the capacitance is varied continuously such that the frequency of the tuning circuit matches the frequency of the incoming radio waves.

Example 8.9

Fig. 8.19 shows a 3.0 μF capacitor connected in series with a parallel combination consisting of a 4.0 μF capacitor, and a 5.0 μF capacitor.

(a) Determine the capacitance of the combination.

(b) If the p.d. between A and B is 20.0 V, calculate

  (i) the charge stored in each capacitor,

  (ii) the p.d. across each capacitor,

  (iii) the energy stored in each capacitor.

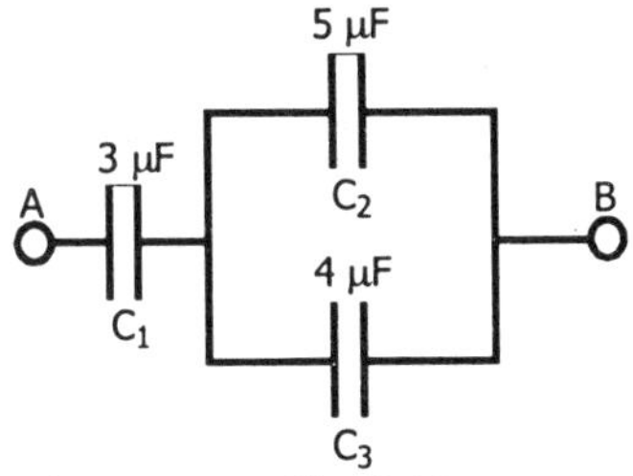

Fig. 8.19

Solution

(a) The capacitance of the parallel combination C_P is $C_P = C_2 + C_3 = 9.0\,\mu F$. This is in series with C_1, thereby giving the effective capacitance C of

$$\frac{1}{C}=\frac{1}{C_1}+\frac{1}{C_P}=\frac{1}{3\,\mu F}+\frac{1}{9\,\mu F}=\frac{4}{9\,\mu F}.\quad \text{Thus, } C = 2.25\,\mu F$$

(b) Knowing the combined capacitance and the potential difference between A and B, we calculate the total charge stored in the combination. This total charge Q is given by $Q = CV = 2.25\times10^{-6}\,F\times 20.0\,V = 45.0\,\mu C$.

(i) the charge in $C_1 = 45.0\,\mu C$.
the capacitors C_2 and C_3 have the same potential across them. The corresponding charges stored Q_2 and Q_3 are obtained from the relations $Q_2 + Q_3 = Q = 45.0\,\mu C$, and $\frac{Q_2}{C_2}=\frac{Q_3}{C_3}$. Substituting the numerical values of C_2 and C_3, we get $Q_2 = 25.0\,\mu C$, and $Q_3 = 20.0\,\mu C$.

(ii) The potential difference across the parallel combination is given by V_P, where $V_P=\frac{Q}{C_P}=\frac{45.0\,\mu C}{9.0\,\mu C}=5.0\,V$. The potential differences across the capacitors are $V_1 = 15.0\,V$, $V_2 = V_3 = 5.0\,V$, where the suffices refer to the capacitors.

(iii) The energy U stored in each capacitor is calculated from $U=\frac{1}{2}QV$. Using the values obtained above, we get

$$U_1=\tfrac{1}{2}\times45.0\times10^{-6}\,C\times15.0\,V=3.375\times10^{-4}\,J$$

$$U_2=\tfrac{1}{2}\times25.0\times10^{-6}\,C\times5.0\,V=6.25\times10^{-5}\,J$$

$$U_3=\tfrac{1}{2}\times20.0\times10^{-6}\,C\times5.0\,V=5.00\times10^{-5}\,J$$

8.13 STATIC ELECTRICITY IN EVERYDAY LIFE

We conclude our study of static electricity by surveying its effects in industry, in the home, and in the classroom. Some of its effects are undesirable, and can sometimes even lead to fatal results. On the other hand, some effects of electrostatics are put to good use.

8.13.1 Electrostatic printing and copying

The primary principle at work in a laser printer is static electricity, built up on the roller of the printer. The core component of the laser printer is the **photoreceptor**, typically a revolving drum or cylinder. This assembly is made out of highly photoconductive material that is discharged by light photons.

Initially, the drum is given a total positive charge by a wire carrying electric current. A tiny beam of laser light is shone on the drum as the latter rotates. This laser light discharges certain points. In this way the laser "draws" the patterns of the images to be printed as a pattern of electric charges, thereby producing what is effectively an **electrostatic image**.

After the pattern is set, the printer coats the drum with positively charged toner a fine, black powder. Since it has a positive charge, the toner clings to the negative discharged areas of the drum, but not to the positively charged background. With the powder pattern affixed, the drum rolls over a sheet of paper, which moves along a belt. Before the paper rolls under the drum, it is given a negative charge by the charged roller. This charge is stronger than the negative charge of the electrostatic image, so the paper can pull the toner powder away. Since it is moving at the same speed as the drum, the paper picks up the image pattern exactly. To keep the paper from clinging to the drum, it is discharged by the corona wire immediately after picking up the toner.

Finally, the printer passes the paper through the fuser, a pair of heated rollers. As the paper passes through these rollers, the loose toner powder melts, fusing with the fibers in the paper. The fuser rolls the finished printed page to the output tray. The fuser also heats up the paper, which is why pages are always hot when they come out of a laser printer or photocopier.

After depositing toner on the paper, the drum surface passes the discharge lamp. This bright light exposes the entire photoreceptor surface, erasing the electrical image. The drum surface then passes the charge corona wire, which reapplies the positive charge. In practice, bringing all the above processes together to give the final production as we know it is quite a technological feat. With the advancement of technology the reprographic industry is today awash with a wide array of printers and photocopiers, some so advanced that they can even print on media other than paper, and do also produce prints in colour.

8.13.2 The van der Graaff generator

About the biggest use of static electricity is in the production of high voltages in what are known as **electrostatic machines**. These machines produce potential differences of the order of millions of volts. The main use of such high voltages is to provide high energies to charged particles (e.g. protons). These highly energetic charged particles can then initiate nuclear reactions. Studies from these have contributed immensely to our current understanding of the nucleus of the atom.

The earliest electrostatic machine developed was the van der Graaff accelerator, commonly called the van der Graaff generator. This machine relies on the accumulation of charges deposited on a metal sphere, by the use of conducting points. **Fig. 8.20** is a schematic diagram of the important features of the van der Graaff generator.

By means of metal points P, charges from a suitable source are deposited on an insulating motor-driven belt at A. Another set of metal points B takes the charge carried on the belt and deposit the charges on the isolated sphere. Since the charge on a metallic conductor resides on its surface, the charge accumulated on this sphere will be distributed on its surface. As the belt runs continuously over a period of time, charges are deposited on the sphere, raising the latter to the high potential. When this high potential is applied to a proton, for example, the proton acquires the very high velocity needed for it to initiate a nuclear reaction. Typical van der Graaff voltages are of the order of 6 MV.

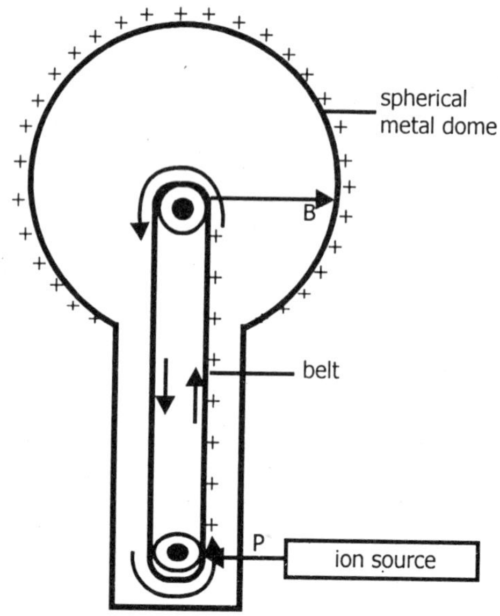

Fig. 8.20 The van der Graaff generator

8.13.3 Undesirable effects

It is a standard demonstration of static electricity that when a comb is run through dry hair, the comb acquires the ability to attract small pieces of paper. Equally so, the newly combed hair tends to accumulate dust in the atmosphere with great ease. This attraction of uncharged bodies by charged bodies forms one of the undesirable effects of static electricity. In many manufacturing industries, static electricity causes huge problems. Examples of such industries include plastics, paper, textiles, printing and electronic industries. These industries have been plagued with static electricity and the related contamination due to attraction of particles on charged surfaces. These problems often cost these industries millions of dollars per year.

In many industries, problems of static electricity are solved by the use of anti-static sprays. The sprays work on the principle that adding surface conductivity to plastics will move them up into the higher conductivity range and prevent the build up of static electricity that is caused by friction. The average anti-static spray is made from a soap-based material that has been diluted in a solvent, such as mild alcohol. A fire retardant is often added to combat flammability of the solvent, after which pressure and a highly volatile liquid such as Freon ® are added. As this spray leaves the nozzle of the aerosol can, the Freons evaporate immediately. A short time after contact with the material, the fire retardant and solvents evaporate leaving a conductive coating on the surface of the material. The plastic has now become conductive and, as long as this coating is not disturbed, it will be difficult to generate static electricity on this material.

In the electronics industry, the effects of static electricity are reduced by grounding the operator. This process entails a direct connection of the operator to ground. If an operator is isolated by standing on a wooden floor or wearing shoes with rubber soles, he/she soon picks up a voltage gradient. Through this isolation, it is possible for an operator to charge to several hundred volts each time he/she handles a piece of charged plastic. As many different pieces of plastics are handled, the operator becomes charged to a higher voltage gradient. In a few extreme cases, it has been found that a continued charging ultimately leads to the operator receiving an electric shock, or sometimes damages a static sensitive device. Personnel grounding equipment become important if the operator sits while working, as is the case in many electronics industries. Anti-static mats are often provided to reduce static effects. These mats provide a direct cable connection to ground, thereby preventing any buildup of static electricity.

EXERCISES

Where necessary, you may use the following constants:

Mass of electron $= 9.1\times10^{-31}$ kg	Speed of light in vacuum $= 3.0\times10^{8}$ m/s
Electron charge $= -1.602\times10^{-19}$ C	Permitivity of free space $= 8.85\times10^{-12}$ F/m
Proton charge $= +1.602\times10^{-19}$ C	$\frac{1}{4\pi\varepsilon_o} = 9.0\times10^{9}$ Nm^2/C^2
Acceleration due to gravity = 9.81 m/s^2	Mass of proton $=1.667\times10^{-27}$ kg

1. Two point charges of +4.0 nC and +2.0 nC are placed 5.0 cm apart in vacuum. Calculate the force between them.

2. Find the ratio of the electrostatic force between the electron and the proton in the hydrogen atom to the gravitational force between them. Assume the two charges are separated by a distance of 1.3×10^{-13} m, and the universal gravitational constant is 6.67×10^{-11} Nm^2/kg^2.

3. Two point charges of 10 nC, and −2 nC are placed 2 cm apart in vacuo. Determine the distance of the line of zero field intensity from either charge.

4. At the corner A of a square ABCD of side 10.0 cm, is placed a charge of +4.0 nC, and at B is placed a charge of +3.0 nC. Determine the magnitude and direction of the force on a +1.5 nC charge placed at the corner C.

5. Two charges attract each other with a force of 1.20×10^{-6} N. Calculate the force that would act between the two charges when their distance apart is
   (i) doubled,
   (ii) halved,
   (iii) tripled.

6. Charges of 5 nC, and −20 nC are placed at points A and B, respectively, a distance of 5 cm apart in vacuo, B being to the right of A. Calculate the force acting on a −10 nC charge placed at a point C in line with A and B, and a distance 2 cm to the right of B.

7. ABC is a right-angled triangle with AB = 6.0 m, and BC = 8.0 m. Charges of +15.0 μC, −20.0 μC, and −20.0 μC are placed at A, B and C, respectively. Calculate the magnitude of the force acting on the charge at C.

8. Two conducting balls each of mass 1.0 g are suspended from a common point by silk threads 1.0 m long. When equal charges are placed on the balls in vacuo, the system comes to rest with each string making an angle of 5° with the vertical. Calculate the charge on each ball.

9. Calculate the magnitude of the electric field in which the electrical force on an electron is equal to the magnitude of its weight.

10. Repeat the calculation in Q9 above for the proton.

11. Three equal charges of 0.5 nC are located at three of the four corners of a square of side 0.10 m in vacuum. Calculate the magnitude of the electric field intensity and potential at the fourth corner.

12. An electron is released from the lower plate of two large parallel plates, separated by a distance of 2.0 cm. The upper plate has a potential of 4.0 kV relative to the lower plate. Determine
   (i) how long it will take for the electron to arrive at the upper plate,
   (ii) the speed of the electron just before striking the upper plate.

13. Suppose that in the helium nucleus there are two protons separated by a distance of 1.50×10^{-15} m. Calculate the electrostatic force between them,

14. Two point charges of 11.0 μC and 5.0 μC are placed 15.0 cm apart in vacuum. Calculate the work done to reduce their separation by 10.0 cm.

15. Two charges of 60 μC and −20 μC are 40 cm apart. A third charge of +30 μC is placed midway between the two charges. Calculate
   (i) the magnitude and direction of the force on the 30 μC charge,
   (ii) the electric field intensity at the midway point.

16. Two charges of 40 μC and 30 μC are 50 cm apart. Find the magnitude of the electric field strength at a point 20 cm from the smaller charge on a line joining the two charges.

17. Two point charges of 40 nC and −90 nC are placed 50 cm apart in vacuo. Determine the distance from either charge at which

(i) the electric field intensity is zero,
(ii) the electric potential is zero.

18. A charge Q is placed on the plates of a parallel-plate capacitor filled with air. The potential difference between the plates is measured to be V_o. A dielectric material of dielectric constant κ is now introduced between the plates, completely filling the space. The potential difference in this instance is measured and found to be V_1.
(a) Which is greater, V_o or V_1? Explain.
(b) In terms of the potential differences V_1 and V_o, deduce an expression for the dielectric constant κ.

19. A 9-MV generator accelerates protons. Determine
(i) the energy of the proton when it reaches its maximum velocity,
(ii) the maximum speed of the proton.

20. A charge of 5.0 nC is placed in a uniform electric field of intensity 7.5×10^4 N/C. Calculate the work done if the charge is moved 40.0 cm
(i) in the direction of the field,
(ii) perpendicular to the field.

21. The plates of a parallel-plate capacitor each of area 0.04 m^2 are 1.0 cm apart. The space between the plates is filled with oil of dielectric constant 20. A potential difference of 220 V is maintained across the plates. Calculate
(i) the capacitance of the capacitor,
(ii) the charge stored in the capacitor,
(iii) the energy stored in the capacitor.

22. The capacitor in Q21 is now connected in series with another capacitor with twice the capacitance, and connected across the same power supply. Calculate the energy stored in each capacitor of the combination.

23. The potential difference in a lightning-flash may be as high as 10^9 V, and the charge passing may be as high as 40 C. Calculate
(i) the energy transformed during a lightning-flash,
(ii) the temperature rise in a lake holding 10^6 kg of water if this energy were to be deposited in the water. Take the specific heat capacity of water as 4.18 kJ/kg/K.

24. You are provided with three capacitors of capacitances 3 μF, 10 μF, and 15 μF.
(a) How may they be connected to produce capacitances of 2 μF, 9 μF, and 12.5 μF?
(b) If the three capacitors are all connected in series with a 100-V battery, determine the charge stored in each capacitor, and the potential difference across it.
(c) If the three capacitors are now connected in parallel, and the combination is powered by the 100-V battery, determine the energy stored in each capacitor.

25. A tuning capacitor has a capacitance that can be varied from 0.1 μF to 5.0 nF, by rotating the plates. The capacitor is charged to a potential difference of 50 V at the maximum value of the capacitance and disconnected from the source. If the plates are now turned to produce minimum capacitance, calculate the potential difference between the plates.

26. A capacitor with vacuum between the plates is connected to a 100-V supply. When the capacitor is immersed in a non-conducting oil of dielectric constant 2.50, the charge on the plates increases by 200 μC. Determine the original capacitance of the capacitor.

27. The plates of a parallel-plate capacitor with vacuum between them are 4.79 mm apart, and each plate carries a charge of magnitude 52 nC. If the electric field between the plates has magnitude $4.80 \times 10^6\ \text{V/m}$, calculate
(i) the potential difference between the plates,
(ii) the capacitance of the capacitor,
(iii) the area of overlap of the plates,
(iv) the energy stored in the capacitor.

28. A spherical capacitor consists of a system of two concentric spheres with radii 12.0 cm, and 15.0 cm. Calculate the capacitance of the capacitor when
(i) the space between the spheres is completely evacuated,
(ii) the space between the plates is filled with oil of dielectric constant 3.5.

9

CHAPTER

Direct Current Electricity

9

CHAPTER

Direct Current Electricity

In the pervious chapter we studied phenomena dealing with charges at rest. We studied the properties of forces between charges, the electric field created around each charge, the potential at any point in the field, and how charges may be stored in capacitors. With this background, we are now in a position to study the effects produced by charges when they are in motion, and the phenomena arising from this motion.

When charges move, they create an **electric current**. When this motion is in one direction only, a **direct current** is produced. When the motion takes place within a conducting loop, starting from one point, going round the loop and ending at the same point, the loop is called an **electric circuit**. In this chapter we study properties of direct currents, and the phenomena dealing with direct current circuits.

9.1 ELECTRIC CURRENT

From electrostatics, we saw that charge can be made to flow from one point to another if there is a potential difference between the two points. Whenever charge flows, an electric current is created. Electric current is defined as

the rate of flow of electric charge.

From the above definition, it follows that if charge Q flows past a point in time t, the current I is given by

$$I = \frac{Q}{t}. \tag{9.1}$$

The unit of current is Coulomb/second, more commonly known as the **ampere** (A), abbreviated **amp** in honour of Andre Marie Ampere (1775 – 1836), one of the pioneers in the study of electricity. From equation (9.1) it is seen that **one amp of current results when a charge of 1 Coulomb flows in a time of 1 second**. We shall learn a more accurate and practical definition of the ampere in the chapters following this one. Common submultiples of the ampere in use are the milliamp (mA), and the microamp (μA), being equal to 10^{-3} A (or 0.001 A), and 10^{-6} A (or 0.000001 A), respectively.

A related parameter for identifying current is the **current density**, defined as

the amount of current flowing per unit area of cross section of the specimen.

In symbols, if we denote the current density by J, then we have

$$J = \frac{I}{A} = \frac{Q}{At},$$

where A is the area of cross section of the specimen through which the current flows. The unit of J is amp per square metre (i.e. A/m^2, or Am^{-2}).

For purposes of electric flow, materials are classified as **conductors**, **semiconductors**, and **insulators**. Insulators do not allow electric current to flow through them easily. Semiconductors have slightly higher electrical conducting properties than insulators. In our study of electrostatics in the previous chapter, we have already studied the behaviour and electrical use of insulators in the construction of capacitors. The conducting properties of semiconductors are studied later on in chapter 15. In this chapter we focus on conductors.

In the class of conductors, we find all the metals. Examples of these are copper, aluminium, iron, sodium, and magnesium. The electronic structure of these elements are such that they have 1, 2, or 3 valence electrons in their outermost shells. Such electrons are so weakly bound to the nuclei that even at normal room temperatures, there is enough energy to overcome the forces holding the electrons in their orbits. When this happens, the electrons become no longer attached to any particular atom, but rather, to all the atoms in the specimen. This is sometimes described by saying that these electrons are 'free' to move about in the metal. In the absence of an external electric field, these free electrons move about rather haphazardly in all directions within the metal, colliding with each other, and with the vibrating atomic centres. The overall velocity in any direction is zero. There is thus no net current in any direction.

When an external electric field is applied to the metal, the overall motions of the valence electrons become aligned against the field direction. This alignment causes a net electric current that can be measured. The net current through the metal can be worked out in terms of the amount of charge flowing through any section of the metal, the area of cross section of the specimen, the type of metal, and the overall speed (or **drift velocity**) with which the valence electrons are moving. We can derive this dependence with reference to **Fig. 9.1**, representing a highly magnified sectional view of a uniform conducting wire carrying a current I.

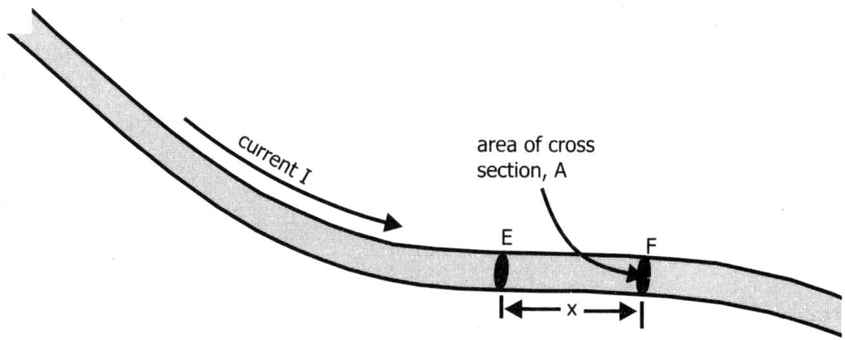

Fig. 9.1 Highly magnified section of a wire carrying current I.

Consider a segment EF of the wire. Free electrons crossing the plane at E go through F, a distance x away. The volume of this segment is Ax. If there are n free electrons per unit volume of the specimen, each moving with drift velocity v_d, the total charge passing through the segment is Q, where

$$Q = nAx$$

The charge flow per unit time between E and F – i.e. the current, is given by

$$I = \frac{Q}{t} = \frac{nAex}{t}, \tag{9.2}$$

where e is the charge on the electron. The ratio of the distance x to the time t is the drift velocity v_d. Substituting this into (9.2) gives

$$I = neAv_d. \tag{9.3}$$

Equation (9.3) relates the current flow to the valency of the metal, its area of cross section, and the drift velocity of the electrons. The current density J is given by

$$J = \frac{I}{A} = nev_d. \tag{9.4}$$

In general, for a charge q, equations (9.3) and (9.4) are modified to read

$$I = nqAv_d, \tag{9.5}$$

and

$$J = \frac{I}{A} = nqv_d. \tag{9.6}$$

In (9.5) and (9.6), n is the number of charge carriers (or mobile charges) per unit volume of the material. The following worked example illustrates how the various parameters can be obtained for a given specimen. From the example, we also learn how the magnitude of the drift velocity of the valence electrons can be calculated.

Example 9.1

A copper rod X of diameter 2.0 mm is joined end-to-end to another copper rod Y of diameter 2.4 mm. The composite wire carries a current of 5.0 A. Given that for copper, the density = 7.8×10^3 kg/m^3, atomic mass = 64.0 g/mole, and valency = 1 (i.e. 1 free electron per atom), calculate

(i) the current density in wire X,

(ii) the value of the drift velocity in wire Y.

Avogadro number = 6.023×10^{23} mol^{-1}; electron charge = 1.6021×10^{-19} C. Ignore any end effects at the junction due to the different diameters.

Solution

We use **Fig. 9.2** to help solve this problem.

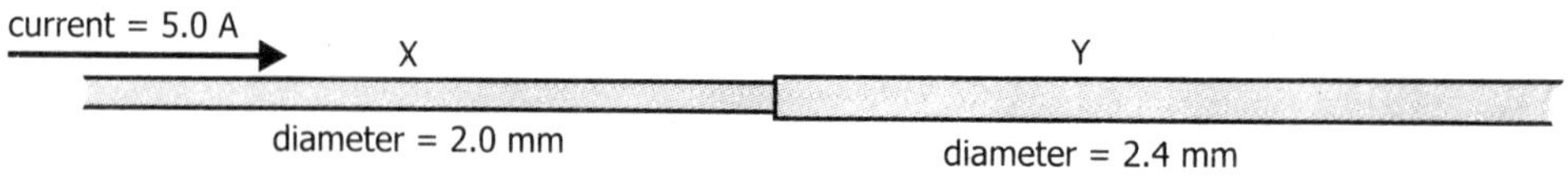

Fig. 9.2

(i) Area of X = $A_X = \pi \dfrac{d_X^2}{4} = \pi \times \dfrac{\left(2.0 \times 10^{-3}\right)^2}{4} = 3.14 \times 10^{-6} m^2.$

Area of Y = $A_Y = \pi \dfrac{d_Y^2}{4} = \pi \times \dfrac{\left(2.4 \times 10^{-3}\right)^2}{4} = 3.77 \times 10^{-6} m^2.$

Current density in X = $\dfrac{I}{A_X} = \dfrac{5.0}{3.14 \times 10^{-6}} = 1.59 \times 10^6 \text{ A/m}^2.$

Current density in Y = $\dfrac{I}{A_Y} = \dfrac{5.0}{3.77 \times 10^{-6}} = 1.33 \times 10^6 \text{ A/m}^2.$

(ii) To find the drift velocity in Y, we need to determine the number of free electrons per unit volume of the metal. This is done as follows: In any given mass m of the metal, the number of atoms is $\frac{m}{M} N_A$, where M is the atomic mass, and N_A is the Avogadro number. The number of free electrons $= \frac{m}{M} N_A \times$ valency. Hence the number of free electrons per unit volume $= \dfrac{m}{V} \cdot \dfrac{N_A}{M} \times \text{valency} = \rho \dfrac{N_A}{M} \times \text{valency},$ where V is the volume of the specimen, and D the density of the metal. Substituting the numerical data for the various quantities, we get

$$n = D \frac{N_A}{M} \times \text{valency} = \frac{7.8 \times 10^3 \text{ kg/m}^3 \times 6.023 \times 10^{23} \text{ mol}^{-1} \times 1}{0.064 \text{ kg/mol}} = 7.34 \times 10^{28} m^{-3}.$$

From equation (9.3), the drift velocity is given by $I = neAv_d$, or $v_d = \dfrac{I}{neA}$.

Substituting the above values of n, and the cross sectional area of Y, we get

$$v_d = \frac{I}{neA} = \frac{5.0 \text{ C/s}}{7.34 \times 10^{28} m^{-3} \times 1.6021 \times 10^{-19} C \times 3.77 \times 10^{-6} m^2} = 1.12 \times 10^{-4} m/s.$$

The extremely small value of the drift velocity gives some idea of the extent to which the motion of each electron is impeded by collisions with other electrons and with the atomic centres in the metal. We are able to record measurable currents in practice due to the extremely large number of these charge carriers passing a given plane at any given time.

9.2 OHM'S LAW

Ohm's law relates the current density through a conductor to the electric field intensity established across the conductor by some external potential difference. The law states:

The current density in a conductor is directly proportional to the electric field intensity across the conductor.

In symbols, if **J** and **E** denote the current density and the electric field intensity, respectively, we have

$$\mathbf{J} = \sigma \mathbf{E}, \tag{9.7}$$

where σ is a constant of proportionality, called the **conductivity** of the material. The conductivity is a property of the material, and varies from one material to another. Obviously, the conductivity of metals is high, and that for insulators is low. This means larger current densities are obtained for the same electric field in a conductor than in an insulator. The unit of conductivity is A/Vm (i.e. amp per volt-metre). This unit is more commonly known as **siemens**. Current density is a vector quantity, the vector being in the same direction as the electric field for positive charges.

A related quantity is **resistivity**. This is a measure of how difficult it is to have charges moving through the material. The resistivity is usually represented by ρ, and is related to the conductivity through the equation

$$\rho = \frac{1}{\sigma}. \tag{9.8}$$

In terms of resistivity, we express Ohm's law as

$$\mathbf{J} = \frac{1}{\rho}\mathbf{E}. \tag{9.9}$$

9.2.1 Resistance

Using equation (9.7) we write down the current I flowing through the conductor in terms of the potential difference V maintained across it as follows:

$$J = \frac{I}{A} = \sigma\frac{V}{L}, \tag{9.10}$$

where A is the area of cross section of the specimen, and L is its length. We thus have

$$I = \frac{\sigma A}{L} V. \tag{9.11}$$

From (9.11), it follows that, for a given specimen (of fixed length and cross section), the current flowing though is directly proportional to the potential difference across it. This provides a more practical and common statement of Ohm's law, that

at constant temperature, the current flowing through a metallic conductor is directly proportional to the potential difference maintained across it.

Equation (9.11) is more commonly written in the equivalent form

$$V = \frac{L}{\sigma A} I. \tag{9.12}$$

This is further expressed as

$$V = RI, \tag{9.13}$$

where R is a constant of proportionality, known as the **resistance** of the specimen. Equation (9.13) represents the more common mathematical expression for Ohm's law. The unit of resistance is Volt/amp, alternatively known as the **Ohm**, symbol Ω. The resistance of a given specimen varies with temperature and with the dimensions of the specimen

(a) Variation with temperature

The resistance of a conducting wire can be explained in terms of collisions among the free electrons and between the free electrons and the vibrating atomic centres. These collisions impede the overall motion of the electrons. When the metal is heated, the collisions become more frequent and violent, thereby decreasing the drift velocity of the electrons. As a result the electrical resistance increases. Hence the resistance of a metal increases as the metal is heated. The resistance R of a metallic conductor at temperature T (in °C) is given approximately by

$$R = R_0\left(1 + \alpha T + \beta T^2\right), \tag{9.14}$$

where α and β are positive constants, and R_o is the resistance at 0 °C. The constant β is normally so small that over small and moderate temperatures, equation (9.14) becomes

$$R = R_0(1 + \alpha T), \tag{9.15}$$

and the constant α is referred to as the **temperature coefficient of resistance**. Since the resistance of the metal increases with temperature, the temperature coefficient of resistance is positive.

Example 9.2

A copper wire has a resistance of 1.05 Ω at a temperature of 20 °C. Given that the temperature coefficient of resistivity of copper is 3.93×10^{-3} /°C, determine the resistance of the wire at 100 °C.

Solution

Data: Resistance at 20 °C, R_{20} = 1.05 Ω; α = 0.00393 /°C.

In this example we use the equation $R = R_0(1 + \alpha T)$, where R_o is the resistance at 0 °C, and R is the resistance at temperature T °C. Given the resistance at 20 °C, we first calculate the resistance at 0 °C.

Thus,

$$R_0 = \frac{R_{20}}{(1 + 20\alpha)} = \frac{1.05\ \Omega}{(1 + 20\ ^{\circ}C \times 0.00393\ /^{\circ}C)} = 0.94\ \Omega.$$

With this value of R_o, we use equation (9.15) to determine the resistance R_{100} at 100 °C. We therefore have $R_{100} = 0.97\ \Omega \times (1 + 0.00393\ /^{\circ}C \times 100\ ^{\circ}C) = 1.35\ \Omega$.

(b) Variation with length and thickness

Comparing (9.12) and (9.13), we express the resistance R of a wire as

$$R = \frac{L}{\sigma A} = \frac{\rho L}{A}, \qquad (9.16)$$

where ρ is the resistivity of the material of the wire. From (9.16), it is seen that resistance depends on the length of the material, its area of cross section, and the substance of which the material is made. Resistivity is a physical property of the material, with units of Ohm-metre, written in symbols as Ωm. **Table 9.1** gives the resistivities of a few substances. It is seen from this table that metals have the lowest resistivities, whilst insulating materials have very high resistivities.

Table 9.1 Resistivity values of some common substances at 20 °C.

Metals & alloys	Resistivity (Ωm)	Other substances	Resistivity (Ωm)
Aluminium	2.75×10^{-8}	Carbon	3.50×10^{-5}
Copper	1.72×10^{-8}	Silicon	2.30×10^{3}
Gold	2.44×10^{-8}	Germanium	0.60
Mercury	95.0×10^{-8}	Amber	5.00×10^{14}
Iron (or steel)	20.0×10^{-8}	Glass	$10^{10} - 10^{14}$
Tungsten	5.25×10^{-8}	Mica	$10^{11} - 10^{15}$
Lead	22.0×10^{-8}	Quartz	7.50×10^{16}
Constantan (alloy of Cu 60%, Ni 40%)	49.0×10^{-8}	Wood	$10^{8} - 10^{11}$
Nichrome	100.0×10^{-8}		

Example 9.3

Calculate the length of fine copper wire with diameter 0.50 mm that will produce a resistance of 2.0 Ω. Resistivity of copper = 17.2 nΩm.

Solution

Data: diameter d of wire = 5.0×10^{-4} m, resistance R = 2.0 Ω
Resistivity of copper $\rho = 1.72 \times 10^{-8}\,\Omega$m.

From equation (9.16), $R = \frac{\rho L}{A}$, or $L = \frac{RA}{\rho}$, where A is the area of cross section.

$A = \frac{\pi d^2}{4} = \frac{\pi \times 25 \times 10^{-8} m^2}{4} = 1.96 \times 10^{-7} m^2$. Thus the required length L of the wire is

$$L = \frac{RA}{\rho} = \frac{2.0\,\Omega \times 1.96 \times 10^{-7} m^2}{1.72 \times 10^{-8}\,\Omega m} = 22.79 \text{ m}.$$

9.2.2 Similarity with heat flow

The equation governing the flow of electricity in conductors is very much similar to the flow of heat through a solid. The rate of flow of electrical charges in a conducting material is given by

$$I = \frac{dQ}{dt} = -\sigma A \frac{dV}{dL}, \tag{9.17}$$

the negative sign expressing the fact that charges flow from a region of higher potential to one of lower potential. Equation (9.17) can be interpreted by saying that the rate of charge flow through a conductor is directly proportional to the area of cross section and the **potential gradient**, the constant of proportionality being the electrical conductivity.

As we saw in Volume 1, the rate of heat conduction in a solid is given by

$$H = \frac{dQ}{dt} = -kA \frac{dT}{dx}. \tag{9.18}$$

In this, $\frac{dQ}{dt}$ represents the rate of heat flow (from a region of higher temperature to one of lower temperature). This shows that the rate of heat conduction is proportional to the area of cross section, and the **temperature gradient**. The constant of proportionality is the thermal conductivity k. That the equations for the flow of electricity and heat in a conductor are similar results from the fact that free electrons are mostly responsible in both modes of conduction.

9.2.3 Resistors

Whilst electrons flow within conductors, their motion is impeded by collisions the electrons make with each other and with the atomic cores. Together with the potential difference, this resistance determines the current through the conductor. This principle is employed to make circuit elements whose role is to control the current flowing in the circuit. These elements are known as **resistors**. The larger the resistance of the resistor, the smaller the current flow through it. **Fig. 9.3** shows the symbols used to represent a resistor. The symbol (a) shows a fixed resistor, whilst (b) shows the representations for a variable resistor.

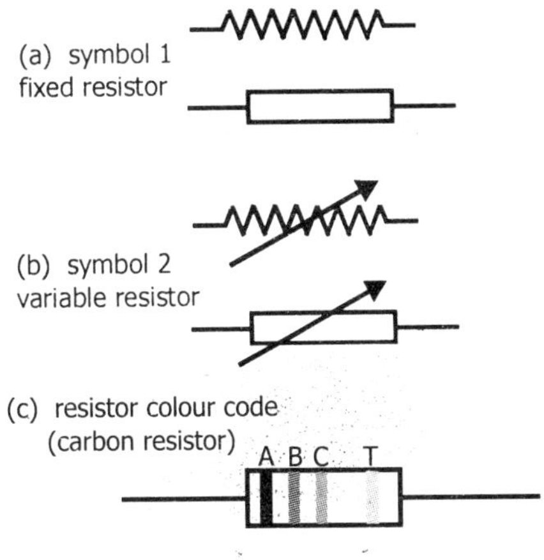

Fig. 9.3 Resistor symbols, (a) and (b); the resistor colour code (c).

The resistance of a standard resistor is normally supplied by the manufacturer. The most common type is the carbon resistor. This comes as a cylinder with different coloured stripes on the barrel, as in (c). The colours are part of a code, known as the **resistor colour code**. This code is used to read off the resistance of the resistor.

Table 9.2 The resistor colour code

A, B, C digits

Black	0	Green	5
Brown	1	Blue	6
Red	2	Violet	7
Orange	3	Grey	8
Yellow	4	white	9
Gold (C only)	-1	Silver (C only)	-2

Tolerance (T) code

Silver	10%
Gold	5%

In (c), the colours are labelled A, B, C, and T. The resistance R is coded in the colours A, B, and C, as follows:

$$R = AB \times 10^{C},\ T. \tag{9.19}$$

T is the tolerance, or error margin; it is normally expressed as a percentage. **Table 9.2** gives the colour code.

Like capacitors, resistors can be arranged as series and parallel combinations. This way it is possible to get a resistance with any desired value.

9.2.4 Resistors in series

Fig. 9.4 shows resistors R_1, R_2, R_3, ..., R_n, all connected in series to a source of voltage supply. The current I in each resistor is the same. The potential differences across the resistors, however, are not necessarily the same. The sum of these potential differences equals to total voltage supply V.

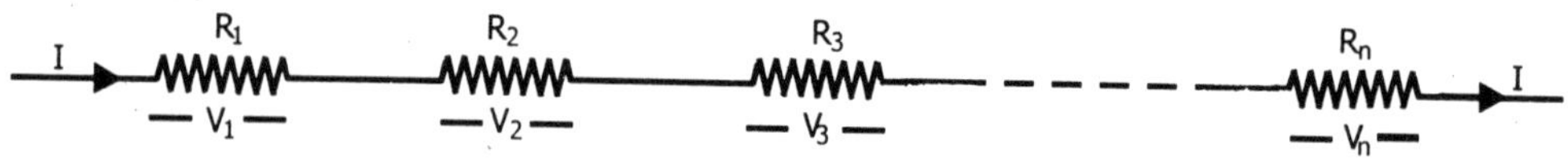

Fig. 9.4 Resistors connected in series.

Thus,

$$V = V_1 + V_2 + V_3 + \ldots + V_n, \tag{9.20}$$

where V is the voltage supply. Using Ohm's law, we express the potential difference across each resistor as the product of the current and the resistance. If R is the overall resistance of the combination, then

$$V = IR = IR_1 + IR_2 + IR_3 + \ldots IR_n.$$

The combined resistance of a set of resistors in series is hence given by

$$R = R_1 + R_2 + R_3 + \ldots + R_n = \sum_{i=1}^{n} R_i. \tag{9.21}$$

9.2.5 Resistors in parallel

Fig. 9.5 shows resistors R_1, R_2, R_3, ..., R_n, connected in parallel to a source of voltage supply. The potential difference V across the resistors is the same, but the currents through them are not the same. The currents add up to the total current I. Thus, we have

$$I = I_1 + I_2 + I_3 + \ldots + I_n.$$

Expressing these currents in terms of the potential difference and the resistances, we have

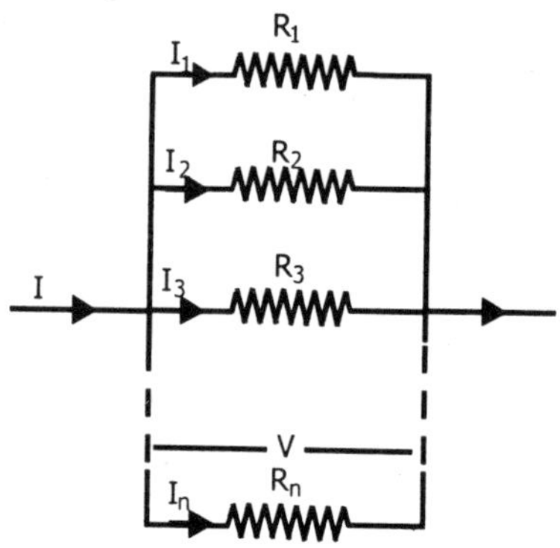

Fig. 9.5 Resistors in parallel.

$$I = \frac{V}{R} = \frac{V}{R_1} + \frac{V}{R_2} + \frac{V}{R_3} + \ldots + \frac{V}{R_n},$$

where R is the effective resistance of the combination. Thus, we have

$$\frac{1}{R} = \frac{1}{R_1} + \frac{1}{R_2} + \frac{1}{R_3} + \ldots + \frac{1}{R_n} = \sum_{i=1}^{n} \frac{1}{R_i}. \tag{9.22}$$

9.3 THE SERIES CIRCUIT

When charges move within a conducting loop, starting from one point, going round the loop and ending at the same point, the loop is called an electric circuit. For charges to move round the loop, there must be a source of potential, or a **cell**.

The simplest circuit is the series circuit, having a single loop containing a cell, and a resistor. **Fig. 9.6** shows a simple series circuit. The figure also shows the common circuit symbols for a cell E, the current and voltage measuring meters and their connections in the circuit, and a switch S. The conventional current direction is taken from the positive terminal to the negative terminal. We note that the electrons in the circuit actually flow from the negative terminal to the positive terminal.

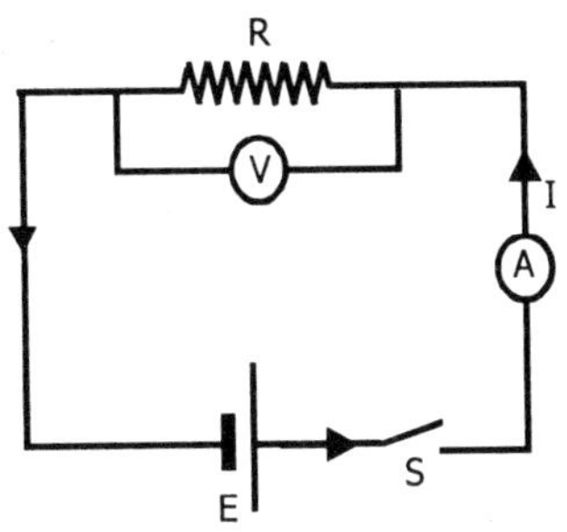

Fig. 9.6 The simple series circuit.

When the switch is closed, current I flows through the circuit. We apply Ohm's law to determine the current flow through the circuit in terms of the potential difference across the resistor, and the resistance R. Hence we write

$$I = \frac{V}{R}. \tag{9.23}$$

Example 9.4

Two resistors each with resistance 2.0 Ω are connected in a circuit to a 6.0 V cell. Calculate the current flowing in the circuit if the resistors are connected

(i) in series,
(ii) in parallel

Solution

Fig. 9.7 shows the two connections.

(i) The combined resistance of the two resistors in series is the sum of the resistances. Hence, the total resistance in the circuit is $R = 2R_1 = 4.0\ \Omega$.. By Ohm's law, the current I is given by

$$I = \frac{E}{R} = \frac{6.0\ V}{4.0\ A} = 1.5\ A.$$

(ii) For the parallel setup, combined resistance is $R = \frac{R_1}{2} = 1.0\ \Omega$.

Thus the current is 6.0 A.

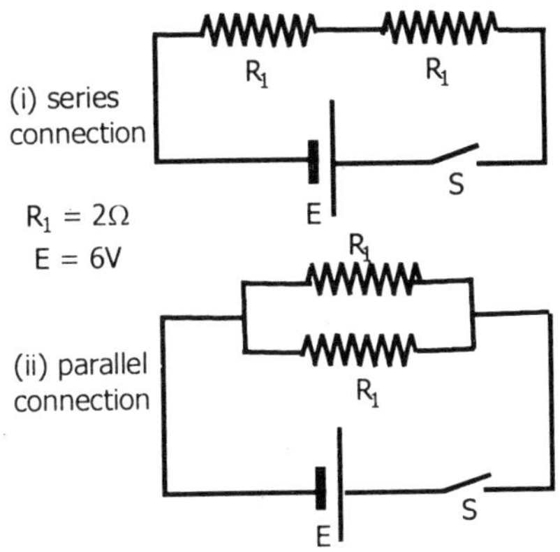

Fig. 9.7

9.4 ELECTROMOTIVE FORCE AND INTERNAL RESISTANCE

The electromotive force in a circuit is the source of electrical potential around the circuit. This potential is usually supplied by an electrolytic, or chemical cell.

The electromotive force (or emf) of a cell is the electrical work done, or energy expended, when a unit positive charge is taken from the positive terminal of the cell to the negative terminal.

The unit of emf is Joules per Coulomb, or Volt. It is important to bear in mind that electromotive force is really **not** a force, but energy. The notion of force tends to arise when we consider that emf is what drives (or 'pushes') charges around the circuit loop.

The **terminal potential difference** across a cell is the potential difference measured across the terminals of the cell. In the series circuit of **Fig. 9.6**, the terminal potential difference is V, the same as that across the resistor. It is found that this terminal potential difference is always less than the emf of the cell by some amount. This difference arises from the fact that some energy is required to drive the charges within the cell. The cell behaves as though it has a resistance within it that opposes the flow of charge through it. This resistance is called the **internal resistance**.

The emf of the cell is equal to the sum of the potential difference across the internal resistance, and that across the external resistor. For the complete circuit, Ohm's law equation would therefore read

$$E = V + Ir, \qquad (9.24)$$

where E is the emf of the cell, I is the current flowing through the circuit, and r is the internal resistance of the cell. Writing V in terms of the current I, and the external resistance R, we get

$$E = I(R + r). \qquad (9.25)$$

Taking the internal resistance into consideration, the circuit diagram for the series circuit can be represented in **Fig. 9.8**, in which the internal resistance r is connected in series with the cell. In the symbol, this series resistance is not normally shown.

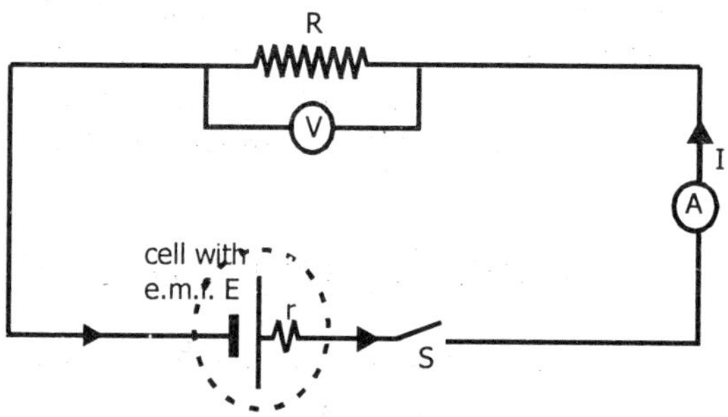

Fig. 9.8 The series circuit with internal resistance.

In any circuit, the overall resistance in the circuit is made up of a series combination of the internal resistance r, and the effective external resistance in the circuit. Internal resistances are normally very small, typical values being between 1 Ω and 5 Ω.

Example 9.5

A cell drives current of 1.2 A around a circuit consisting of two resistors each of resistance 2.0 Ω, connected in series. When the same resistors are connected in parallel, the current measured is 3.0 A. Determine the emf and internal resistance of the cell.

Solution

The effective resistance R_S of the two resistors in series is 4.0 Ω. When the two resistors are connected in parallel, the effective resistance R_P of the parallel combination is 1.0 Ω.

The Ohm's law equations for the two combinations are as follows:
For the series combination,

$$E = I(r + R_S) = 1.2 \times (4 + r). \qquad (1)$$

For the parallel combination,

$$E = I(r + R_P) = 3.0 \times (1 + r). \qquad (2)$$

Solving equations (1) and (2) above simultaneously, we get E = 6.4 V, and r = 1.5 Ω.

Example 9.6

A battery charger sets up a potential difference of 13.6 V at the terminals of a battery with an emf of 12.0 V, and internal resistance 0.95 Ω. Calculate
(i) the charging current,
(ii) the rate of conversion of electrical energy to chemical energy,
(iii) the rate of heat production in the battery.

Solution

(i) During the charging process, we find the charging current using Ohm's law for the complete circuit. Thus, the charging current I is given by $V = E + Ir$, where V is the potential difference of the charger, E and r being the emf, and internal resistance, respectively, of the battery being charged. Hence, $I = \frac{V - E}{r} = \frac{13.6 - 12.0}{0.95} = 1.68$ A.

(ii) The total power in the circuit is made up of two parts: the power used for charging, and that lost through the internal resistance of the battery being charged. The rate of conversion of electrical energy to chemical energy is P, where

$$P = EI = 12.0\ \text{V} \times 1.68\ \text{A} = 20.16\ \text{W}.$$

(iii) The energy lost through the internal resistance of the battery as heat, is the reason why batteries warm up slightly when they are being charged. The power P_r dissipated across the internal resistance is $P_r = I^2 r = 1.68^2 \times 0.95 = 2.68$ W.

9.5 COMBINATIONS OF CELLS

Cells can be connected in series, in opposition, or in parallel. We discuss below the effective emf and internal resistance of these combinations.

9.5.1 Cells in series

Fig. 9.9 shows a number of cells E_1, E_2, E_3, ..., E_n, connected in series. The positive terminal of one cell is connected to the negative plate of the one next to it. The cells all send currents in the same direction, and the internal resistances are all in series.

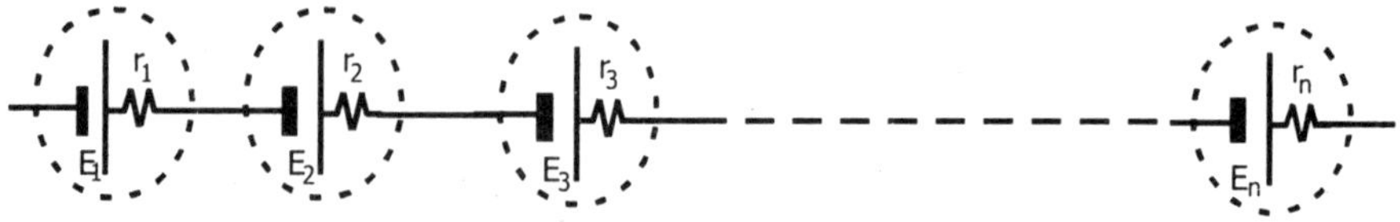

Fig. 9.9 Cells connected in series.

If E and r are the emf and internal resistance, respectively, of the combination, then

$$E = E_1 + E_2 + E_3 + ... + E_n = \sum_{i=1}^{n} E_i, \quad (9.26)$$

and

$$r = r_1 + r_2 + r_3 + ... + r_n = \sum_{i=1}^{n} r_i. \quad (9.27)$$

9.5.2 Cells in opposition

The opposition arrangement is similar to the series setup, with the difference that the positive terminals of some cells are connected to each other, thus sending currents in opposite direction – hence the term 'opposition'. **Fig. 9.10** shows cells in opposition.

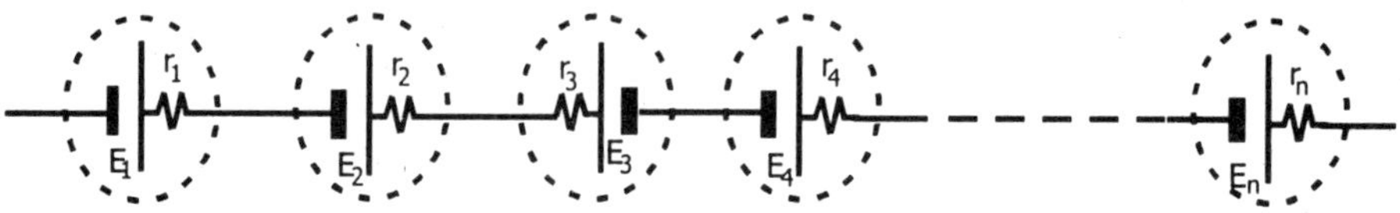

Fig. 9.10 Cells connected in opposition.

The emf E, and the internal resistance r of the combination are given respectively by

$$E = E_1 + E_2 - E_3 + E_4 + ... + E_n = \sum_{i=1}^{n} E_i, \quad (9.28)$$

and

$$r = r_1 + r_2 + r_3 + ... + r_n = \sum_{i=1}^{n} r_i. \quad (9.29)$$

Note the difference in the summations represented by equations (9.26) and (9.28). In (9.28), the summation takes the orientations of the cells into consideration. As resistances have no polarities, the internal resistances add up even for cells in opposition.

9.5.3 Cells in parallel

When cells are connected in parallel, all their positive terminals are connected to one point, whilst their negative terminals connected to another point. The two points then serve as the terminals of the combination. If the cells are not identical (i.e. not having the same emf and internal resistance), current loops will exist between any pair of cells, making the analysis very much complicated. The simplest case is when the cells are identical. For a number of identical cells each with emf E,

$$\text{combined e.m.f.} = E \tag{9.30}$$

The internal resistance is the equivalent of a number of identical resistors connected in parallel.

$$\text{effective internal resistance} = \frac{r}{n} \tag{9.31}$$

9.6 ELECROLYSIS

Electrolysis is the process by which an electric current is passed through a chemical substance in order to bring about a chemical change. The chemical change is one in which the substance gains or loses electrons – **oxidation**, or **reduction**. The process is carried out in an **electrolytic cell**, an apparatus consisting of positive and negative electrodes held apart and dipped into a solution containing positive and negative ions. **Fig. 9.11** is a schematic diagram of a simple electrolytic cell using sodium chloride as the **electrolyte** – i.e. substance that conducts electricity as a result of positive and negative ions.

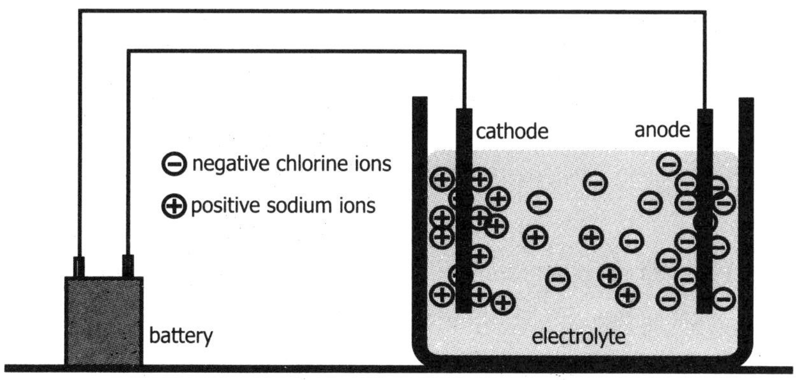

Fig. 9.11 Electrolysis used to break down common salt (sodium chloride). The negative chlorine ions are collected at the anode, whilst the positive sodium ions travel to the cathode.

The substance to be transformed may form the electrode, may constitute the solution, or may be dissolved in the solution. Electrons from the power supply enter through the negatively charged electrode (or cathode); positively charged ions in the electrolyte travel to this electrode, combine with the electrons, and are transformed to neutral elements or molecules. The negatively charged ions of the solution travel to the anode, give up their electrons, and are transformed into neutral atoms or molecules. If the substance to be transformed is the electrode, the electrode slowly dissolves as the current passes through the solution. An example of this occurs when copper electrodes are used in an electrolyte of copper sulphate solution. The copper of one electrode (the anode) is dissolved, and deposited on to the other electrode (the cathode).

9.6.1 Faraday's laws of electrolysis

Electrolysis was studied extensively by Michael Faraday (1791 – 1867), leading to the formulation of two laws, known as the laws of electrolysis. These are as follows:

1. ***The mass of element deposited in electrolysis is directly proportional to the quantity of electricity used***, and

2. ***the mass of different elements deposited by the same quantity of electricity is directly proportional to their atomic masses***.

When a steady current I flows through the electrolyte for a time t, the total charge Q transferred is given by $Q = \text{current} \times \text{time} = It$.

If m is the mass deposited, from Faraday's laws, we write

$$m = zIt, \tag{9.32}$$

where z is a constant, known as the **electrochemical equivalent**, of the ion. The electrochemical equivalent of an ion is defined as the mass of ion liberated by the passage of unit quantity of electricity. The unit quantity of electricity is defined as one that deposits one gram-equivalent of the element or compound. The unit is known as the **Faraday**. Its numerical value has been the subject of many detailed studies; measurements have found it to be around 96,500 coulombs. Thus one Faraday of charge passing through an electrolyte will liberate exactly 1 gram-atom of element or molecule.

9.6.2 Uses of electrolysis

The major areas of application of electrolysis are in battery and fuel cell technology, and in metallurgy. There are many spontaneous chemical reactions that liberate electrical energy, when allowed to proceed under special conditions. A process of this type involves the direct conversion of chemical energy into electrical energy. A device in which this is brought about is the electrical cell, or battery. Today's technology has seen improvements on this general principle to produce cells ranging from the simple dry cells, to rechargeable cells.

One application of electrolysis in metallurgy is in **electroplating**, also known as **electrodeposition**. This process involves coating the surface of a metal using an electric current. With this method, metals can be coated by other metals to protect against chemical attack. An example is the coating of iron with copper to prevent corrosion. Using electrodeposition, plastics can also be coated with thin metallic layer to make them conducting. In electroplating, the specimen to be plated is arranged to be one of the electrodes in a suitable electrolytic cell. When the current is turned on, metal ions in the electrolyte travel to the cathode, pick up electrons from the latter to form the metal atoms and attach to the plate.

9.7 ELECTRICAL ENERGY AND POWER

From our study of electrostatics, we saw that the work done, or energy expended U, when a charge Q is placed at a point in an electric field is given by

$$U = QV,$$

where V is the potential at the point. For charges in motion, we express the charge in terms of the current I, and the time t, for which the charge flows. Thus,

$$U = VIt. \tag{9.33}$$

If the charges flow through a resistor of resistance R, we can further express the current in terms of the potential difference V across the resistor, and its resistance R.
Hence,

$$U = VIt = I^2Rt = \frac{V^2}{R}t. \tag{9.34}$$

This energy appears in the form of heat. As a result, it is often referred to as **energy dissipation**.

The resistor slowly warms up as the current flows through it. This is the heating effect of a current. In some cases the heating is desirable, whilst it is a nuisance in other cases. Examples of cases in which the heating is desirable include toasters, household electric irons, lamp filaments, cookers, electric furnaces, and ovens. The heating is not desirable in other appliances like fans, transformers, electric motors, and electric generators. The electrical energy dissipated per unit time, or Power P, is given by

$$P = \frac{U}{t} = VI = I^2R = \frac{V^2}{R}. \tag{9.35}$$

Example 9.7

A car radio draws a current of 3.5 A when operating off a 12-volt car battery. Calculate
(i) the rate at which the radio consumes energy,
(ii) the resistance of the radio circuit.

Solution

(i) Rate of energy consumption by the car radio is its power dissipation. From equation (9.34) above, the power P is given by

$$P = IV = 3.5\text{ A} \times 12.0\text{ V} = 42.0\text{ watts.}$$

(ii) Using the power value obtained above, we calculate the resistance R as

$$P = I^2R,\ \text{or } R = \frac{P}{I^2} = \frac{42.0\text{ J/s}}{3.5^2\text{ A}^2} = 3.42\ \Omega.$$

By far the biggest source of electrical energy comes from electric power stations. In the generation of electric power, energy in some primary form is converted finally into electrical energy. The starting form of energy may be chemical, as in a coal-fired or oil-fired power station. This chemical energy is converted into heat energy, used to produce steam. The steam is then sent at high velocity to turn turbine blades, finally producing electricity. In hydro powered stations, the mechanical energy of the water at the bottom of a waterfall is used to turn turbine blades, thus producing electricity. In either of these cases, the electricity produced is in the form known as **alternating current**, the details of which, and its conversion into direct current, we shall study in later chapters.

The electrical energy from a power generating station is sold to consumers in units, where 1 unit of electricity is the kilowatt-hour, defined as

the energy delivered when an appliance rated at 1 kilowatt (1 kW) runs continuously for 1 hour.

Thus, 1 unit = 1kWh = 1000 J/s x 3600 s = 3.60×10^6 J.

Example 9.8

An electric motor rated at 500 W is made to run continuously for 30 days. If 1 unit of electricity costs $0.25, calculate the cost of running this motor for the period.

Solution

We calculate the total energy U consumed for the period.

$$U = \text{Power rating} \times \text{time (in seconds)} = 500\text{ W} \times 30 \times 24 \times 3600\text{ s} = 1.296 \times 10^9\text{ Ws.}$$

Converting this to kWh, we get $U = \dfrac{1.296 \times 10^9}{3.6 \times 10^6} = 360$ kWh.

The cost of running the motor is 360 kWh x $0.25/kWh = $ 90.00.

9.8 RESISTANCE NETWORK CIRCUITS

Ohm's law is very useful when applied to single circuit loops. Most practical circuits are of the type known as **network circuits**, and consist of a complicated mesh of circuit **loops**, with some loops sharing common elements. The common elements might be resistors, cells, or simply conducting wires. **Fig. 9.12** is an example of a simple resistor network circuit.

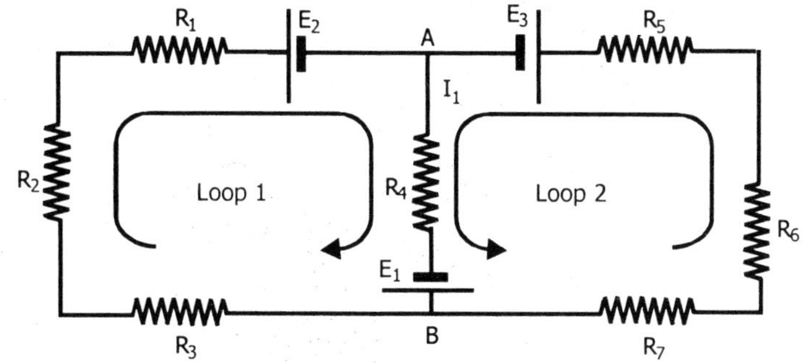

Fig. 9.12 A simple resistance network circuit.

We study network circuits in terms of **loops** and **junctions**, where

- a circuit loop is any closed path, and
- a junction is the point where elements from different loops meet.

In the circuit of **Fig. 9.12**, points A and B are junctions, and two loops are shown. For purposes of clarity, the third loop is not shown. This loop constitutes the outside circuit. We analyse the network circuit using two rules, known as **Kirchhoff's rules**. There are two of them: the current (or junction) rule, and the voltage (or loop) rule.

Kirchhoff's junction (or current) rule states that

the algebraic sum of the currents at any junction is zero.

At any junction there are currents flowing **into**, and **out of**, the junction. The junction rule stipulates that **the total current flowing into a junction is equal to the total current flowing out of the junction.**

Kirchhoff's loop (or voltage) rule states that

in any closed loop, the sum of the potential differences across the resistors is equal to the algebraic sum of the emfs in that loop.

The application of Kirchhoff's rules is illustrated by the following numerical example.

Example 9.9

In the circuit of **Fig. 9.12**, find the currents flowing in each arm, given that $R_1 = R_2 = R_6 = 4\ \Omega$, $R_3 = R_7 = 2\ \Omega$, $R_4 = R_5 = 1\ \Omega$, $E_1 = 12.0$ V, $E_2 = 8.0$ V, and $E_3 = 6.0$ V. Ignore the internal resistances of the cells.

Solution

Fig. 9.13 is a reproduction of **Fig. 9.12**, with the component values stated. The currents, I_1, I_2, and I_3 are assumed to be flowing in the directions shown.

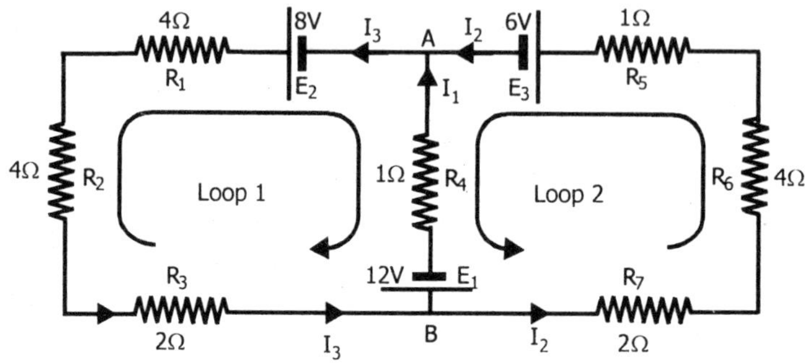

Fig. 9.13

At the junctions A and B, we apply the current (or junction) rule to get

$$I_1 + I_2 - I_3 = 0 \qquad (1)$$

In any closed loop above, the cells are either in series, or in opposition. We use Kirchhoff's loop rule for the two loops shown, in the directions of the arrows. Kirchhoff's voltage rule equations applied to this circuit are as follows:

Loop 1: $\quad -I_1R_4 - I_3R_3 - I_3R_2 - I_3R_1 = E_1 - E_2,$

giving

$$-I_1 - 7I_3 = 12 - 8 = 4 \qquad (2)$$

Loop 2: $\quad I_1R_4 + I_2R_7 + I_2R_6 + I_2R_5 = E_1 - E_3,$

or

$$I_1 + 7I_2 = 12 - 6 = 6 \qquad (3)$$

Solving equations (1), (2), and (3) simultaneously, we get I_1 = -10.00 A, I_2 = 2.28 A, and I_3 = -7.82 A. The negative signs of the currents indicate that the actual current flow in the arm is opposite to the direction assumed earlier.

9.9 THE WHEATSTONE BRIDGE

The Wheatstone bridge provides an experimental arrangement that uses resistance networks to determine the ratio of resistances as well as the resistance of an unknown resistor. **Fig. 9.14** is the circuit diagram for the Wheatstone bridge.

The bridge circuit consists of four resistances P, Q, R, and S, connected as shown. When the switch keys K_1 and K_2 are closed, current I from the battery divides into I_1 and I_2. At junction B, current I_g flows through the galvanometer (or milliammeter), with resistance r. We apply Kirchhoff's rules to the circuit loops indicated as follows:

Loop 1:

$$I_1Q - I_gr - I_2S = 0, \tag{9.36}$$

Loop 2:

$$(I_2 - I_g)R - (I_1 + I_g)P - I_gr = 0. \tag{9.37}$$

In the practical application, the resistances in the circuit are slowly varied to a point such that no current flows through the galvanometer. The bridge is then said to be **balanced**. This is described as the **balance point**, for which $I_g = 0$. Equations (9.36) and (9.37) are hence rearranged to give

$$\frac{P}{Q} = \frac{R}{S}. \tag{9.38}$$

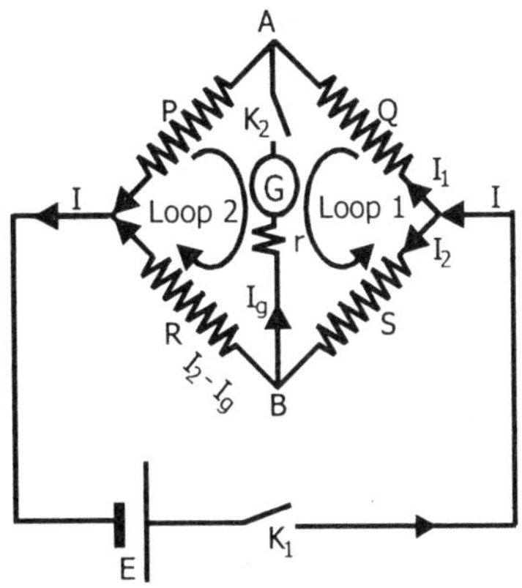

Fig. 9.14 Theoretical circuit for the Wheatstone bridge.

The practical Wheatstone bridge circuit is shown schematically in **Fig. 9.15**. On a wooden or metal plank are mounted a metre rule, metal connecting strips and connecting blocks, with gaps as shown in the arrangement. Resistors R and S are placed in the gaps provided. A uniform resistance wire runs over the metre rule, and soldered to the metal strips at the ends A and B of the rule. One terminal of the galvanometer G is connected to

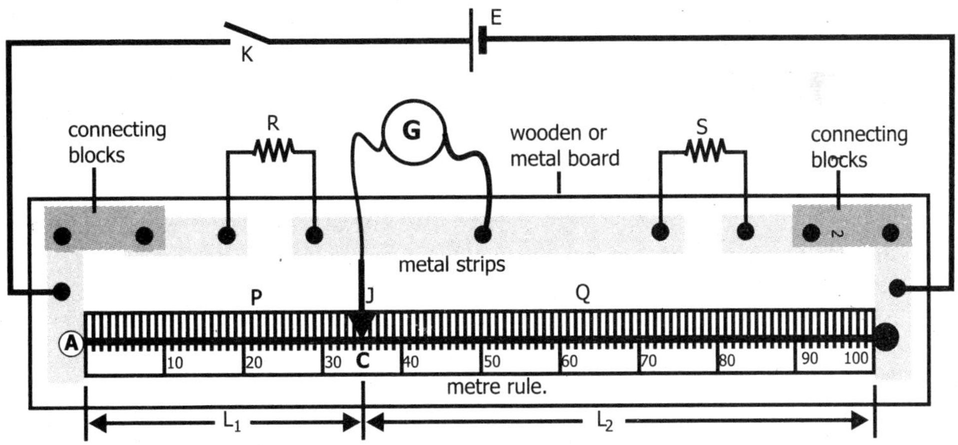

Fig. 9.15 The practical Wheatstone bridge circuit.

the mid-point of the central strip; the other terminal of the galvanometer is connected to a sliding contact, known as the jockey J. When the key is closed current flows through the resistance network. Contact on the wire is varied slowly until a point is reached when no current flows through the galvanometer. At this balance point,

$$\frac{R}{S} = \frac{P}{Q},$$

where P and Q are the resistances of the segments AC, and CB, respectively, of the wire. Expressing these latter resistances in terms of the lengths of the segments, we get

$$\frac{R}{S} = \frac{L_1}{L_2}. \tag{9.39}$$

Thus the ratio of the resistances is the same as the ratio of balance lengths.

The main use of the Wheatstone bridge lies in the applications of equation (9.38). One use of the bridge is to compare resistances. In this application, the resistors are placed in the positions R and S, in the circuit, and balance of obtained. The ratio of the balance lengths is then the same as the ratio of the resistances.

The same bridge can be used to find the value of an unknown resistor, by comparing it with a standard (i.e. known) resistor. From the ratio of the balance lengths, the unknown resistance can be determined.

9.10 THE POTENTIOMETER

The potentiometer is used to compare and measure potential differences. In its outlook, the potentiometer is very similar to the resistance bridge, both consisting of a length of metre wire passing over a metre rule. **Fig. 9.16** shows the practical potentiometer circuit used for comparing the electromotive forces of two cells.

The cell E drives current through the main circuit, in series with the wire AB, dropping a constant voltage across it. Another source of potential difference V (e.g. a battery) is now connected with its positive side at A, and the negative side to the jockey through the galvanometer, a shown. The sliding contact with the wire is moved along the wire until the galvanometer reading is zero. If L is the balance length, then

$$V \alpha L.$$

Often, for purposes of accuracy, a resistor is placed in the circuit, with value such that the balance point is obtained somewhere around the middle of the wire.

If two sources of emf E_1 and E_2 are to be compared, they are connected in turn, and the corresponding balance lengths L_1 and L_2 determined. The ratio, E_1/E_2 is given by

$$\frac{E_1}{E_2} = \frac{L_1}{L_2}. \qquad (9.40)$$

9.10.1 Measurement of unknown emf

The same circuit of **Fig. 9.16** can be used to determine an unknown emf. This is done by first calibrating the potentiometer wire. In calibrating the wire, a standard cell with known emf E_S is balanced on the wire. Standard cells have very accurately known emf, sometimes to five or more places of decimal. If the balance length obtained with the standard cell is L_S, then the voltage drop per unit length of the potentiometer wire is E_s/L_s. Using the standardised potentiometer, the next step is to obtain balance with the unknown emf, and hence determine the unknown voltage.

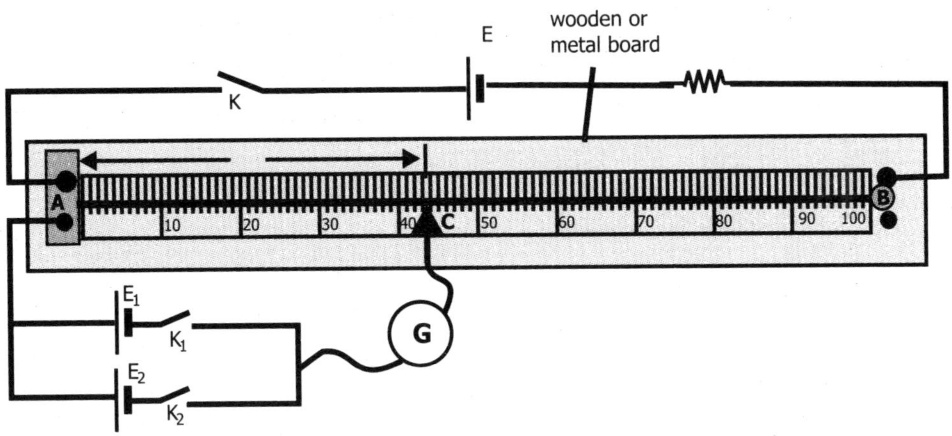

Fig. 9.16 Potentiometer circuit. Comparison of emfs E and E .

9.10.2 Experimental verification of Ohm's law

Often the problem of experimentally verifying Ohm's law connecting potential difference with current is complicated by the demand that a moving coil voltmeter should not be used to measure the potential difference across the resistor. This is because the calibration of the voltmeter assumes Ohm's law is true. In verifying Ohm's law therefore, we need to use a voltage measurement that does not assume Ohm's law. The potentiometer circuit shown in **Fig. 9.17** is often employed.

The current I through the resistor R is varied using the variable resistor, or rheostat. The corresponding p.d. across the resistor is measured by balancing on the calibrated potentiometer. A plot of the variation of I vs. V is linear, verifying Ohm's law. The resistance of the resistor is obtained from the slope b/a of the line.

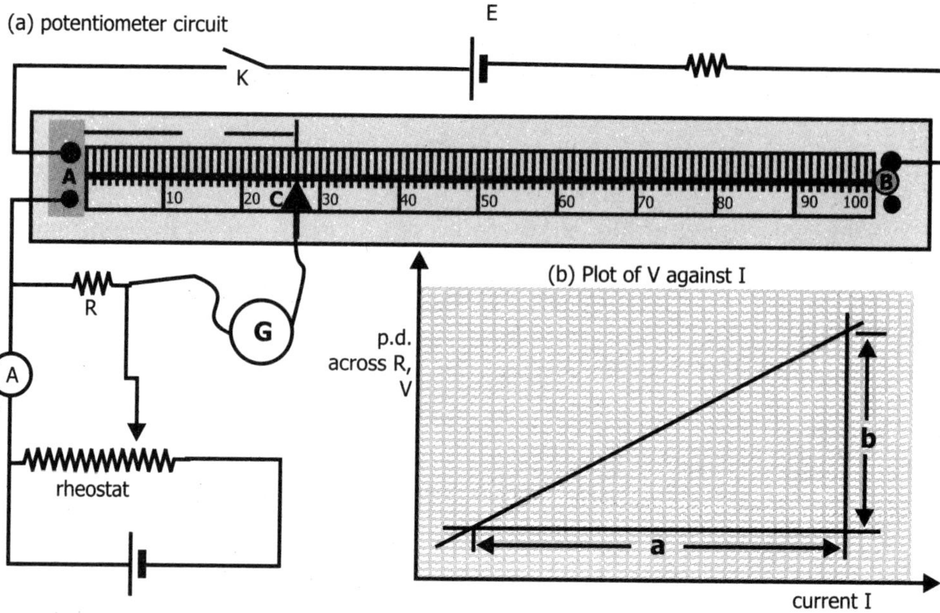

Fig. 9.17 Verifying Ohm's law. (a) circuit using calibrated potentiometer, (b) plot of V vs. I.

EXERCISES

Where necessary, use values in **Table 9.1**

1. The heating coil of a household iron is 4.0 m long. The wire is made of nichrome, with resistance 15.0 Ω. Calculate the diameter of the wire.

2. Manganin wire of diameter 0.5 mm is used to make a resistor of resistor 10.0 Ω. Calculate the length of wire to the nearest cm.

3. Calculate the current that would be drawn by a resistor of 40.0 Ω, running off a 240-Volt power supply.

4. Resistances of 2.0 Ω, 3.0 Ω, 4.0 Ω, and 5.0 Ω are all connected in series with a battery of 24.0 V. Neglecting the internal resistance of the battery, calculate
   (i) the total resistance in the circuit,
   (ii) the current flowing in the circuit.

5. Three resistors or 5.0 Ω, 10.0 Ω, and 20.0 Ω are connected in parallel across a 12-volt supply. Calculate
   (i) the combined resistance,

(ii) the current flowing through each resistor,
(iii) the total current in the circuit,
(iv) the electrical power dissipated through each resistor.

6. For the resistor combination represented by **Fig. 9.18**, determine
(i) the effective resistance between A and B,
(ii) the current flow through AB if a voltage source of 24 V is connected across AB.

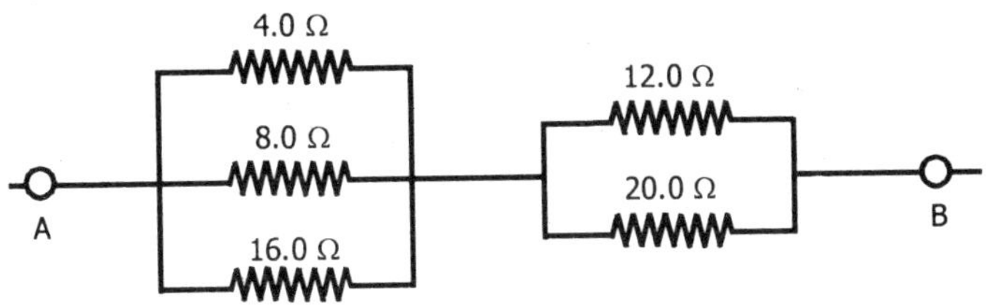

Fig. 9.18

7. Given 10 cells each of emf 1.5 V, and internal resistance 0.20 Ω, determine the current that would flow through an external resistor of resistance 2.5 Ω, when the cells are
(i) all in parallel,
(ii) all in series,
(iii) four in opposition to the rest.

8. A city hall maintains a bank of 30 electric lamps (connected in parallel), each having a resistance of 200 Ω when hot, connected to the terminals of a 240 V supply. Find the cost of operating these lamps for 8 hours, if one unit of electricity costs $0.25.

9. In the circuit of **Fig. 9.19**, a source of 12.0 V is connected across AB, and the resistances are as shown. Calculate
(i) the current through each resistor,
(ii) the power dissipated in each resistor.

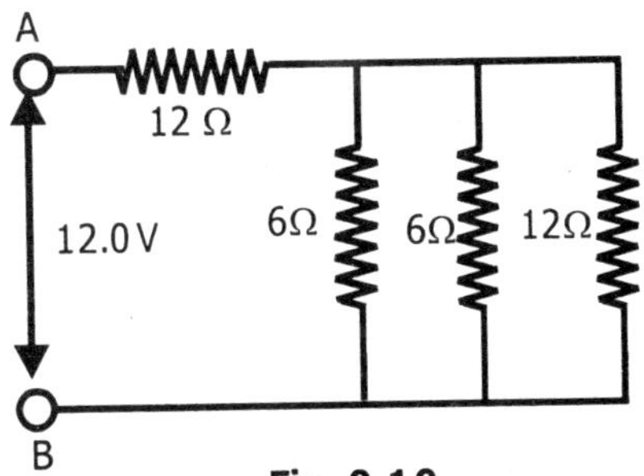

Fig. 9.19

10. The electric machinery in a milling factory is known to develop 210 kW of power. The power consumption for lighting and heating in this factory is about 90 kW, split equally between them. Calculate
(i) the total electrical energy consumption in the factory in one month of 20 working days at an average 8 hours per day.
(ii) the cost save in Summer (3 months), when the factory turns off the heating, given that 1 unit of electricity costs $0.25.

11. A 6.0-volt rechargeable battery with internal resistance 0.6 Ω is to be charged from a supply of 15.0 volts. Given that the maximum charging rate that the battery can withstand is 5.0 A, calculate the resistance that must be connected in series with the battery to achieve this charging current. Draw the diagram of the charging circuit.

12. A battery charger applies a potential difference of 14.0 V at the terminals of a battery of emf 12 V, and internal resistance 0.92 Ω. Determine
    (i) the charging current,
    (ii) the rate of electrical energy conversion into chemical energy,
    (iii) the rate of heat energy production in the battery,
    (iv) how long it would take to charge a dead battery if it is rated at 35 amp-hour.

13. Four identical lamps, rated at 1.5V, 500 mA are to be connected
    (a) in series with, and
    (b) in parallel across
    a battery in normal operation. Determine the power dissipated in each combination.

14. A potentiometer circuit is made up of a 2.0-volt battery in connected across a resistance wire of length 2.0 m. A cell of emf 1.5 V is connected so as to get balance on the wire.
    (a) Draw a circuit diagram for this connection.
    (b) Determine the position of the balance point from either end of the wire.

15. In a Wheatstone bridge setup with 2.0 m resistance wire, an unknown resistor X on the left side, produces balance with a standard resistor Y of resistance 12 Ω, on the right. If the balance point is 134.50 cm from the left end of the wire, find the unknown resistance.

16. For the circuit shown in **Fig. 9.20**, determine
    (i) the currents in each arm,
    (ii) the potential difference across AB.

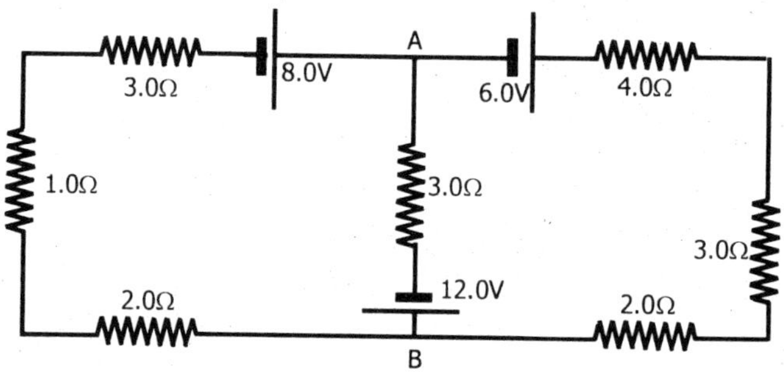

Fig. 9.20

17. The electrical work done in taking a positive charge round any closed circuit is zero. By applying this statement to the circuit of **Fig. 9.20**, derive expressions for Kirchhoff's voltage rule for each loop.

18. **Fig. 9.21** shows the circuit diagram for charging a 12-volt car battery that has internal resistance of 0.8 Ω. The charger delivers a potential difference of 21.0 V, at a current of 2.0 A, through a protective resistor R. Calculate
   (i) the value of R,
   (ii) the rate of electrical energy conversion to chemical energy,
   (iii) the rate of heating of the battery.

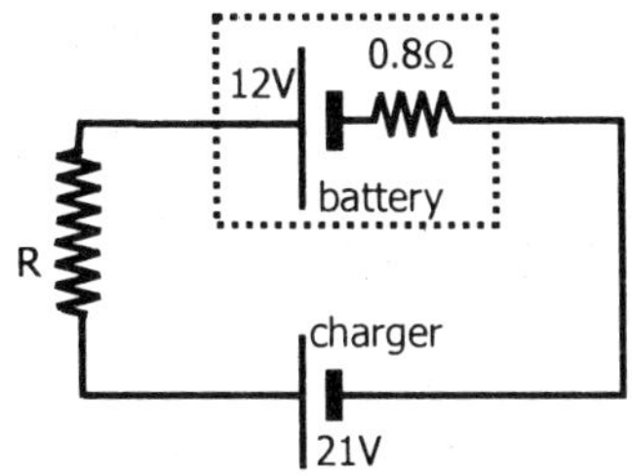

Fig. 9.21

19. A battery of emf 24 V is connected in the circuit of **Fig. 9.22**. S is a switch that can be connected to one of four positions, X, Y, Z, and U. Each resistor has resistance 4 Ω. Find the ammeter reading when S is connected
   (i) to X,
   (ii) to Y,
   (iii) to Z,
   (iv) to U.

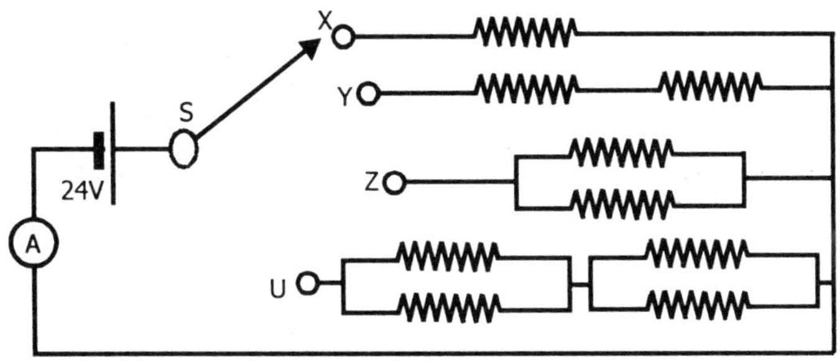

Fig. 9.22

20. In the circuit of **Fig. 9.22**, calculate the power dissipated in each resistor when the switch is connected to the points X, Y, Z, and U.

21. A metre bridge was used to compare the resistance of a coil of wire with a standard 10 Ω resistor. Balance lengths of 40.6 cm and 59.4 cm were obtained when the coil was maintained at 0 °C and 100 °C, respectively. Determine the temperature at which the balance length for the coil would be 38.0 cm.

22. By using a Wheatstone bridge circuit, a coil of wire was found to have a resistance of 5 Ω in melting ice. When the coil was heated to 100 °C, a hundred-ohm resistor had to be connected in parallel with the coil in order to keep the bridge balanced at the same point. Calculate the temperature coefficient of resistance of the coil.

23. State Kirchhoff's rules for electrical network circuits. For the circuit shown in **Fig. 9.23**, determine

(i) the readings indicated by the meters,

(ii) the potential at X,

(iii) the potential difference across the 100 Ω resistor.

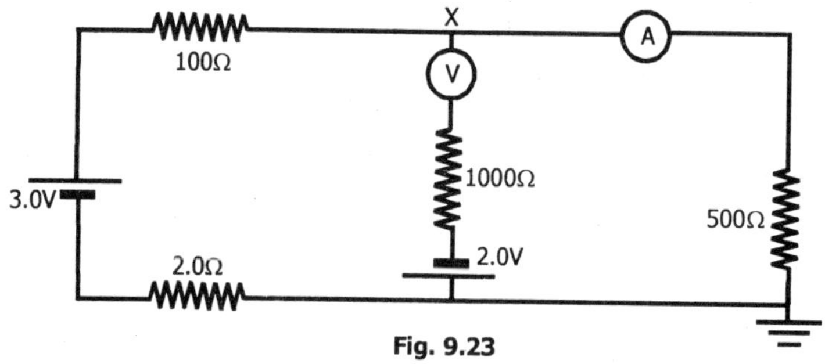

Fig. 9.23

24. In the circuit represented by **Fig. 9.24**, the cells have negligible internal resistance. Find

(i) the resistance to current flow in the whole circuit,

(ii) the reading of the meter,

(iii) the potential difference across the 20 Ω resistor.

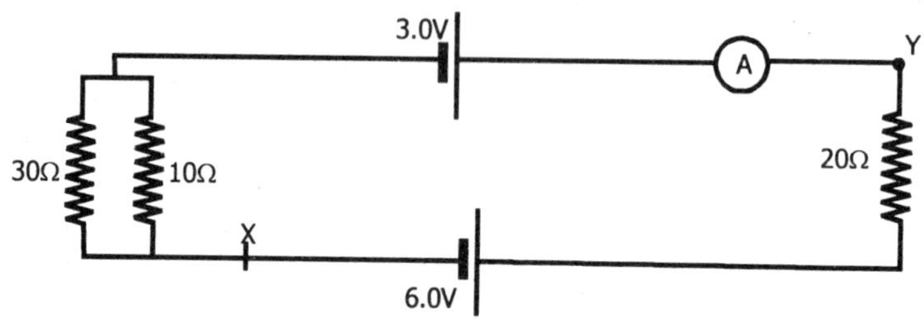

Fig. 9.24

25. In the circuit represented by **Fig. 9.25**, two resistors of 12 Ω and 20 Ω can be connected either in series or in parallel using the switch S joined to either of positions a or b. When S is connected to a, the ammeter reads 1.108 A, and reads 4.50 A when S is in position b. Calculate

(i) the effective external resistances when S is connected to a, and to b,

(ii) the internal resistance of the voltage source,

(iii) the power delivered in both connections.

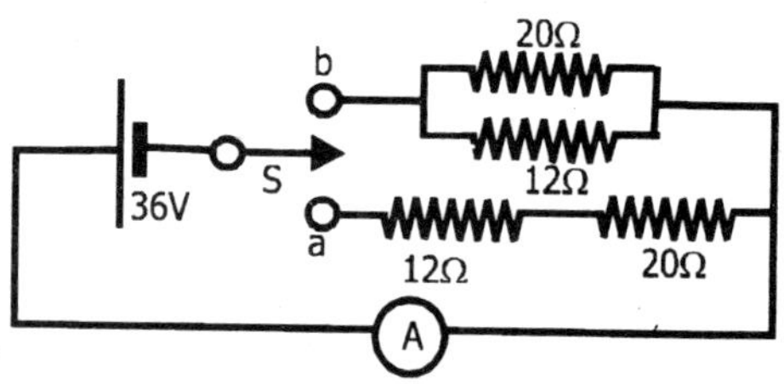

Fig. 9.25

26. The terminal potential difference of a battery is measured to be 12.50 V when the battery delivers a current of 5.0 A, and 12.2 V when a current of 15 A is delivered. Determine the emf and internal resistance of the battery.

27. For the network represented by **Fig. 9.26**, calculate

(i) the total current delivered by the power supply,

(ii) the current flowing in the 4Ω resistor.

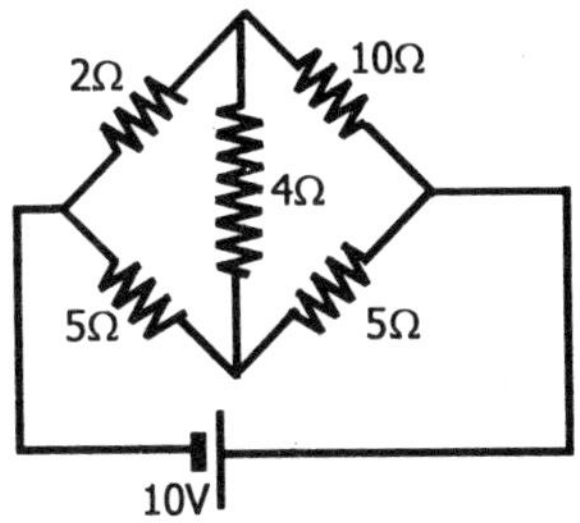

Fig. 9.26

28. In the circuit represented by **Fig. 9.27** calculate the currents in the two cells when

(i) the switch S is open, and

(ii) the switch S is closed.

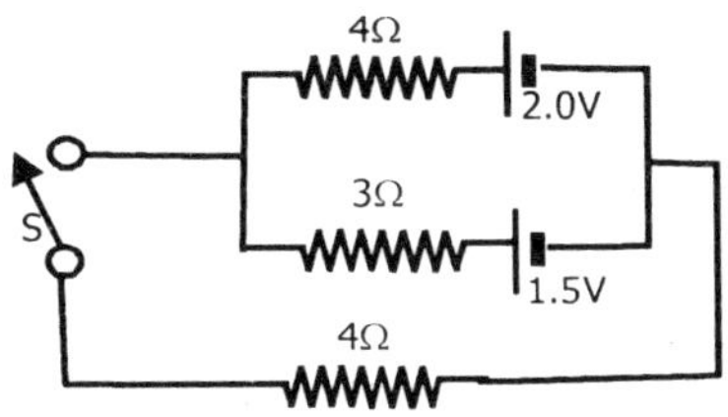

Fig. 9.27

29. In the circuit of **Fig. 9.28**, the p.d. measured between A and B is 3.0 V, and that measured between B and C is 12.0 V. The battery is assumed to have negligible internal resistance.

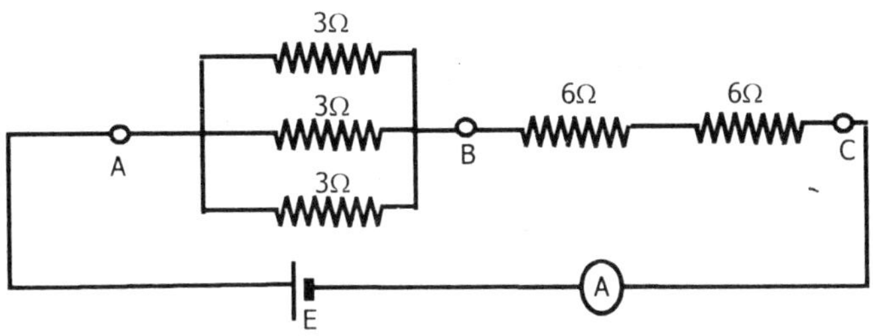

Fig. 9.28

Determine

(i) the total resistance of the combination.

(ii) the emf E of the cell.

(iii) the reading on the ammeter.

30. Calculate the current taken from the battery in the circuit of **Fig. 9.29**, and the terminal potential difference of the battery, if it consists of ten alkaline cells in series, each with emf 1.5 V, and internal resistance 0.04 Ω.

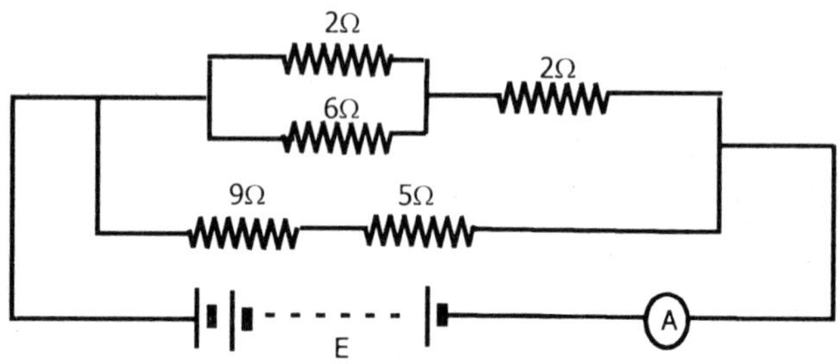

Fig. 9.29

10

CHAPTER

The Magnetic Field

10
CHAPTER

The Magnetic Field

The study of magnetism started with the discovery of the mineral called **lodestone**, which was found to attract iron and other magnetic materials. Today, much is known of this mineral – also called magnetic iron ore, or iron oxide. From these early discoveries, interest grew in the study of the property of magnetism. By the end of the nineteenth century, scientists had tested all the known elements and compounds for their magnetic properties. As a result, these materials were grouped into categories based on their magnetic behaviour. Theories have been put forward to account for the magnetic behaviour of materials.

With the discovery of electricity, and studies ensuing from that discovery, it was soon realised that a steady current through a conducting wire creates a magnetic field around the wire. This means the wire acquires the property of attracting magnetic materials brought close to it. This phenomenon is often described as 'the magnetic effect of a steady current', just as we studied the heating effect of a current in the last chapter.

The effects and uses of magnetism, and its relationship with electricity are used in many technologies. These technologies are very diverse in scope, covering applications in music recording, transformers, motors and generators, moving coil instruments, and many more. In this chapter we study the principles of magnetism – from permanent magnets, to the magnetic fields of steady currents. We also study magnetic forces on current wires and moving charges.

10.1 PERMANENT MAGNETS

Permanent magnets come in many shapes and sizes, ranging from the rather small bar magnets, to the large horse-shoe magnets. We visualise a magnetic field by imaginary lines of force, which start at the north pole N of the magnet, and end at the south pole S, as shown in **Fig. 10.1**. The closer the lines of force are at any point, the stronger the magnetic field at that point. Again, in common with the electrostatic field, the number of lines of force per unit area at a point gives the strength of the field at that point. One of the earliest findings of magnetic interaction between permanent magnets is that

Like poles repel each other, and unlike poles attract.

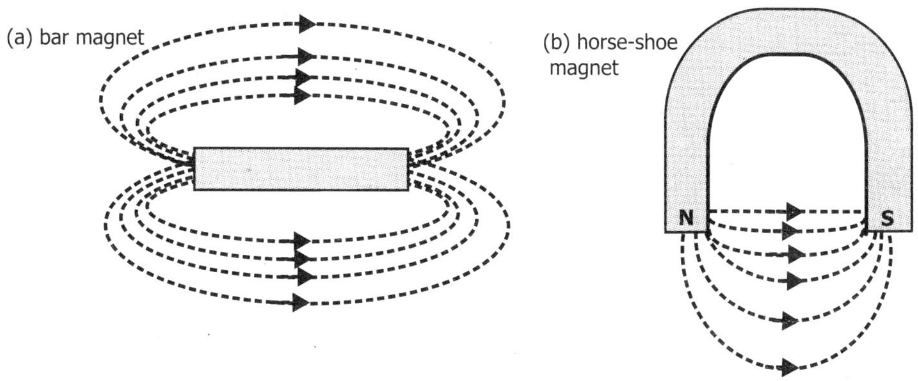

Fig. 10.1 Magnetic lines of force around (a) a bar magnet, and (b) around a horse-shoe magnet.

10.2 INVERSE-SQUARE LAW

Early interest in the study of magnetism focused on the force between two magnetic poles. From these studies, it was found that

the force between two magnetic poles is proportional to the magnitude of each pole, and inversely proportional to the square of their distance apart.

This law, known as the **inverse-square law**, is very similar to the inverse-square law of Coulomb, dealing with the force between two electrostatic point charges. Like the electrostatic force, the magnetic force between two magnetic poles is directed along the line joining the two poles. The force is repulsive if the poles are of the same type (e.g. two North poles), and attractive if the poles are different. A mathematical expression for the force **F** between two magnetic poles as

$$\mathbf{F} = \frac{\mu_o}{4\pi}\frac{q_1 q_2}{r^2}\hat{\mathbf{r}}, \qquad (10.1)$$

where q_1 and q_2 represent the pole strengths of the two magnets. The constant μ_o is known as the **permeability of free space**. We shall discuss more on the permeability of free space shortly in this chapter.

The strengths of two magnets may be compared by measuring the forces they exert on a third magnet. This method, however, does not give a unit in terms of which the magnetic pole strength can be expressed. Using the inverse-square law, this problem was solved by setting $\frac{\mu_o}{4\pi} = 1$.

Unit magnetic pole strength is that existing between two magnetic poles 1 m apart, that will produce a force of 1 Newton.

With the connection between steady currents and magnetic fields, the pole strength is now defined in terms of a steady current flowing through unit length of conducting wire.

10.3 THE EARTH'S MAGNETIC FIELD

Studies have shown that the earth behaves very much like a bar magnet, tilted 11° from the axis of rotation of the earth.

Fig. 10.2 shows the magnetic lines of force around the earth created by the earth's magnetic field. A freely pivoted bar magnet comes to rest pointing in the direction of the north pole. Thus, the South pole of the earth's magnetic field is around the geographic North pole. That a freely suspended magnet points North belies the principle of the magnetic compass – about the oldest navigation instrument. The magnetic field around the earth is used by marine animals to navigate the sea.

The magnetic equator is a great circle around the earth, equidistant from the two magnetic poles, and the magnetic latitude is measured from the magnetic equator, in the same way as the geographic latitude is measured from the geographic equator.

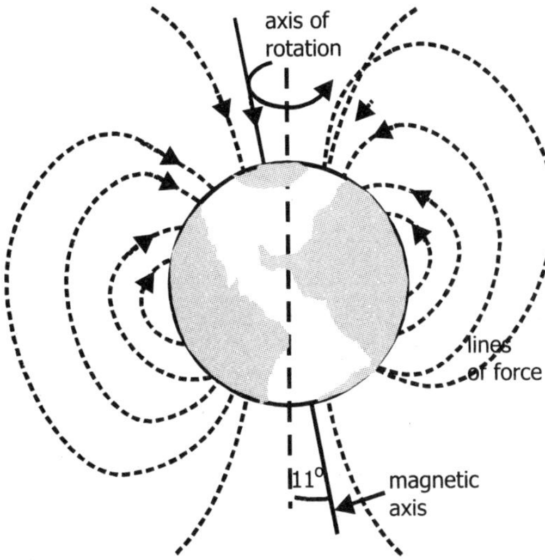

Fig. 10.2 Lines of force in the earth's magnetic field.

10.4 MAGNETIC MATERIALS

By the end of the nineteenth century, all the known elements and many compounds had been tested, and found to have some magnetic property. The magnetic properties are largely determined by the behaviour of the negatively charged electrons that orbit the nuclei. Materials exhibit one of three types of magnetism: **diamagnetism**, **paramagnetism**, and **ferromagnetism**.

The most common type of magnetism shown by materials is diamagnetism. Diamagnetic materials exhibit a weak repulsion by both poles of a magnet. For most atoms and molecules, the total magnetic moment of all the orbital electrons is zero. When these materials are placed in an external magnetic field, a small magnetic moment is induced by the interaction of the field with the atomic electrons. This induced moment is opposite to the applied magnetic field.

Some materials (e.g. chromium) show weak induced magnetization when brought near a magnet. These materials are said to be paramagnetic. This magnetism is due to incomplete electron shells, giving the atom a net magnetic moment. In the absence of an external magnetic field these atomic magnetic fields are randomly ordered, giving a net zero magnetic field. When an external magnetic field is applied, the 'atomic magnets' will react to the field to produce net overall magnetism for the material. The magnetization disappears when the external magnetic field is removed.

Ferromagnetic behaviour is exhibited by only three elements – iron, cobalt, and nickel. These materials have the ability of remaining permanently magnetised even when the external field is removed. At the atomic level, ferromagnetism is explained by the fact that the spins of some of the electrons in adjacent atoms become coupled, thereby creating local magnetic domains that are all in the same direction. When an external magnetic field is applied to the material, the number of domains aligned with the field increases. A very strong permanent magnetisation results.

Fig. 10.3 depicts the magnetic domains in (a) an unmagnetised ferromagnetic material, and (b) a magnetised ferromagnet. In an unmagnetised ferromagnetic material, the magnetic domains are randomly oriented in all directions; the net magnetic field of zero. When the material is magnetised, the domains get aligned in the field direction.

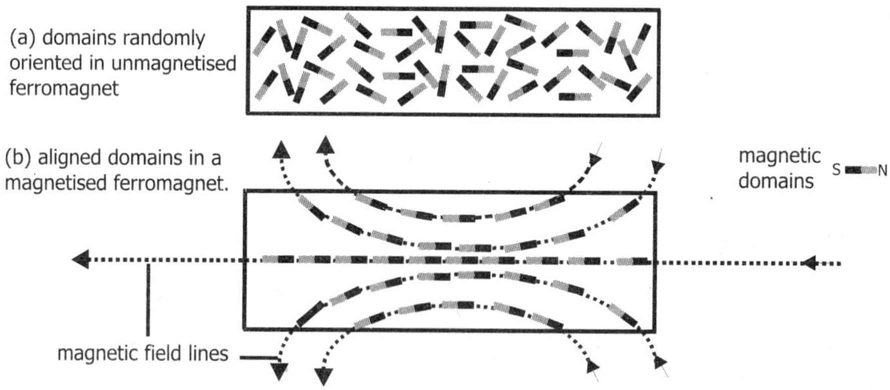

Fig. 10.3 Magnetic domains of a ferromagnet. The unmagnetised material has the domains randomly oriented as in (a). In the magnetised sample, the domains are aligned along the field direction, as in (b).

10.5 MAGNETIC UNITS

Magnetic fields are represented mathematically by two quantities: the **magnetic flux density**, usually symbolised by **B**, and the **magnetic field strength**, or **magnetic field intensity**, denoted by **H**. The magnetic field intensity might be thought of as the magnetic field resulting from the flow of currents in wires, whilst the magnetic flux density is the total magnetic field including contributions from the magnetic properties of the materials in the field. These contributions become appreciable especially for ferromagnetic materials. When a current flows in a wire wrapped around a soft iron cylinder, the magnetizing field is quite weak. However, the actual average magnetic flux density **B**, within the iron may be thousands of times stronger, because of enhancements produced by the alignment of the millions of the tiny atomic magnets in the material. The magnetic flux density is expressed in weber/m^2 (or tesla, abbreviated T), whilst the magnetic field intensity is in amp/m

10.6 MAGNETIC FIELD OF A STEADY CURRENT

The connection between electricity and magnetism had been suspected even in the early days of the discovery of magnetism. In 1820, Hans Christian Orsted showed that an electric current flowing in a wire produces its own magnetic field. Ampere repeated Orsted's experiments, and was able to express the magnetic forces between current carrying conductors in terms of simple mathematical relationships. He also demonstrated that a current flowing through a loop of wire produced a magnetic dipole, almost the same as that existing in a permanent magnet. This led Ampere to suggest that magnetism in any form is caused by circulating currents at the atomic or molecular level.

The magnetic field due to a steady current in a conducting wire at a point distance r from the wire is given by the **Biot-Savart law**, so-named after two French physicists who formulated it, Jean-Baptiste Biot, and Felix Savart. We can understand the statement of the law with reference to **Fig. 10.4** (a).

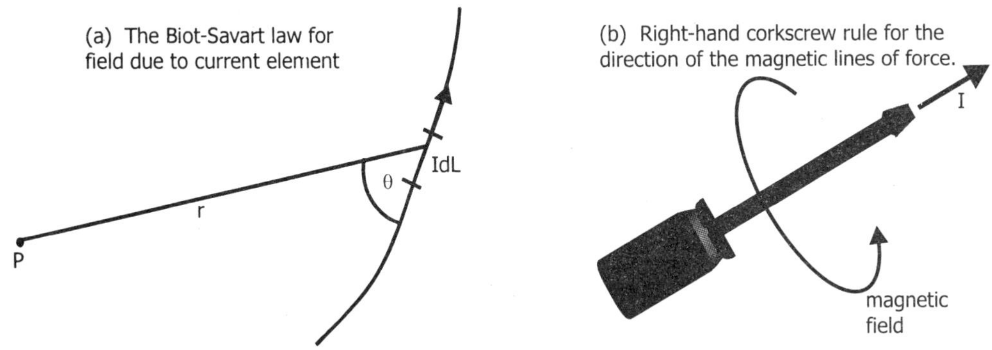

Fig. 10.4 Magnetic field of a current-carrying wire showing (a) the Biot-Savart law, and (b) the cork-screw rule showing the direction of the magnetic field.

Consider a steady current I flowing through a wire. We take the wire to consist of a large number of elements, each producing a magnetic field at any given point P. The overall magnetic field at P is resultant of the fields due to all such elements. We define a quantity known as the **current element** as follows:

For any segment of the wire, the current element is the product of the current through the wire and the length of the segment.

The magnetic field $d\mathbf{B}$ at P due to the current element $Id\mathbf{L}$ of the wire is given by

$$d\mathbf{B} = \frac{\mu_o}{4\pi} \cdot \frac{Id\mathbf{L} \sin\theta}{r^2} \hat{\mathbf{u}}, \tag{10.2}$$

where μ_o is the **permeability of free space**, with value $4\pi \times 10^{-7}$ henrys/metre. The unit vector, $\hat{\mathbf{u}}$ is perpendicular to both the current element and the vector **r**, in a right-hand rotation from Id**L** to **r**. In the diagram, if both the current and **r** are in the plane of the page, the magnetic field **dB** will be perpendicular to the two, going into the page. The direction is given by the **right-hand corkscrew rule**:

If the current is in the same direction as the direction of advance of the screw, then the magnetic lines of force are in the direction of the screw rotation.

This statement is demonstrated pictorially in **Fig. 10.4** (b). In the sections following we use the Biot-Savart law to derive expressions for the magnetic field due to currents in simple symmetrical cases.

10.6.1 Field due to a long straight wire

As our first application of the Biot-Savart law, we find the magnetic field at a point P, a perpendicular distance a from a long straight wire carrying current I. Using **Fig. 10.5**, we select a current element and determine the field due to this element, sum over the whole length of the wire to find the overall magnetic field at P due to the wire.

By the Biot-Savart law, the field **dB** at P due to the chosen current element has magnitude given by

$$dB = \frac{\mu_0}{4\pi}\frac{IdL\sin\theta}{r^2}.$$

From the geometry of the figure, we have

$r = a\sec\phi$, or $r^2 = a^2\sec^2\phi$, and $L = a\tan\phi$,

giving $dL = a\sec^2\phi d\phi$.

Substituting into the Biot-Savart law gives

$$dB = \frac{\mu_0}{4\pi}\frac{Ia\sec^2\phi d\phi.\cos\phi}{a^2\sec^2\phi} = \frac{\mu_0 I}{4\pi a}\cos\phi d\phi.$$

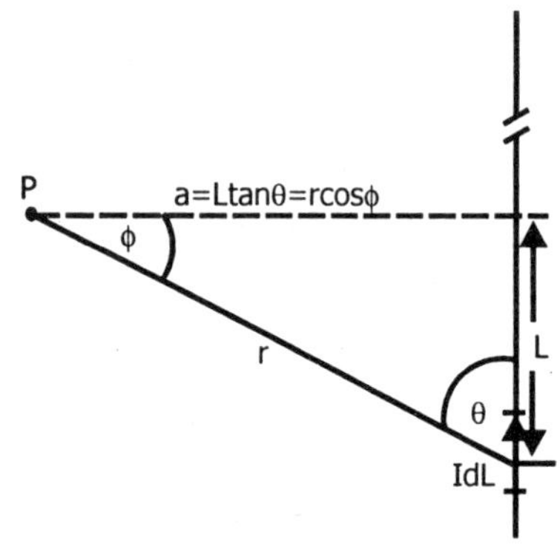

Fig. 10.5 Field at a perpendicular distance a from a long straight wire.

The field at P is given by

$$B = \int dB = \frac{\mu_0 I}{4\pi a}\int_{-\pi/2}^{\pi/2}\cos\phi d\phi = \frac{\mu_0 I}{2\pi a}. \qquad (10.3)$$

In this derivation, we have used the assumption that for an infinitely long wire, a<<L. Thus the angle ϕ varies from a minimum of $-\pi/2$ to a maximum of $+\pi/2$.

Example 10.1

Two long straight vertical wires X and Y, are placed 100 mm apart, with Y placed to the right of wire X. X carries a current of 2.0 A, whilst Y carries a current of 4.0 A, the currents being in opposite directions. Calculate

(i) the resultant magnetic field midway between the wires,
(ii) the resultant magnetic field in a plane 150 mm to the right of wire X,
(iii) the distance from X of the line of neutral points in the resultant magnetic field.

Solution

We use **Fig. 10.6** to solve this problem. In this figure, the wires are assumed to lie in the plane of the page. The directions of the magnetic field due to each wire are indicated as either coming out (arrowhead), or going into (arrow tail) of the page. The suffixes of the magnetic fields refer to the wires.

(i) For the plane M containing points that are midway between the wires, the resultant magnetic field B_M is given by $B_M = B_X + B_Y$, in direction out of the page.

$$B_X = \frac{\mu_0 I_X}{2\pi a}$$

$$= \frac{4\pi \times 10^{-7} H/m \times 2.0A}{2\pi \times 5.0 \times 10^{-2} m} = 8.0 \times 10^{-6} \text{ tesla.}$$

Similarly,

$$B_Y = \frac{\mu_0 I_Y}{2\pi a} = \frac{4\pi \times 10^{-7} H/m \times 4.0A}{2\pi \times 5.0 \times 10^{-2} m} = 1.60 \times 10^{-5} \text{ tesla.}$$

Thus, the magnetic field at M = 2.4×10^{-5} tesla.

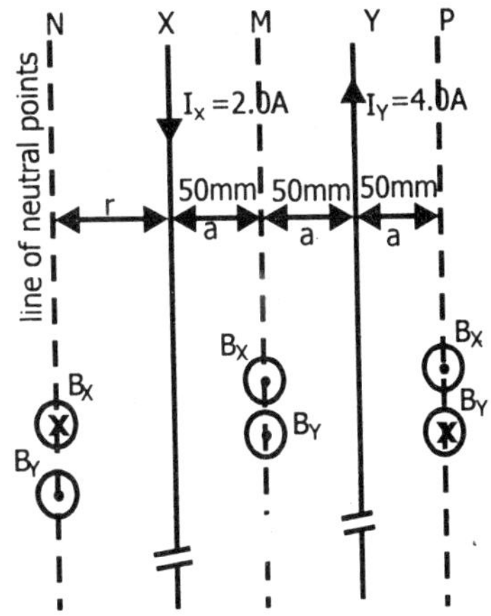

Fig. 10.6 Magnetic field around two parallel wires, with currents in opposite directions. The fields are in the same direction between the wires, and in opposite directions on either side.

(ii) At points on the line P, 150 mm from X, the contributing magnetic fields are in opposite directions. The resultant field B_P is given by $B_P = B_Y - B_X$, into the page.

For this side,

$$B_X = \frac{\mu_0 I_X}{2\pi \times 3a} = \frac{4\pi \times 10^{-7} H/m \times 2.0A}{2\pi \times 3 \times 5.0 \times 10^{-2} m}$$

$$= 2.67 \times 10^{-6} \text{ tesla,}$$

and

$$B_Y = \frac{\mu_0 I_Y}{2\pi a} = \frac{4\pi \times 10^{-7} H/m \times 4.0A}{2\pi \times 5.0 \times 10^{-2} m}$$

$$= 1.60 \times 10^{-5} \text{ tesla.}$$

The resultant magnetic field is given by

$B_P = B_Y - B_X = 1.33 \times 10^{-5}$ tesla, into the page.

(iii) The line of neutral points N, is that for which the resultant magnetic field is zero – i.e. the field due to X is equal and opposite to that due to Y. From the currents given, this line falls to the left of X. At this line, $B_N = B_Y - B_X = 0$. If N is a distance r from X, we write

$$B_X = \frac{\mu_o I_X}{2\pi r} = B_Y = \frac{\mu_o I_Y}{2\pi(r+2a)}, \text{ from which } r = \frac{2aI_X}{I_Y - I_X}.$$

Substituting the values for the currents and the separation of the wires, we get $r = 100$ mm. Thus the line of neutral points is 100 mm to the left of X.

10.6.2 Field at the centre of a circular coil

Fig. 10.7 shows a circular coil with N turns of wound wire, carrying a steady current I. The mean radius of the coil is a. We use Biot-Savart law to find an expression for the magnetic field at the centre O of the coil.

The magnetic field dB due to a current element IdL of the wire is given by

$$dB = \frac{\mu_o}{4\pi}\frac{IdL\sin\theta}{r^2}.$$

The resultant field at the centre O is

$$B = \int dB = \frac{\mu_o I}{4\pi a^2}\int \sin\theta dL$$

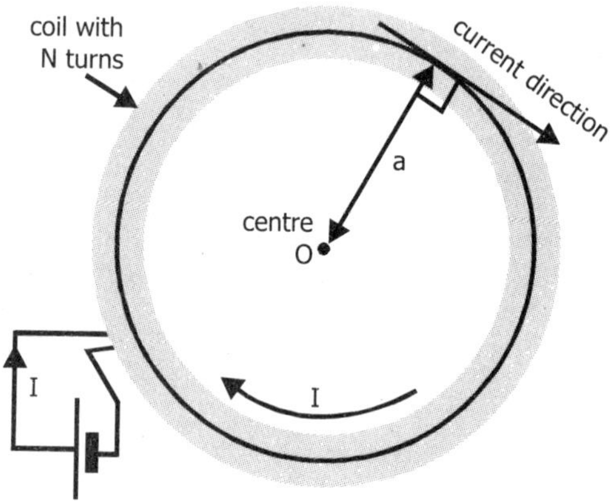

Fig. 10.7 Field at the centre O, of a circular coil of mean radius R, having N turns of wire.

At any point on the loop, the current direction is along the tangent. Hence $\theta = 90^o$, giving $\sin\theta = 1$. Also, the integral with respect to L is the total length of the wire, equal to the product of the circumference of the loop and the number of turns of coil = $2\pi aN$.

The field at the centre of a circular coil of radius a with N turns carrying current I is given by

$$B = \frac{\mu_o NI}{2a}. \qquad (10.4)$$

With the given current direction, and the coil being in the plane of the page, the magnetic field is directed into the page.

10.6.3 Field along the axis of a circular coil

A related example of the Biot-Savart law is in finding the field along the axis of a circular coil of radius a, with N turns of wire, carrying current I. We refer to **Fig. 10.8.** Consider a point P on the axis of the coil, a distance x from the centre of the coil.

The magnetic field $d\mathbf{B}'$ due to a current element $I d\mathbf{L}$ of the wire given by

$$d\mathbf{B}' = \frac{\mu_o}{4\pi}\frac{I d\mathbf{L}\sin\theta}{r^2} = \frac{\mu_o}{4\pi}\frac{I d\mathbf{L}\sin\theta}{(a^2+x^2)}\hat{\mathbf{u}}, \qquad (10.5)$$

As we saw in the previous example, $\sin\theta = 1$, since $\theta = 90°$. The field **dB′** is at right angles to both r and the plane of the coil, and has two components: a horizontal component **dB′**$\sin\phi$, and a vertical component, **dB′**$\cos\phi$. Over each turn of the coil, the vertical components cancel out, leaving the horizontal component as the effective field.

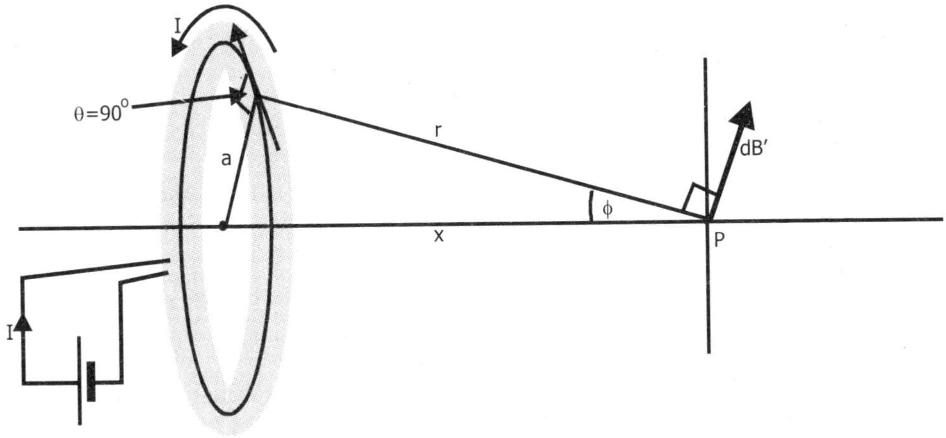

Fig. 10.8 Magnetic field at a point, P on the axis of a circular coil

Thus the effective field dB at P due to any loop is horizontal, and given in magnitude by

$$dB = dB'\sin\phi = \frac{\mu_o I}{4\pi}\frac{dL}{\left(a^2 + x^2\right)}\sin\phi.$$

For all current elements of the loop, the resultant field B at P is of magnitude given by

$$B = \int dB = \frac{\mu_o I}{4\pi}\frac{\sin\phi}{\left(a^2 + x^2\right)}\int dL.$$

From the diagram, $\sin\phi = \dfrac{a}{\sqrt{a^2 + x^2}}$, and the total length of the wire = $2\pi aN$.

Thus,

$$B = \frac{\mu_o I}{4\pi}\frac{2\pi Na^2}{\left(a^2 + x^2\right)^{3/2}} = \frac{\mu_o NIa^2}{2\left(a^2 + x^2\right)^{3/2}}. \qquad (10.6)$$

It is seen from the above that the magnetic field along the axis has its greatest value at the centre of the coil. At large distances from the centre (i.e. $x >> a$), the magnetic field falls off approximately as $1/x^3$. It is left as an exercise for the student to show that equation (10.4) represents a special case of (10.6), for which $x = 0$.

Example 10.2

A large circular coil with 200 turns has diameter 1.0 m, and carries a current of 20 A.

(a) Determine the magnitude of the magnetic field at the centre of the coil.
(b) Find the magnitude of the magnetic field at a point on the axis 40 cm from the centre of the coil.
(c) Find the point along the axis for which the magnitude of the field is one-half that at the centre of the coil.

Solution

(a) From equation (10.4), the magnetic field at the centre of the coil has

$$B = \frac{\mu_0 NI}{2a} = \frac{4\pi\times10^{-7}\ \text{H/m}\times200\times20\ \text{A}}{2\times0.50\ \text{m}} = 5.027\times10^{-3}\ \text{T}.$$

(b) For any point along the axis of the coil, we use equation (10.6) to calculate the magnitude of the magnetic field. Thus, for the point 0.40 m from the centre, the field B is given by

$$B = \frac{\mu_0 NIa}{2(a^2+x^2)^{3/2}} = \frac{4\pi\times10^{-7}\ \text{H/m}\times200\times20\ \text{A}\times0.50\ \text{m}}{2\times(0.50^2\text{m}^2+0.40^2\text{m}^2)^{3/2}} = 2.393\times10^{-3}\ \text{T}.$$

(c) For the point at which the field is one-half that at the centre, we find the distance x along the axis from the expression $B = \frac{\mu_0 NIa}{2(a^2+x^2)^{3/2}} = \frac{1}{2}\frac{\mu_0 NIa}{2(a^2+0^2)^{3/2}}$.

From this, we get $x = 0.38$ m.

10.6.4 Electromagnets

It is sometimes desirable to produce a high and constant magnetic field at points in space, a magnetic field whose magnitude can be controlled as required. Also, it is often required that the magnetic field be turned on and off when needed. Such controlled magnetic fields are produced by **electromagnets**, two of the common types being the **solenoid**, and **Helmholtz coils** - shown in **Fig. 10.9**.

The cylindrical solenoid is made up of a very large number of turns of wire closely wound on a cylindrical former. The number of turns of the coil is so large that there is at any one point, several layers of windings, each may be taken as a circular coil. The field lines are parallel inside the solenoid, but diverge on either side of the solenoid, as shown in **Fig. 10.9** (a) and (c).

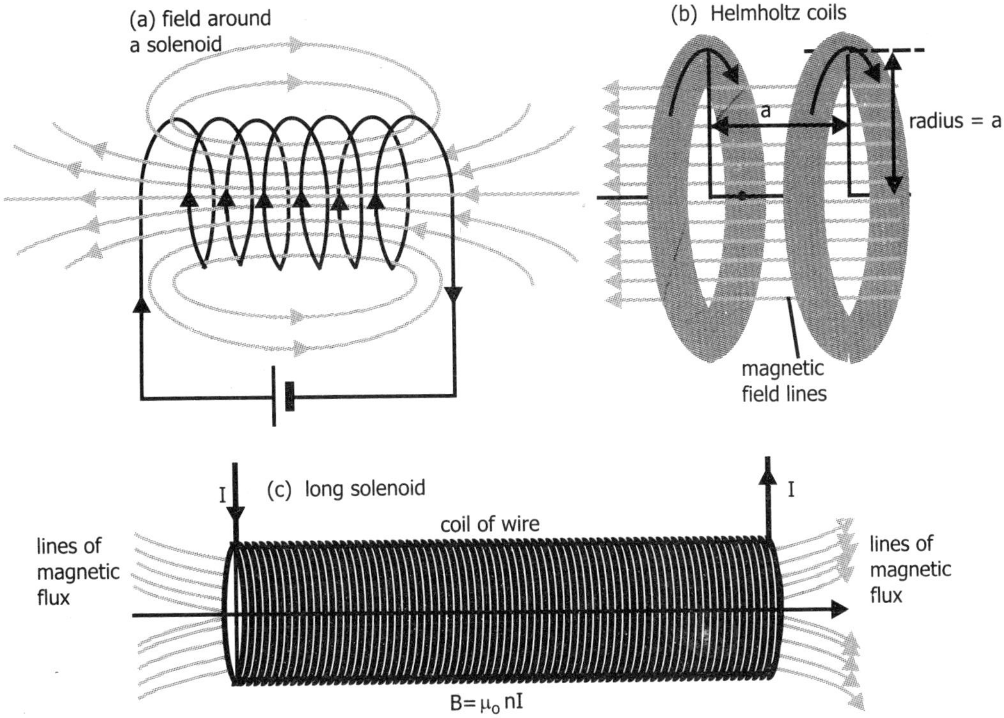

Fig. 10.9 Electromagnets: (a) field around a cylindrical solenoid, (b) Helmholtz coils, (c) long solenoid.

For a long solenoid, (i.e. one for which the length is very large compared to the radius), the magnetic field at the centre of the solenoid has a constant value B, given by

$$B = \mu_0 nI, \qquad (10.7)$$

where n is the number of turns per unit length of the coil.

Helmholtz coils consist of two thin circular coils of the same radius a, with the same (large) number of turns N. The coils are held coaxially, and parallel to each other, with their planes separated by a distance equal to the radius. The currents through the coils flow in the same direction. The magnetic field at any point between the coils is made up of contributions from both coils. For points nearer either coil the field due to that coil is strong, whilst the contribution due to the other coil is weak. The result is a constant magnetic field anywhere along the common axis of the two coils. The magnetic field on the axis at any point distance x from one coil is of magnitude given by

$$B = \frac{\mu_0 NIa^2}{2}\left\{\left[a^2 + x^2\right]^{-3/2} + \left[a^2 + (a-x)^2\right]^{-3/2}\right\},$$

where I is the current through the coils. At the middle P of the coils, x = a/2, yielding a field of magnitude

$$B = \left(\frac{4}{5}\right)^{3/2} \frac{\mu_o NI}{a}. \tag{10.8}$$

Electromagnets provide a way of finding values of the magnetic quantities, H and B. If a solenoidal coil with n turns per metre carries current I has air or vacuo within it, this coil will have a magnetising force H, given by

$$H = nI. \tag{10.9}$$

This gives the units of H as amp-turns per metre, or simply A/m, since turns is a pure number. From equation (10.7), the magnetic flux density arising from this is

$$B = \mu_o H. \tag{10.10}$$

If a magnetic material is now placed in the field, the flux density becomes

$$B = \mu H, \tag{10.11}$$

where μ is the permeability of the material. Similar to the case in electrostatics, we have

$$\mu = \mu_o \mu_r, \tag{10.12}$$

where μ_r is the **relative permeability** of the material.

The magnetic permeability of free space is connected to the electric permitivity of free space through the equation

$$\frac{1}{\mu_o \varepsilon_o} = \left(\frac{1}{4\pi \times 10^{-7}}\right)\left(\frac{1}{8.85 \times 10^{-12}}\right) = 8.99 \times 10^{16} m^2/s^2.$$

This is numerically equal to the square of the speed of light in free space.

Thus, we write

$$\frac{1}{\mu_o \varepsilon_o} = c^2. \tag{10.13}$$

Equation (10.13) comes not as a coincidence, but from the fact that electromagnetic waves consist of vibrating electric and magnetic fields. Hence the motion of these waves through free space must be related to the electric and magnetic constants of free space.

Example 10.3

A long solenoid has 2000 turns of wire wound around a cylindrical former of length 50.0 cm, and diameter 2.0 cm.

(a) Calculate the magnitude of the magnetic field at the centre of the solenoid when it carries a current of 20 A.

(b) It is desired to use a pair of Helmholtz coils with the same diameter as the solenoid, to provide the same magnitude of the magnetic field for the same current. Calculate the number of turns of each coil.

Solution

(a) We first find the number of turns per metre n of the solenoid. From the given data,

$$n = \frac{\text{number of turns of coil}}{\text{length of coil}} = \frac{2000\text{ turns}}{0.50\text{ m}} = 4000\text{ turns per metre.}$$

From equation (10.7), the magnetic field at the centre of the coil has magnitude

$$B = \mu_0 nI = 4\pi\times10^{-7}\,\text{H/m}\times4000\,\text{m}^{-1}\times20\text{ A} = 0.1\text{ T.}$$

(b) For the Helmholtz coils, we have the following data: radius a = 0.01 m, I = 20 A, and B = 0.1 T (from above calculation). Using equation (10.8), the magnetic field at the centre has magnitude B given by $B = \left(\frac{4}{5}\right)^{3/2}\frac{\mu_0 NI}{a}$, giving $N = \left(\frac{5}{4}\right)^{3/2}\frac{Ba}{\mu_0 I}$.

Substituting the numerical values of B, a, and I, we get

$$N = \left(\frac{5}{4}\right)^{3/2}\frac{Ba}{\mu_0 I} = \frac{5^{3/2}}{8}\,\frac{0.1\text{ T}\times0.01\text{ m}}{4\pi\times10^{-7}\,\text{H/m}\times20\text{A}} = 55.6\text{ turns,}$$

showing that far less number of turns are needed for producing a certain magnetic field using Helmholtz coils than is required for a solenoid.

10.7 FORCE ON A MOVING CHARGE IN A MAGNETIC FIELD

In our study of electrostatics, we saw that a charged particle in an electric field experiences a force in the direction of the field, or against the field, depending on the sign of the charge. Similarly, a charged particle in a magnetic field would experience a force. In the case of the magnetic field, however, the charge must be moving. The force acting on the charged particle results from the interaction of the external magnetic field and the magnetic field created by the charge. Since the latter exists only when there is a current (i.e. motion of charges), it follows that the force exists only when the charge is moving.

For a charge q moving with velocity v in a magnetic field of flux density B, the force acting on the charge is given by the expression

$$\mathbf{F} = q\mathbf{v}\mathbf{B}\sin\theta\,\hat{\mathbf{u}}, \qquad (10.14)$$

where θ is the angle between the velocity vector and the magnetic field, and $\hat{\mathbf{u}}$ is a unit vector perpendicular to both **v** and **B**. The direction of the force is therefore perpendicular to both the direction of motion of the charged particle, and to the magnetic field. The force is zero when the particle moves in the direction of the field. The direction of the force is often expressed by **Fleming's left-hand** rule, illustrated here in **Fig. 10.10**.

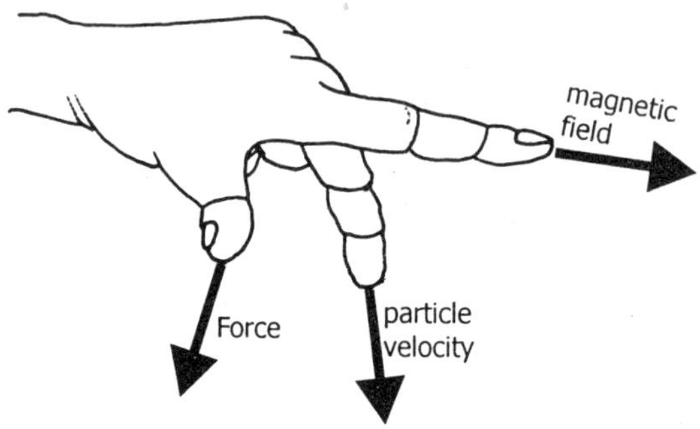

Fig. 10.10 Illustration of Fleming's left-hand rule.

This rule states that

if we hold the first finger, the second finger, and the thumb of the left hand at right angles to each other, such that the first finger points in the direction of the magnetic field, the second finger in the direction of the velocity, then the thumb will point in the direction of the force.

Example 10.4

A proton of mass 1.667×10^{-27} kg, and charge 1.6021×10^{-19} C, moves with velocity 6.0×10^{6} m/s into a magnetic field. The magnetic field has flux density 0.5 T, in a direction making an angle of 45° to the direction of motion of the proton. Calculate

(i) the magnitude of the force on the proton,

(ii) the acceleration of the proton, assumed constant.

Solution

(i) The force acting on the proton has magnitude given by equation (10.9).

$$F = Bqv\sin\theta = 0.5\text{ T}\times(1.602\times10^{-19}\text{ C})\times(6.0\times10^{6}\text{ m/s})\times\sin45^{\circ},$$
$$= 3.4\times10^{-13}\text{N}.$$

(ii) To get the acceleration of the proton, we set the above force equal to the product of the mass m, and acceleration a of the proton.
Thus, we have

$$F = ma\text{, or } a = \frac{F}{m} = \frac{3.4\times10^{-13}\text{N}}{1.667\times10^{-27}\text{kg}} = 2.0395\times10^{14}\text{ m/s}^2.$$

10.8 MOTION OF CHARGES IN ELECTRIC AND MAGNETIC FIELDS

Fig. 10.11 illustrates particles with charge q, moving with velocity v into (a) an electric field, and (b) a magnetic field. The force $\mathbf{F_e}$ on the charge q in an electric field **E**, is given by $\mathbf{F_e} = q\mathbf{E}$. This force is parallel to the field. Equation (10.9) gives the force acting on a charge moving in a magnetic field with velocity **v** at and angle θ to the field direction.

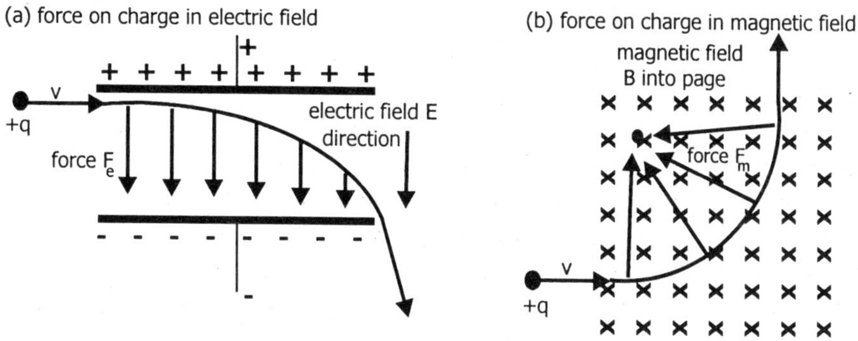

Fig. 10.11 Force on a positive moving charge in (a) an electric field, and (b) a magnetic field. The magnetic force is perpendicular to both the magnetic field B, and the velocity vector. The path is circular. The force due to the electric field is the direction of the field. In both fields, the particle exits at a tangent to the path in the field.

The force on the charged particle in the magnetic field is perpendicular to both the magnetic field and the direction of motion of the particle. The path in the field is circular. For steady motion in the magnetic field, we find the radius of this circular path by setting the magnitude of the magnetic force equal to the centripetal force on the particle. Thus,

$$Bqv = m\frac{v^2}{r}, \tag{10.15}$$

or

$$r = \frac{mv}{Bq}. \tag{10.16}$$

We note the following:

1. In the electric field, the force on the charged particle causes a displacement in the direction of the field. There is thus work done by the field. The speed, and hence the kinetic energy, of the charged particle is increased by the action of the field.
2. The magnetic field also causes a displacement of the charged particle. By contrast, however, the displacement produced in the magnetic field has no component in the direction of the magnetic field, as the latter is at right angles to the plane of motion.

Example 10.5

A proton accelerated through a potential difference of 3.0 kV enters a magnetic field of flux density 0.1 T, in a plane perpendicular to the motion of the proton. Calculate

(i) the speed of the proton on entering the magnetic field,
(ii) the radius of the circular path that the particle describes in the magnetic field.

Mass of proton = 1.667×10^{-27} kg, charge on proton = $+1.602 \times 10^{-19}$ C.

Solution

(i) We find the velocity v of the proton by equating its gain in kinetic energy to the electrostatic energy. Thus, $\frac{1}{2}mv^2 = qV$, where q is the charge on the proton of mass m, and V is the potential applied.

This gives $v = \sqrt{\frac{2eV}{m}} = \sqrt{\frac{2 \times 1.6021 \times 10^{-19}\,\text{C} \times 3000\,\text{V}}{1.667 \times 10^{-27}\,\text{kg}}} = 7.50 \times 10^5$ m/s.

(ii) Using equation (10.16), we calculate the radius of the circular path in the magnetic field. Thus, $r = \frac{mv}{Bq} = \frac{1.667 \times 10^{-27}\,\text{kg} \times 750000\,\text{m/s}}{0.1\,\text{T} \times 1.602 \times 10^{-19}\,\text{C}} = 7.80$ cm.

Much of our current knowledge of subatomic particles and the atom has been obtained from experiments involving the action of electric and magnetic fields on charged particles. We study a few of these applications in the following sections.

10.8.1 The velocity selector

The **velocity selector** selects charged particles with the same velocity. **Fig. 10.12** is a sectional view of the velocity selector. Particles with charge +q, travelling with a range of velocities, enter through a collimating slit S_1 into an area containing both electric and magnetic fields. A second collimating slit S_2 is placed in line with S_1. The electric field acts between a pair of electrodes connected to the terminals of a high voltage supply, whilst the magnetic field is provided by an electromagnet. The two fields are such that the electric and magnetic forces on the particles are in opposite directions. Particles go through both collimating slits only if the electric and magnetic forces on them are equal.

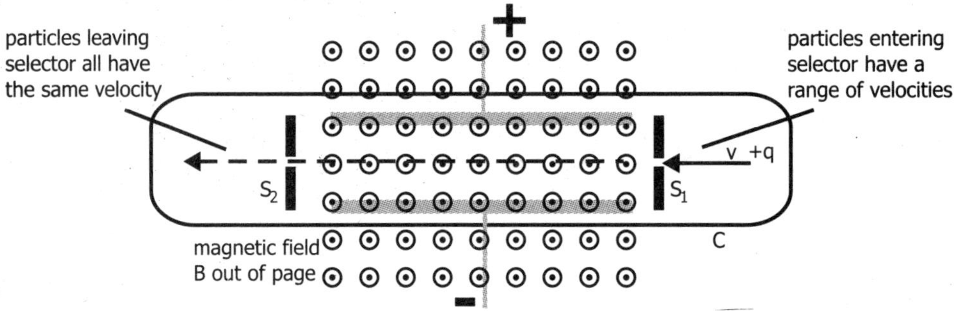

Fig. 10.12 The velocity selector. Only particles with same velocity v = E/B exit from the second collimator.

Thus,

$$qvB = qE, \tag{10.17}$$

where E is the electric field between the plates, a distance d apart, with potential difference V. B is the magnetic flux density between the plates, and v is the common velocity of the charged particles. Particles going through the slits have velocity given by

$$v = \frac{E}{B}. \tag{10.18}$$

The instrument can therefore be used to select particles with a fixed velocity, determined by the magnitudes of the electric and magnetic fields applied.

Example 10.6

Protons accelerated through a potential difference of 6.0 kV enter a velocity selector with electrodes connected to 500 V d.c. supply. The electrodes are 1.0 cm apart in vacuo. Calculate the magnitude of the magnetic field that would cause no displacement in the path of the protons. For the proton, mass = 1.667×10^{-27} kg, charge = $+1.602 \times 10^{-19}$ C.

Solution

As in the previous example, we find the entry velocity v by equating the kinetic energy of the charged particle to the electrostatic energy used in accelerating it.

Thus, $$v = \sqrt{\frac{2eV}{m}} = \sqrt{\frac{2 \times 1.602 \times 10^{-19}\,\text{C} \times 6000\,\text{V}}{1.667 \times 10^{-27}\,\text{kg}}} = 1.07 \times 10^{6}\ \text{m/s}.$$

For the particles to pass through undeviated, the force due to the magnetic field must be equal and opposite to that due to the electric field. Using equation (10.13), the magnitude of the magnetic field B is $B = \frac{E}{v}$, where E is the electric field intensity, given by

$$E = \frac{V}{d} = \frac{500\ \text{V}}{0.010\ \text{m}} = 50000\ \text{V/m}.$$

Substituting the values of E and v, we get

$$B = \frac{E}{v} = \frac{50000\ \text{V/m}}{1.07 \times 10^{6}\ \text{m/s}} = 4.70 \times 10^{-2}\ \text{T} = 47\ \text{mT}.$$

10.8.2 The mass spectrometer

The mass spectrometer was first developed by Bainbridge and used in the early experiments on atomic masses. A beam of charged particles from a velocity selector is subjected to a magnetic field at right angles to their velocity vector. The particles then describe circular paths with radius given in terms of the mass of the particles, their speed, and the magnetic flux density. **Fig. 10.13** is a schematic view of a section through the spectrometer.

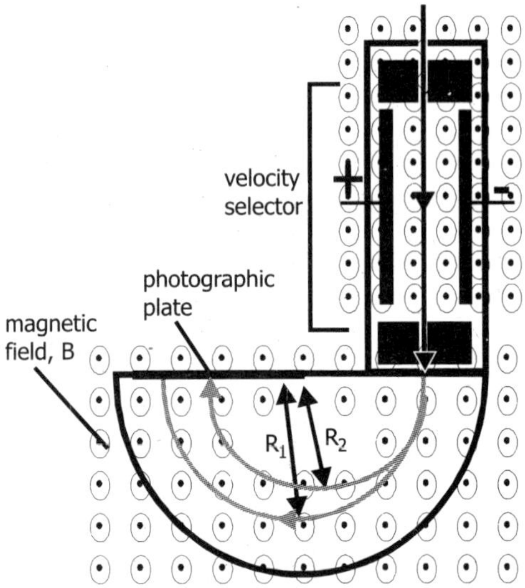

Fig. 10.13 The mass spectrometer. Two particles of the same charge but different masses fall on different points on the photographic plate.

Charged particles enter the spectrometer from the velocity selector. The magnetic field B exerts a force on the particles, causing them to describe circular paths of radius depending on their masses. The particles fall on a photographic plate, making impressions on the plate. From these spots, the radii of the circular orbits can be determined. For a particle of mass m, charge q, describing a circular path of radius R, we have

$$\frac{mv^2}{R} = Bqv$$

or

$$m = \frac{BqR}{v}, \tag{10.19}$$

from which the mass can be obtained.

10.8.3 The cyclotron

In some aspects of nuclear structure and nuclear reaction studies, charged particles with high energies are needed. The cyclotron is one of the machines developed to achieve such high energies. The cyclotron relies on the constant time taken by a charged particle to complete a circular path in a uniform magnetic field. Between the poles of a strong electromagnet, an evacuated space contains two pairs of semicircular electrodes called **dees**, on account of their shape. The magnetic field is applied at right angles to the plane of the particle beam. Charged particles from a source near the centre move in a circle because of the force on them exerted by the magnetic field.

An alternating voltage of high frequency changes polarity just as the particle gets to the gap between the dees. This way the charged particles are attracted to one Dee as they are simultaneously repelled from the other, with the result that the energy of the particle increases each time the particle moves from one dee to the other. Due to their increasing energy and momentum, the ions move in circles of progressively larger radii. They still take the same time in each dee. At each gap, the energy is increased by qV, where q is the charge on the particle, and V is the alternating voltage. The particles complete many revolutions until the radius of the semicircle has increased to a point at which they strike a target at the outside edge of the dee. **Example 10.7** illustrates the principles of operation of the cyclotron.

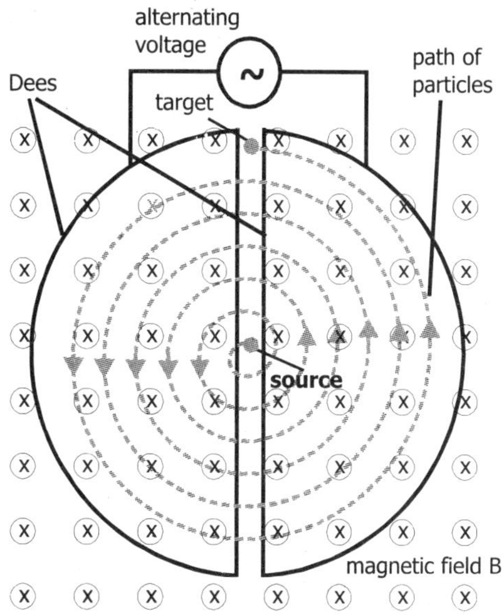

Fig. 10.14 The Cyclotron

Example 10.7

The electromagnet in a cyclotron usually produces a very high magnetic field. In a typical case, the field used to accelerate protons is found to be 4.0 T, and the kinetic energy achieved is 4.8×10^{-13} J.

Calculate

(i) the velocity of the protons,
(ii) the radius of the circular orbit of the protons,
(iii) the angular speed of the protons,
(iv) The kinetic energy of the particle exiting the Dee.

Solution

(i) We find the velocity of the protons from the equation $E_k = \frac{1}{2}mv^2$, where E_k is the kinetic energy of the particle. This gives

$$v = \sqrt{\frac{2E_k}{m}} = \sqrt{\frac{2 \times 4.8 \times 10^{-13}\ J}{1.667 \times 10^{-27}\ kg}} = 2.4 \times 10^7\ m/s.$$

(ii) To get the radius of the orbit, we equate the magnetic force on the proton to the mechanical force keeping it in the circular path. If R is the radius of the orbit, we write $Bqv = \frac{mv^2}{R}$, from which $R = \frac{mv}{Bq}$. Thus, we have

$$R = \frac{mv}{Bq} = \frac{1.667 \times 10^{-27} kg \times 2.4 \times 10^7 m/s}{4.8\ T \times 1.602 \times 10^{-19} C} = 0.052 \text{ m (or 5.20 cm)}.$$

(iii) The angular speed ω is obtained by dividing the velocity by the radius, both of which are obtained above. Thus $\omega = \frac{v}{R} = \frac{2.40 \times 10^7 m/s}{0.052\,m} = 4.61 \times 10^8 s^{-1}$.

(iv) The exiting kinetic energy K.E. is given by

$$K.E. = \frac{1}{2}mv^2 = \frac{1}{2}m\omega^2R^2 = \frac{1}{2} \times 1.667 \times 10^{-27} \times (4.61)^2 \times 10^{16} \times 0.052^2 = 5.0 \times 10^{-13} J$$

10.9 FORCE ON CURRENT WIRE IN A MAGNETIC FIELD

From the previous section a charge q, moving in a magnetic field with speed v, experiences a force, $F = qvB\sin\theta$, where B is the magnetic flux density, and θ is the angle between the velocity and magnetic field vectors. Since current in a wire results from the motion of charges in the wire, this expression can ultimately lead us to find the magnetic force on a conductor of length L, placed in a magnetic field.
Thus, we have

$$F = qvB\sin\theta = \left(\frac{q}{t}\right)(vt)B\sin\theta,$$

where t is the time taken for the mobile charges to move through the length L of the wire, with speed v. The magnetic force on the wire then becomes

$$F = ILB\sin\theta. \tag{10.20}$$

The direction of the force is given by Fleming's left-hand rule (see **Fig. 10.10**):

If the First finger points in the direction of the Field, the seCond finger in the direction of the Current, the thuMb will be in the direction of the Motion (or force).

Notice the capitalized letters in the above statement. This is designed to assist you in remembering the assignment of the fingers and the thumb. From equation (10.20), it follows that for a given current and magnetic field, maximum magnetic force is experienced when the wire is at right angles to the magnetic field direction, and zero when the wire is parallel to the magnetic field.

Example 10.8

A wire of length 20.0 cm and mass 50.0 g, is suspended in air in a region with magnetic field of 0.50 T at right angles to the wire. Determine the current that would have to flow through the wire to just prevent it from falling to the ground.

Solution

Fig. 10.15 shows the wire in the magnetic field, and the forces acting. Using the left-hand rule, it is seen that the magnetic field is in the direction shown (i.e. into the page), in order to produce a magnetic force F upwards to oppose the weight W of the wire.

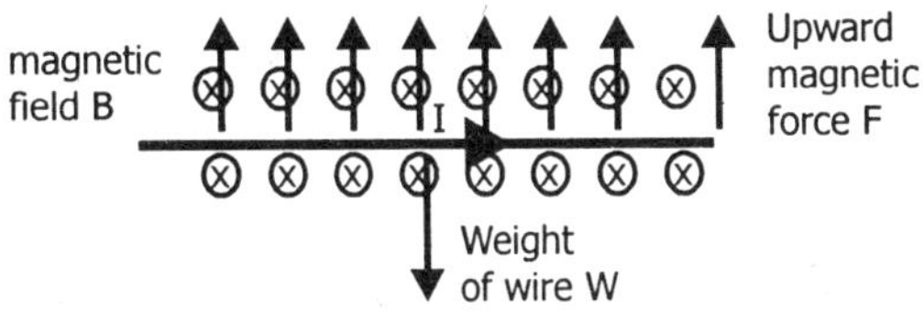

Fig. 10.15 Magnetic field and forces on wire

From (10.20), $F = ILB\sin 90^{o} = ILB$

To just prevent the wire from falling to the ground, this magnetic force must be equal to the weight of the wire. Thus, $ILB = mg$.

This gives the current I as $I = \dfrac{mg}{LB} = \dfrac{0.050\,kg \times 9.81\,m/s^2}{0.20\,m \times 0.50\,T} = 4.9\,A.$

10.10 FORCE BETWEEN CURRENT WIRES

This force can be considered to result from the fact that each wire lies in the magnetic field created by the other wire. A steady current through a conductor produces a magnetic field. Two current-carrying wires lying side by side would each produce magnetic fields that would interact with each other, producing a force. Using a combination of the right-hand screw rule and the left-hand rule, it can be seen that the force between the two wires would be attractive (i.e. the wires tending to move toward each other) when the currents are in the same direction, and repulsive when the currents are in opposite directions. **Fig. 10.16** shows two current wires lying side by side. The magnetic field B_1 created by current I_2 in wire 2 at the position of wire 1 is $B_1 = \dfrac{\mu_0 I_2}{2\pi a}$. Similarly, the field at the location of wire 2 is $B_2 = \dfrac{\mu_0 I_1}{2\pi a}$. The force F per

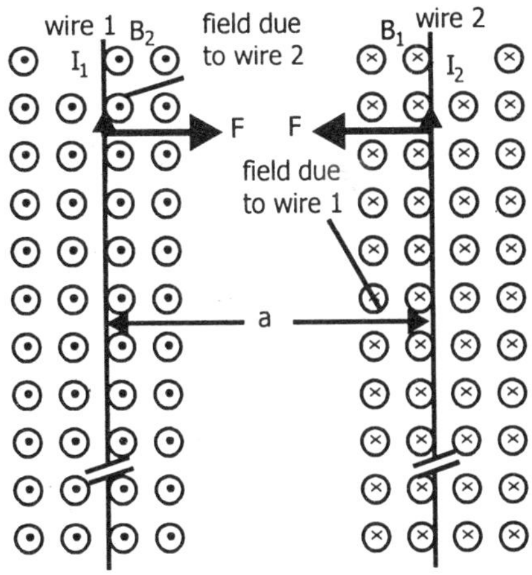

Fig. 10.16 Force between two parallel current wires. The force is attractive with currents in same direction.

unit length of the wires is

$$F = \frac{\mu_0 I_1 I_2}{2\pi a}. \qquad (10.21)$$

For two long straight parallel wires carrying current I, the force per unit length is then

$$F = \frac{\mu_0 I^2}{2\pi a}. \qquad (10.22)$$

If a current of 1 A flows through both wires, placed 1.0 m apart in vacuo, substituting the value of μ_o, we get the force per unit length to be 2.0×10^{-7} N. This provides an absolute definition of the ampere as follows:

One ampere is the value of that constant current which, flowing through two infinitely long straight wires placed 1.0 m apart in vacuo, would exert a force of 2.0×10^{-7} Newton per metre of their length.

Example 10.9

A light rectangular wire loop, measuring 25 cm by 10 cm lies next to a long straight wire carrying a current of 20 A. Determine the magnitude and direction of the force acting on the loop when a current of 30 A flows through it.

Solution

Fig. 10.17 shows the wire and the loop lying side by side. The forces on the sides of the loop result from the magnetic field created by the long wire. The forces on the sides ST and RU are equal in magnitude and opposite in direction, thereby producing a net force of zero in the direction parallel to the wire. The forces on the sides RS and TU are also in opposite directions, with the force on RS upwards, whilst that on TU is in the downward direction. Thus the net force on the loop is the difference between the two vertical forces. Force F_1 on RS is given by

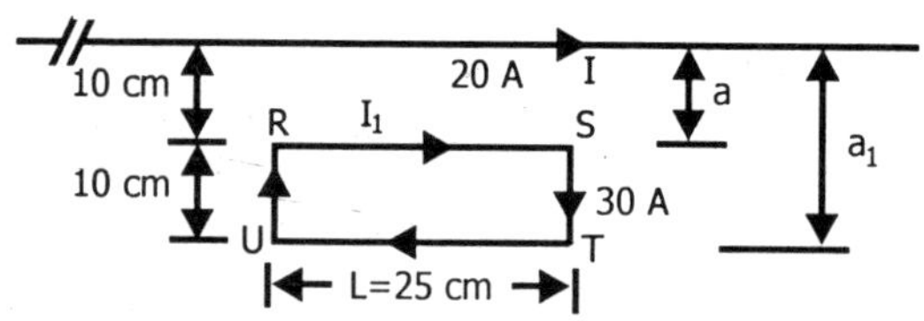

Fig. 10.17

$F_1 = \frac{\mu_0 I I_1 L}{2\pi a} = \frac{4\pi \times 10^{-7} H/m \times 20A \times 30A \times 0.25m}{2\pi \times 0.10} = 3.00 \times 10^{-4}$ N upward. Force F_2 on TU is given by $F_2 = \frac{\mu_0 I I_1 L}{2\pi a_1} = \frac{4\pi \times 10^{-7} H/m \times 20A \times 30A \times 0.25m}{2\pi \times 0.20} = 1.50 \times 10^{-4}$ N downward. The net force on the loop is hence 1.50×10^{-4} N upward toward the wire.

10.11 SOME APPLICATIONS OF MAGNETIC FORCES AND FIELDS

There are many devices which operate either as a result of the magnetic field produced through the passage of an electric current, or by the force created when two magnetic fields interact with each other. We briefly consider a few of these devices.

10.11.1 The relay

Many motion gadgets operate using the **relay**. Relays are used extensively for the local or remote control of circuits. Electric motors in inaccessible places, for instance, can be operated by the use of relays. Relays are designed to connect or disconnect circuits as desired. **Fig. 10.18** is a schematic diagram showing the main parts of a relay. In this example, the relay is used to connect the circuit at X with the battery E, thereby providing an output voltage across the resistor R. This output can be used to, for example, turn on a siren, or switch a lamp on.

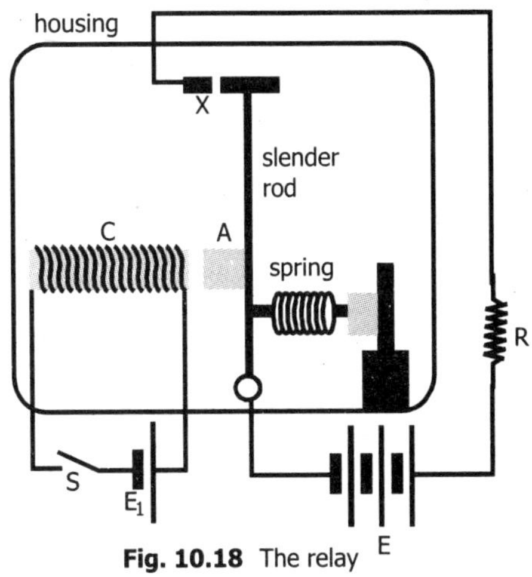

Fig. 10.18 The relay

Inside the housing of the relay is a coil C wrapped around an iron former. The coil is connected externally to the battery E_1, through the switch S. Upon closing S, the current in the coil produces a magnetic field that attracts a soft iron armature A. The armature is mounted on a slender rod, and a spring at the back pulls the connection at X off when S is open. Although there are more complicated relays on the market nowadays, the principles of operation have remained the same. In some versions, S takes the form of a radio frequency receiver, so the circuit can be connected remotely.

10.11.2 The door bell

The door bell, or electric bell works in similar to the relay. **Fig. 10.19** shows the schematic arrangement. The electromagnet is switched on and off in quick succession, driving the spring-loaded hammer to strike the gong G. Pressing the doorknob closes the switch S. The flow of current magnetises the coil. The spring-loaded hammer H is then pushed forcibly to strike the gong, breaking the circuit as it does so. The hammer springs back after striking the gong, thereby magnetising the coil once again, and the cycle is repeated while S remains closed.

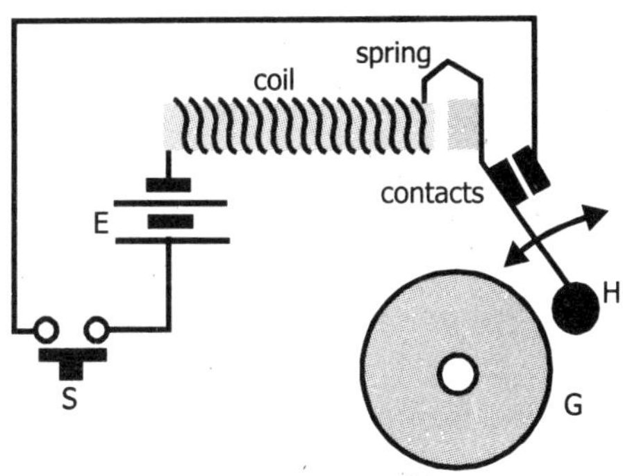

Fig. 10.19 The door bell

10.11.3 Magnetic recording

Today's computing, music, and video industry is awash with varieties of devices based on magnetic recording. In these, in formation is stored in magnetic material such as iron oxide, formed as a thin film on a nonmagnetic backing, such as plastic. The direction of magnetisation of small localised areas of the magnetic material represents the stored information. The information is stored on, and accessed from the magnetic material by means of small electromagnets – the read and write heads. Audio tapes, video tapes, and computer floppy disks are among the widely used examples of magnetic recording media. They belong to the class of devices known as **moving magnetic surface memory**.

10.11.4 The electromagnetic pump

In the metal industry, the **electromagnetic pump** is used to keep molten metal moving. A magnetic field is applied over a portion of the system. In the diagram – **Fig. 10.20** – the field is shown in the vertical direction in the plane of the page. The combination of the magnetic field and electric current provides the force that pushes the molten metal along the containment.

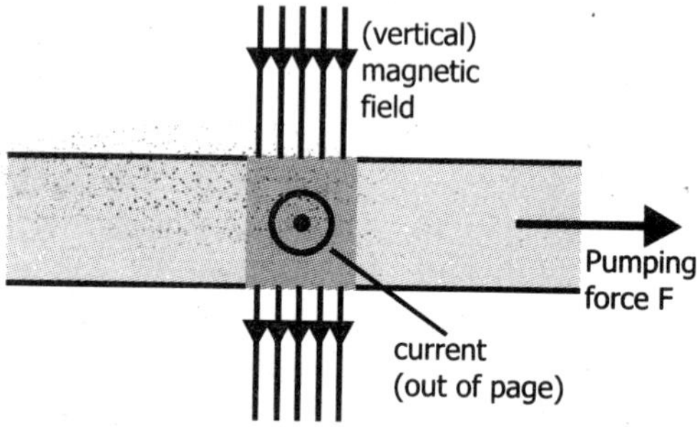

Fig. 10.20 The electromagnetic pump

One advantage of the electromagnetic pump is that it has no moving parts that would break down due to the high temperatures of the molten metal. Although first developed and applied mainly in the nuclear reactor industry, the electromagnetic pump has found other useful areas of application, such as the movement of blood. In artificial kidney machines, the heart needs to be assisted by some external pump. Mechanical pumps tend to damage blood cells moving through them, since these machines have moving parts. In such situations, the electromagnetic pump offers a good alternative.

10.12 TORQUE ON A COIL IN A MAGNETIC FIELD

From the above study, a force always acts on a current element when placed in a magnetic field. If the wire were free to move, this force would lead to motion in a direction perpendicular to both the magnetic field and the current direction.

Consider for simplicity, a rectangular wire loop PQRS, length a, and width b, pivoted about an axis XY, as shown in **Fig. 10.21**. Since the currents flowing in opposite sides of the coil are equal and in opposite directions, the force on opposite sides are equal in magnitude but opposite in direction. This gives the total force

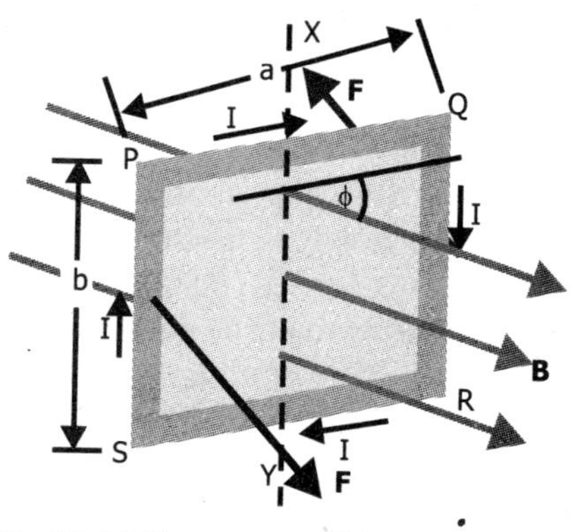

Fig. 10.21 Torque on a coil in a magnetic field

around the whole loop as zero. There is, however, a torque acting as the coil is pivoted about the axis XY. This torque is the product of the force (on SP or QR) and the perpendicular distance between the sides in question. Thus, the couple, or torque C about the line XY is given by

$$C = Fa \sin \phi = BIba \sin \phi.$$

Bearing in mind that the product ab = area of the coil, we have

$$C = BAI \sin \phi,$$

where A= ab is the area of the coil. If the coil has N turns of wire, the total area through which the magnetic field threads is NA. The torque acting on such a coil becomes

$$C = BANI \sin \phi. \qquad (10.23)$$

Whilst equation (10.23) is derived using a rectangular coil, the equation holds nonetheless for any coil of wire. This is so because the final expression does not contain the dimensions of the coil, but rather the area of the coil. Hence we can apply the equation to a circular coil by using the area for the circular coil. The equation also shows that the torque is greatest when the plane of the coil is at right angles to the magnetic field, and zero when the field is parallel to the plane of the coil.

10.13 MOVING COIL INSTRUMENTS

A range of instruments for measuring current, voltage, and resistance use the torque on a coil in a magnetic field. These are commonly known as **moving coil instruments**. A schematic diagram showing the features of such devices is shown in **Fig. 10.22**.

A coil wrapped around a soft iron block is freely pivoted between the poles (N and S) of a magnet. Connection is made with the terminals of the instrument. When current flows through the coil, a torque results. This torque tends to rotate the coil. A return spring opposes the complete rotation of the coil. A pointer attached to the spring moves over an angular scale. For any current I flowing through the coil, the pointer moves through an angle representing a balance between the torque due to the magnetic field, and the mechanical torque of the spring. From equation (10.23), the torque C set up due to the magnetic field when a current I flows through the coil is given by $C = BANI \sin \phi$, where B is the magnetic field of the permanent magnet, A the area of the coil, and N is the number of turns of the coil. The angle ϕ is the angle between the direction of the magnetic field and the plane of the coil. In these instruments, the poles of the magnet are shaped such that the magnetic field is at all times perpendicular to the plane of the coil. The expression "radial magnetic field" is often used to describe this situation. Thus, $\phi=90^{\circ}$.

If k is the torsion constant of the spring, the opposing couple τ in the spring is

$$\tau = k\theta, \tag{10.24}$$

where θ is the angle of twist of the spring. Equating the two couples, we have

$$BANI = k\theta.$$

This gives

$$I = \frac{k}{BAN}\theta. \tag{10.25}$$

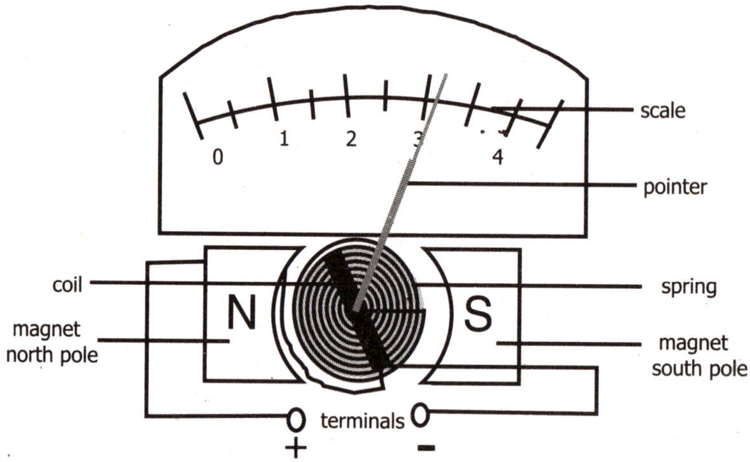

Fig. 10.22 The moving coil galvanometer (or milliammeter).

From (10.25), it is seen that for a given spring, magnet, and coil, the current flowing through the coil is directly proportional to the angular deflection of the pointer. The instrument can hence be calibrated directly to read the current, rather than the angle.

The galvanometer described above is also called a **milliammeter**, as it can only measure currents in the milliamp range. Large currents of the order of amperes can burn the pivoted coil. However, it is possible to measure currents of the order of amperes with the use of a **shunt** resistor connected in parallel with the coil. The value of the shunt resistor is such that most of the current passes through the shunt, leaving only the small part that can be accommodated by the coil to pass through the latter. This way, it is possible to convert the instrument into an ammeter.

To measure voltage we place a resistor in series with the coil. Using Ohm's law, the potential difference across the resistor and coil is then given by

$$V = IR = \frac{kR}{BAN}\theta, \tag{10.26}$$

where R is the total resistance of the coil and the external resistor. Thus the p.d. is proportional to the angular deflection. As the voltage measurement relies on the application of Ohm's law, this explains why the moving coil instrument is not used in experiments to verify Ohm's law.

EXERCISES

$\mu_0 = 4\pi\times10^{-7}\,H/m$; $\varepsilon_0 = 8.85\times10^{-12}\,F/m$.

electron charge, $e = -1.602\times10^{-19}\,C$ Proton mass $= 1.667\times10^{-27}\,kg$

1. The current in a conductor is directed westward. Draw a diagram showing the lines of magnetic flux around the conductor

2. Calculate the magnetic flux passing through an area of 0.10 m^2 in a region where the magnetic flux density is 0.25 T normal to the area.

3. Calculate the magnetic flux density at a point 5.0 cm away from a long straight wire carrying a steady current of 10 A.

4. A horizontal power line carries a current of 100 A in a north-south direction. Determine
   (i) the magnitude of the magnetic field at a point 1.0 m below the wire,
   (ii) the direction of the magnetic field at the point.

5. A wire is in the shape in **Fig. 10.23**, consisting of a combination of a circular loop of radius a, ana a long straight section. A current I flows through the wire. Show that the minimum magnetic field at the centre of the coil has magnitude $\frac{3\mu_0 I}{2a}$, where μ_0 is the permeability of free space.

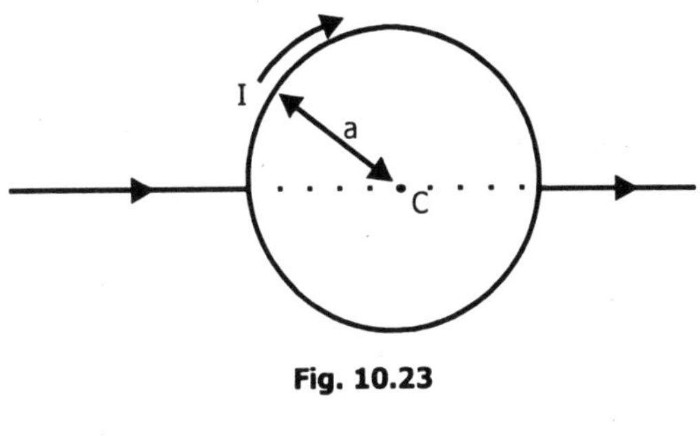

Fig. 10.23

6. A straight wire of length 0.10 m, has a mass of 7.0×10^{-3} kg. The wire lies in a uniform magnetic field of flux density 0.5 T perpendicular to its length. Calculate the current that would have to flow through the wire such that the magnetic force on the wire will be equal and opposite to its weight.

7. In **Fig. 10.24**, R and S represent the sections of two long straight current-carrying wires, placed a distance 2d apart. Both wires carry the same current I, with the current through R coming out of the page, while that through S goes into the page, both currents being perpendicular to the plane of the page.

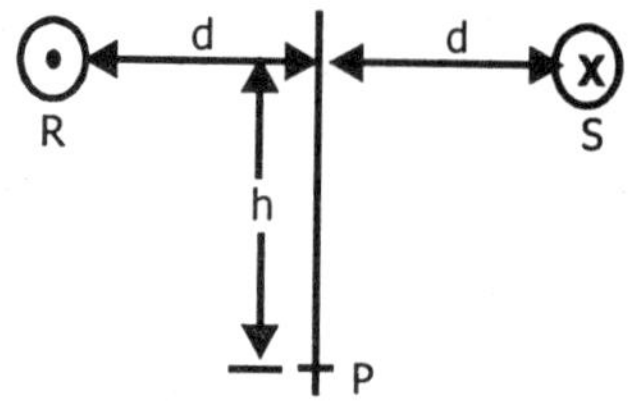

Fig. 10.24

(a) Show that the resultant magnetic field B at the point P on the perpendicular bisector of the line joining the two wires is given by $B = \dfrac{\mu_0 Id}{\pi(h^2 + d^2)}$.

(b) If the two wires are 20 cm apart, and carry a current 25 A, calculate the magnitude of the magnetic field at P if P is 10 cm from the line joining their centres.

8. A light wire frame is in the form of a semi-circle of diameter 10 cm, joined at either end by straight portions in line with the diameter. Calculate the magnetic field at the centre of the frame, when a steady current of 20 A flows through the frame.

9. A solenoid of length 50 cm, has 5000 turns of conducting wire. The solenoid has a cross sectional area of 4.0 cm^2, and a steady current of 4.0 A flows through the coil. Calculate

(i) the flux density inside the solenoid,

(ii) the total flux inside the solenoid.

10. A solenoid of length 0.60 m has 1500 turns of conducting wire. If a current of 5.0 A flows through the coil, find the magnitude of the magnetic flux density at the centre of the solenoid.

11. A solenoid has a length of 50 cm and contains 4000 turns of conducting wire. The diameter of the solenoid is 4.0 cm. What current would be needed for the solenoid to produce a magnetic flux density of 0.5 T at its centre?

12. The maximum current that can flow through a given wire is found to be 20 A. This wire is used to construct a pair of Helmholtz coils each of mean diameter 10.0 cm. Determine the number of turns of each coil that would produce a magnetic field of 0.1 T on the axis at a point midway between the coils.

13. If the wire in Exercise 12 were to be wrapped around a hollow cylinder of length 50 cm, and the same diameter of 10.0 cm, calculate the total number of turns that would produce the same magnetic field inside the cylinder as before, with the same current as before.

14. Using Biot-Savart law, deduce an expression for the magnetic field at any point on the axis of a circular coil having N turns of mean radius a, carrying current I.

(a) Using the above expression, deduce the magnetic field at the centre of a pair of Helmholtz coils

(b) If the coil has diameter 24.0 cm, with 2000 turns, and a current of 20 A flows through it, determine the magnitude of the magnetic flux density at the centre of the Helmholtz coils.

15. A wire having a linear density of 0.025 kg/m lies in a horizontal plane that contains a magnetic field of flux density 0.025 T perpendicular to the wire. A current I flowing through the wire prevents the wire from falling to the ground. Calculate this current.

16. Calculate the magnetic field at the centre of a long air-cooled solenoid having 5000 turns per metre, carrying a current of 10 A. If an iron core with relative permeability 300 is now inserted in the solenoid, calculate the new value of the magnetic field.

17. A magnetic field is generated by a pair of Helmholtz coils of diameter 10 cm. Each coil has 200 turns of wire carrying current 10 A.
(a) Calculate the magnetic field at the centre of the pair.
(b) If a rod carrying a current of 5 A is inserted between the coils at right angles to the common axis, calculate the force acting per unit length of the rod.

18. A circular coil of radius 10 cm has 200 turns of wire through which a current of 10 A flows. The coil is placed in a uniform magnetic field of 0.5 T in the positive x-direction.
(a) Determine the positions of minimum and maximum torque on the coil.
(b) Calculate the maximum and minimum torques on the coil
(c) Determine the angle between the plane of the coil to the magnetic field such that the torque is one-quarter the maximum torque in magnitude.

19. Two horizontal parallel wires each 50 cm long are arranged in vacuo, one above the other, and connected in series so that the current flowing through one is in opposite direction to that flowing through the other. The lower conductor is fixed, while the upper conductor is free to move vertically in guides under its own weight, the conductors always remaining parallel. The current flowing through the conductors is 35 A.
(a) Draw a diagram showing the magnetic field and the forces on the two wires.
(b) If the upper conductor has a mass of 1.50 g, calculate the distance apart between the two conductors at equilibrium.

20. A square wire loop of side 0.50 m, lies with its plane parallel to a magnetic field with flux density 0.30 T. The wire carries a current of 20 A.
(a) Draw a force diagram showing the magnetic force acting on each side of the wire.
(b) Calculate the magnitude of the magnetic force on each side.

21. At a certain location, the horizontal component of the earth's magnetic field is 2.5×10^{-5} T, due north. A positively charged particle of mass 5.0×10^{-27} kg moving due east, has just the right speed for the magnetic force to balance its. Calculate the speed of the particle, given that its charge is 3.2×10^{-19} C.

22. A proton accelerated through a potential difference of 3.0 MV enters a magnetic field of flux density 0.5 T, in a plane perpendicular to the motion of the proton. Calculate the radius of the circular path that the proton will describe in the magnetic field.

23. A proton beam travelling at a speed of 6.5×10^5 m/s enters at right angles to a uniform magnetic field with flux density of magnitude 0.50 T. If the proton experiences a force of 4.0×10^{-8} N, determine the angle between the magnetic field and the velocity of the protons.

24. An alpha particle of mass 6.668×10^{-27} kg, and charge 3.20×10^{-19} C, enters a magnetic field of flux density 0.75 T in a direction at right angles to the field. If the speed of the alpha particles is 4.5×10^5 m/s, determine the radius of the circular path described by the particle in the field.

25. A charged particle enters a velocity selector that is constructed using an electric field of intensity 5.65×10^3 N/C, and a magnetic field of flux density 0.121 T. On switching off the electric field, the particle is found to describe a circular path of radius 0.030 m. Calculate the charge-to-mass ratio of the particle.

26. In a mass spectrometer, charged particles with a selected velocity enter a region with a uniform magnetic field perpendicular to the velocity vector. The particles then describe a circular orbit. In an experiment using singly ionised atoms travelling with velocity 2.50×10^5 m/s, in a magnetic field of flux density 0.13 T, the radius of the circular path is measured to be 20.0 cm. Calculate
(i) the charge-to-mass ratio of the ions,
(ii) the mass of each ion.

27. Protons are accelerated in a potential difference of 3.0 MV, and enter a region where they are subjected to a magnetic field of 1.0 T at right angles to their direction of travel. Calculate the radius of the circular path they describe.

11

CHAPTER

Electromagnetic Induction

11 CHAPTER

Electromagnetic Induction

Interactions between moving charges and magnetic fields long convinced early workers of the definite connection between the electric field and the magnetic field. While it was easy to produce magnetic fields from steady currents and moving charges, the reverse effect – producing a steady current from a magnetic field – took quite some time to be realised in practice. Faraday showed that there is indeed an electric field resulting from a magnetic field, but that this electricity only existed while the magnetic field was changing. The results of early experiments with this phenomenon established the connection between electric and magnetic fields and forces, and the branch of study now known as **electromagnetism**, was born. Today, electricity and magnetism are no longer treated as separate subjects, but rather parts of the same entity. Devices based on these interactions are constantly being developed, and range from high voltage power generation to modern telephone communication, and computer technologies.

In this chapter, we study the principles underlying electromagnetism – the laws governing the generation, and the effects of electromagnetism. We end the chapter with a survey of devices based on electromagnetism.

11.1 INDUCED ELECTROMOTIVE FORCE

An electric field (or electromotive force) created as a result of a magnetic field occurs only when the magnetic field is changing. This emf is referred to as **induced emf**. To induce an emf in a circuit, there must be a changing magnetic field threading the circuit. The early experiments used permanent magnets that were moved toward or away from a coil of wire. In these setups, the emf was induced only when there was relative motion between the coil and the magnet. As a result, this emf was originally described as **motional induced** emf. In another method of changing the magnetic field, the current flowing in a nearby coil is changed. **Fig. 11.1** (a) and (b) illustrate these two methods. Today, because of the different modes of inducing an emf, we refer to this emf simply as induced emf.

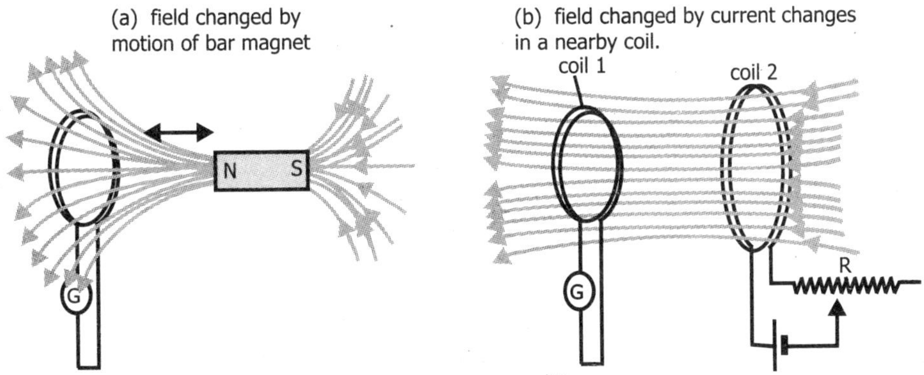

Fig. 11.1 Induced emf in a coil brought about by (a) motion of a magnet into toward and away from coil, and (b) current changes in a nearby coil, effected by the variable resistor R.

The situation depicted in (a) is often called "motional induced emf", and it was the established method of inducing emf. Here, the magnetic field threading the coil is changed whenever there is relative motion between the coil and the permanent magnet. This relative motion is achieved if (i) the magnet is moved with respect to a stationary coil, (ii) the coil is moved with respect to a stationary magnet, and (iii) both magnet and coil are moved toward or away from each other. In (b), current is made to flow in a coil (coil 2) connected to a battery and a variable resistor. A steady current in the coil produces a magnetic field at the location of coil 1. An emf is induced in coil 1 whenever the magnetic field through the coil changes. This change can be brought about either by relative motion between the two coils, with a constant current in coil 2, or by changing the current in coil 2 (using the variable resistor R), or both.

11.2 LAWS OF ELECTROMAGNETIC INDUCTION

Experiments such as the ones depicted in **Fig. 11.1** above have shown that the induced emf depends on how fast the magnetic flux is changed. In (a), this has to do with how fast the magnet is thrust toward, or away from the coil. In (b), keeping the two coils a fixed distance apart, the induced emf is found to depend on how fast the current in coil 2 is changed. These observations are embodied in two laws, known as the laws of electromagnetic induction.

11.2.1 Faraday's law

This law deals with the magnitude of the induced emf. It states that

the magnitude of the induced electromotive force is directly proportional to the rate of change of magnetic flux.

From the above statement, we can express Faraday's law in symbols as

$$|\varepsilon| \propto \frac{d\Phi}{dt}, \tag{11.1}$$

where ε is the induced emf brought about by a flux change $d\Phi$ in a coil in time dt.

11.2.2 Lenz's law

Lenz's law deals with the direction of the induced emf. It states that

the induced electromotive force is in such a direction as to oppose the change that gives rise to it.

11.2.3 The combined law and flux linkage

Combining Faraday's and Lenz's laws, we get the combined law of electromagnetic induction, which allows us to determine both the magnitude and the direction of the induced emf resulting from a given flux change.

In symbols, this is expressed as follows:

$$\varepsilon = -\frac{d\Phi}{dt}. \tag{11.2}$$

In equation (11.2), the negative sign represents the inclusion of Lenz's law. The proportionality expressed by Faraday's law is also present, with a proportionality constant of unity.

In the application (11.2), one concept we take into account is that if a coil has N turns, and the magnetic flux changes by an amount $d\phi$, the total flux change $d\Phi$ in the whole coil is given by

$$d\Phi = N d\phi. \tag{11.3}$$

The quantity $Nd\phi$ is called the **magnetic flux linkage**.

Example 11.1

A circular coil of mean diameter 4.0 cm, has 100 turns of wire, with total resistance 15 Ω. The coil lies with its plane perpendicular to a uniform magnetic field of 500 mT, provided by an electromagnet. The current in the electromagnet is slowly switched off over 5 ms. Calculate

(i) the magnetic flux linkage in the coil before and after the current in the electromagnet is switched off,
(ii) the current flow in the coil when the electromagnet is just switched off,
(iii) the charge flow through the coil.

Solution

(i) Data: diameter of coil d = 0.04 m, number of turns of coil N = 100 turns.
resistance of wire R = 15 Ω, magnetic flux density B = 0.5 T.

The magnetic flux linkage Φ at any time is given by $\Phi = N\phi$, where ϕ is the flux through one turn of the coil. Also, since the magnetic field is perpendicular to the plane of the coil, the flux linking a turn of the coil at any time is equal to the product of the flux density and the area A of the coil. Thus, the flux linkage Φ_1 with the electromagnet on is given by $\Phi_1 = N\phi = NBA$. Using the given data,

$$\Phi_1 = NBA = NB\pi\frac{d^2}{4} = 100 \times 0.5\ \text{weber/m}^2 \times \pi \times \frac{0.04^2\,\text{m}^2}{4} = 0.063\ \text{weber}.$$

When the electromagnet is switched off, the magnetic flux density is zero. Thus the flux when the magnet is switched off is zero.

(ii) From equation (11.2), the induced emf in the coil caused by the change in magnetic flux is given by $\varepsilon = -\frac{d\Phi}{dt}$. The current flow I in the coil therefore becomes

$$I = \frac{\varepsilon}{R} = -\frac{1}{15}\frac{(\Phi_1 - 0)}{t} = -\frac{0.063\ T}{15\ \Omega} \times \frac{1}{0.005\ s} = 0.84\ A.$$

(iii) The charge flow Q through the coil is $Q = It = 0.84A \times 0.005s = 4.2 \times 10^{-3}C.$

Example 11.2

A circular coil of cross sectional area $2.0\ cm^2$, having 200 turns of wire, rotates at 240 r.p.m. (revolutions per minute), in a uniform magnetic field of 0.5 T. If the wire has a total resistance of 5 Ω, calculate

(i) the maximum value of the emf induced at the ends of the coil,
(ii) the maximum current flow between the ends.

Solution

Data: area of coil $A = 2.0 \times 10^{-4}\ m^2$, number of turns N = 200 turns
Frequency of rotation f = 4.0 /s, magnetic flux density B = 0.5 T
Resistance of coil R = 5.0 Ω.

(i) Maximum flux linkage is obtained when the magnetic flux linking the coil is a maximum. This occurs when the plane of the coil is at right angles to the direction of the magnetic field. The maximum flux linkage in the coil is given by $\Phi_{max} = NBA = 200 \times 0.5\ weber/m^2 \times 2.0 \times 10^{-4} m^2 = 0.02$ weber. The minimum flux linkage occurs when the plane of the coil is parallel to the magnetic field. In this momentary position, the area of coil exposed to the field is zero. Hence the minimum magnitude of the flux linkage is zero. The magnitude of the maximum flux change is hence 0.02 weber. In rotation, the coil takes one-quarter of a period to go from the point where its plane is perpendicular to the field to the point where its plane is aligned to the field. The induced emf therefore has a maximum value given by

$$\varepsilon_{max} = \frac{(\Phi_{max} - \Phi_{min})}{t} = \frac{(0.02 - 0)weber}{0.25 \times 0.25\ s} = 0.32\ V.$$

(ii) The maximum current I_{max} is given by $I_{max} = \frac{\varepsilon_{max}}{R} = \frac{0.32\ V}{5.0\Omega} = 64\ mA.$

11.2.4 The search coil

The search coil is a device that uses induced emf to measure the magnetic flux at a point in a magnetic field. **Fig. 11.2** is a schematic diagram of how the search coils is used.

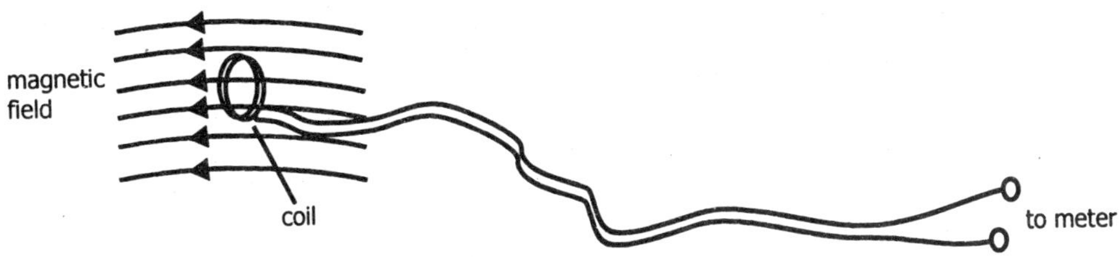

Fig. 11.2 Measurement of flux using the search coil

A small coil is joined to a suitable charge measuring meter. The coil is inserted in a magnetic field, and oriented such that maximum magnetic flux passes through the coil, as indicated by maximum reading on the meter. The coil is then removed to a field-free region, and the charge flow through the coil is measured.

The solution in **Example 11.1** provides the theory behind the measurement. If R is the total resistance of the coil and connecting wire, we can write an expression for the charge flow through the coil when it is moved from a region of maximum flux to a field-free region. This charge flow is given by

$$R\frac{dQ}{dt} = -\frac{d\Phi}{dt},$$

from which $d\Phi = RdQ$. If the charge measured is Q, the flux Φ at the point is given by

$$\Phi = RQ.$$

If the resistance of the coil and adjoining cable is arranged to be 1.0 Ω, the charge measured would be numerically equal to the magnetic flux.

11.3 MUTUAL AND SELF INDUCTION

From our study of electromagnetic induction so far, we conclude that two current-carrying coils lying side by side will interact with each other through their magnetic fields if their currents are varied. This interaction is described by the concept of **mutual induction** between the coils. Also, a single coil carrying a current has a magnetic field. This magnetic field would change if the current in the coil changes; this latter change in magnetic field would induce an emf in the same coil. This kind of interaction involving magnetic field changes in a single coil with itself is referred to as **self induction**.

11.3.1 Mutual induction

Consider two current-carrying coils, labelled coil 1 and coil 2, lying close to each other, as shown in **Fig. 11.3**. Current flow in one coil creates a magnetic field that would interact with the magnetic field created by current flow in the other coil. A change of current in coil 1, for instance, will produce an induced emf in coil 2, and vice versa.

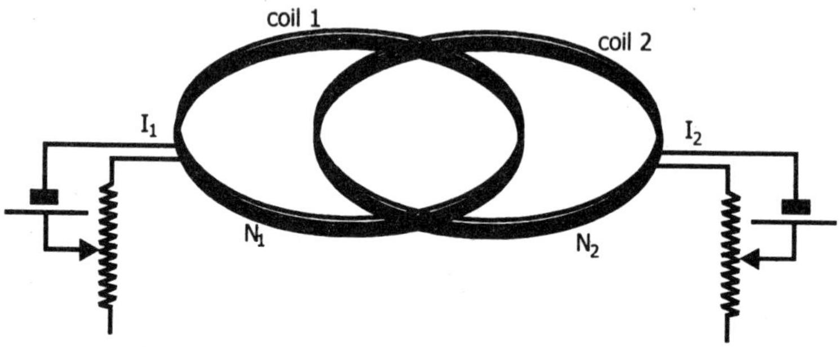

Fig. 11.3 Mutual induction between two coils

Currents I_1, and I_2, respectively flow in the coils 1 and 2. A change in of current in coil 1 would lead to a flux change in coil 2, which we represent as Φ_{21}, read as "flux change in coil 2 due to current change in coil 1". Similarly, we can have and a flux change of Φ_{12} in coil 1 due to current change in coil 2. The corresponding electromagnetic energy per unit time in each coil due to these flux changes are:

$$U_{21} = -\mathcal{E}_2 I_2 = -I_2 N_2 \Phi_{21}, \quad (11.4)$$

and

$$U_{12} = -\mathcal{E}_1 I_1 = -I_1 N_1 \Phi_{12}, \quad (11.5)$$

where U_{21} is the electromagnetic energy in coil 2 resulting from current change in coil 1, and U_{12} is the energy in coil 1 due to current change in coil 2. By energy conservation, we equate the left-hand sides of (11.4), and (11.5). Thus,

$$I_2 N_2 \Phi_{21} = I_1 N_1 \Phi_{12},$$

giving

$$\frac{N_2 \Phi_{21}}{I_1} = \frac{N_1 \Phi_{12}}{I_2} = M \text{ (a constant)}. \quad (11.6)$$

From equation (11.6), the flux linkage in coil 2 per unit current change in coil 1 is equal to the flux linkage in coil 1 per unit current change in coil 2. The constant M represents the degree of electromagnetic interaction between the two coils. It is known as the **mutual inductance**. We can define the mutual inductance between the two coils as

the magnetic flux linkage in one coil per unit current change in the other coil.

A current change I between two coupled coils leads to a flux change given by

$$\Phi = \frac{MI}{N}, \tag{11.7}$$

where M is the mutual inductance between the coils, and N is the number of turns of the chosen coil. In terms of the mutual inductance, the induced emf becomes

$$\varepsilon = -N\frac{d\Phi}{dt} = -M\frac{dI}{dt}. \tag{11.8}$$

The units of M are weber/amp. This unit is also known as the **henry**, abbreviated as H.

11.3.2 Self induction

Current changes in a coil produce changes in the magnetic flux linkage, and hence an induced emf in the same coil. In other words, the coil interacts with itself. This electromagnetic interaction of a coil with itself is known as **self-induction**, and represented by L. From analysis similar to the one above for mutual induction, we have

$$L = \frac{N\Phi}{I}, \tag{11.9}$$

where N is the number of turns, Φ the flux change, and I is current change in the coil. The induced emf is expressed in terms of the self-inductance as

$$\varepsilon = -N\frac{d\Phi}{dt} = -L\frac{dI}{dt}. \tag{11.10}$$

Like mutual inductance, the unit of self inductance is the Henry. From equation (11.10), the self inductance of a circuit can be defined as

the magnitude of the flux linkage per unit current change in the same coil.

When two coils are coupled, the mutual induction between them can be obtained from the self inductances of the coils, and the degree of electromagnetic coupling between them. For two coils with inductances L_1 and L_2, coupled together, the mutual inductance M between them is given by

$$M = k\sqrt{L_1 L_2}, \tag{11.11}$$

where k is the **coefficient of coupling**, representing the percentage of the magnetic flux from one coil that goes through the other coil, with $0 \le k \le 100$.

11.4 THE INDUCTOR

An inductor is any device or any part of a circuit that is designed to introduce a known value of inductance. The inductor is usually a coil of wire wound sometimes on a soft iron former, the latter greatly increasing the magnetic flux. As a circuit component, the inductor has many uses, many of which we shall study later in connection with alternating currents. Inductors come in two forms. In one of these, the value of the induction is fixed. This is the fixed or standard inductor. In the other type, it is designed that all or part of the coil can be connected in the circuit as desired. Thus the inductance can be changed, making this the variable type. The standard symbols for the inductor are shown in **Fig. 11.4**.

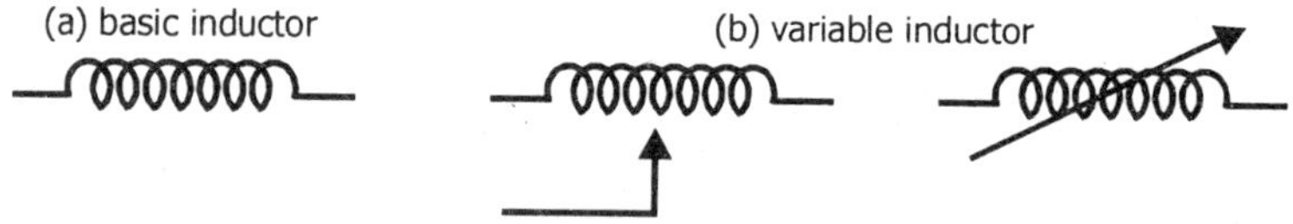

Fig. 11.4 Circuit symbols for the inductor

Just as resistors and capacitors can be connected in series and in parallel, it is also possible to have series and parallel combinations of inductors. In analysing these combinations, however, we make some assumptions for purposes of simplicity.

1. That the inductors are pure inductors. Since inductors are coils of wire, there is always some resistance, albeit small in many cases. The assumption of pure inductors implies that the effect coil resistance to current changes can be neglected.
2. That the inductors are **non-interacting**. This implies that the inductors are so far apart that we can assume the mutual inductance between the coils to be zero. Thus we only need to combine the self inductances. The analysis becomes a lot more complicated if the mutual inductances are to be considered.

11.4.1 Inductances in series

Fig. 11.5 shows a number of inductors connected in series across a direct current supply.

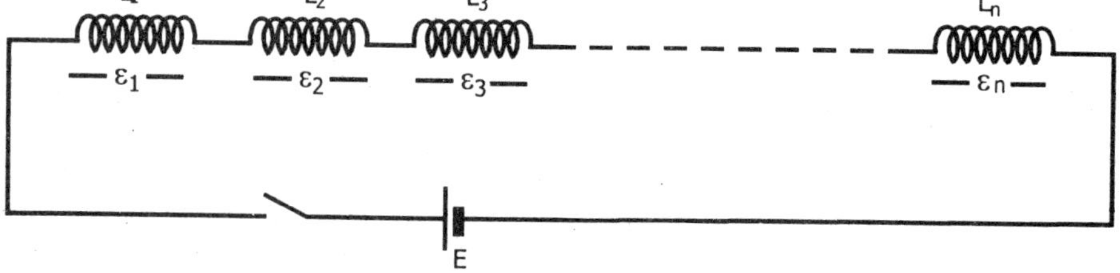

Fig. 11.5 Inductors in series

When the key is closed, emf ε_1 is induced in inductor L_1, ε_2 is induced in L_2, etc. The sum of these induced emfs is equal to the emf E of the source.
Hence,

$$E = \varepsilon_1 + \varepsilon_2 + \varepsilon_3 + \ldots\ldots\ldots + \varepsilon_n.$$

The current changes in the coils, however, are the same. Thus, we write

$$E = -L\frac{dI}{dt} = -L_1\frac{dI}{dt} - L_2\frac{dI}{dt} - L_3\frac{dI}{dt} - \ldots\ldots - L_n\frac{dI}{dt},$$

where L is the inductance of the combination. The overall inductance is hence given by

$$L = L_1 + L_2 + L_3 + \ldots + L_n = \sum_{i=1}^{n} L_i. \qquad (11.12)$$

11.4.2 Inductances in parallel

Fig. 11.6 shows a number of inductors connected in parallel. When the switch key is closed, current starts to flow in the circuit, bringing about a change in magnetic field in the inductors. The current in each conductor at any instant of time depends on its inductance. For a current change dI from the battery, we have

$$dI = dI_1 + dI_2 + dI_3 + \ldots\ldots + dI_n.$$

The induced emf across each conductor, however, are all the same, since the inductors are connected in parallel across a common supply. Thus,

$$\varepsilon_1 = \varepsilon_2 = \varepsilon_3 = \ldots\ldots = \varepsilon_n = E,$$

and

$$\frac{dI_1}{dt} = -\frac{E}{L_1}, \quad \frac{dI_2}{dt} = -\frac{E}{L_2}, \quad \frac{dI_3}{dt} = -\frac{E}{L_3}, \ldots, \quad \frac{dI_n}{dt} = -\frac{E}{L_n}.$$

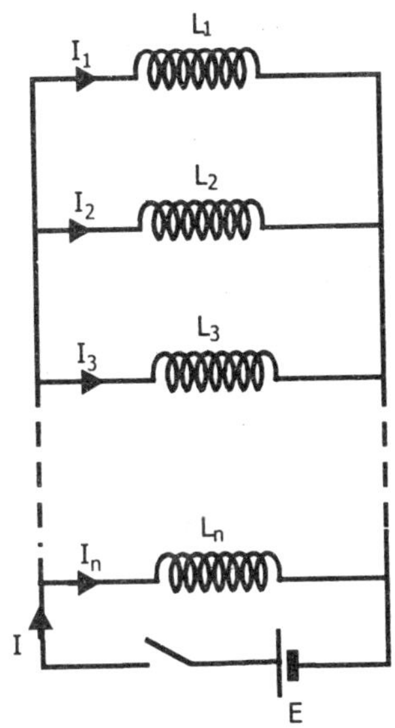

Fig. 11.6 Inductors in parallel

If L is the effective inductance of the parallel combination, we have $\frac{dI}{dt} = -\frac{E}{L}$.

The effective inductance therefore becomes

$$\frac{1}{L} = \frac{1}{L_1} + \frac{1}{L_2} + \frac{1}{L_3} + \ldots\ldots\ldots + \frac{1}{L_n},$$

or

$$\frac{1}{L} = \sum_{i=1}^{n} \frac{1}{L_i}. \qquad (11.13)$$

11.5 ELECTROMAGNETIC ENERGY IN AN INDUCTOR

In order to have current flowing in an inductor, some energy is required. Consequently, an inductor carrying a current has a store of energy. This energy is stored in the magnetic field, just as the energy stored in a capacitor resides in the electric field.

Consider an inductor connected to a supply voltage, as shown in **Fig. 11.7**. When the switch S is closed, the current rises steadily from zero to a final value I. Over an infinitesimal time, the current change dI, brings about an energy change dU in the inductor, given by

$$dU = LIdI.$$

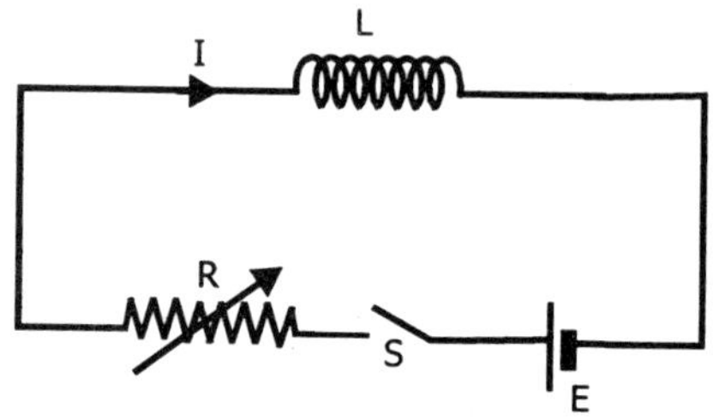

Fig. 11.7 Energy stored in an inductor.

Thus, the total energy U supplied when the current changes from zero value to a final value I, is given by

$$U = \frac{1}{2}LI^2. \quad (11.14)$$

11.6 ELECTROMAGNETIC DEVICES

Not long after the first demonstrations of electromagnetic induction, it was realised that the principles could be applied to many practical devices. Devices that operate using these principles are called electromagnetic devices.

Electromagnetic devices fall into a number of categories. One group contains devices that produce a force output. These devices employ the interaction between the magnetic field of a permanent magnet, and the field created by current flowing through a conductor. The interaction then produces a force that can be used to drive an external load. The electric motor falls into this category. In the other group of electromagnetic devices, a voltage output results. This output voltage can be produced from the electromagnetic interaction between two coils, one of which has a varying current. An emf is induced in the other coil. An example in this group is the transformer. Alternatively, a voltage output can be achieved by rotating a coil in the magnetic field of a permanent magnet. The alternating current generator is an example of this. Today, there are many more devices on the market, and more devices and applications are being constantly designed and developed. We limit our brief survey here to the most established devices.

11.6.1 The electric motor

In the electric motor, an electric current flows through a coil, resulting in a mechanical energy output. Examples of electric motors can be found in the driving mechanism of the tape recorder and video cassette players, the starters and windscreen wipers of motor vehicles, and electric fans in all kinds of machinery. Motors come in a large range of sizes depending on the use.

The electric motor uses the interaction between an electric field and a magnet. The steady current creates a magnetic field around a coil. This magnetic field interacts with a fixed magnetic field to produce rotation of the coil. The fixed magnetic field can be an electromagnet or a permanent magnet. **Fig. 11.8** is a schematic representation of the motor, that uses fixed permanent magnet.

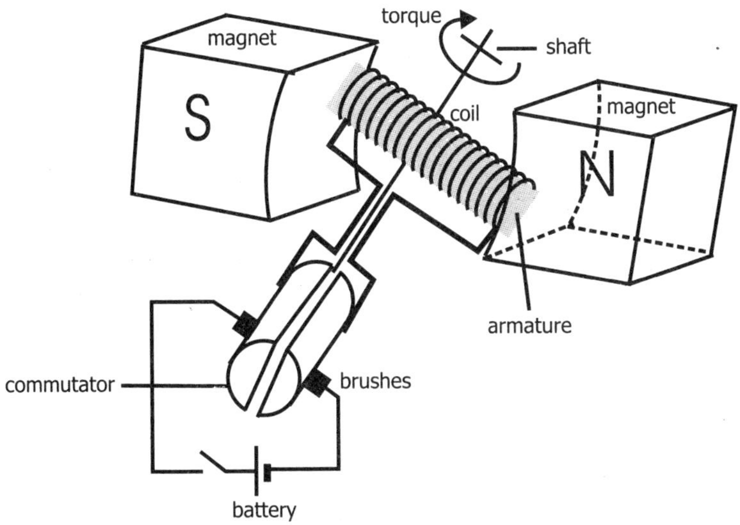

Fig. 11.8 The electric motor

In most motors, the magnet stands still, and the coil carrying the current moves. The moving part of the motor is made of a soft iron core called **armature**, around which the coil is wound. A shaft passing through the armature allows it to rotate between the poles of the magnet. A small drum-shaped arrangement, called the **commutator** is also connected to the coil. Electric current passes through carbon **brushes** to the coils. One end of the armature shaft is linked to the machinery to be turned, the machinery making up the external **load**.

As we learnt in the last chapter, a couple acts on a current-carrying coil that is placed in a magnetic field. In the motor, when the current is switched on, the couple turns the armature and coil in a direction determined by the current flow in the coil, and the magnetic field. This couple is in the same sense until the coil turns through 180°, at which point the current direction is reversed by the commutator. This way, the torque on the coil maintains the same of direction of rotation at all times.

In most motors we find today, the fixed magnet is also an electromagnet, provided by current flowing in a separate coil known as the **field coil**. Two modes of connection are in use for the field coil. In one, the field coil is connected in series with the moveable coil. This is the series connection, and in it the current reduces as the external load to be driven increases, and vice versa. In the parallel connection, the field coil is connected in parallel with the moveable coil. The current in this coil remains constant even when the load changes. The series connection is preferred in small motors where the external load remains practically constant. Motors used in heavy industry are designed to carry varying external loads. The parallel connection is preferred for these.

As the coil rotates, it induces an emf in the external circuit. By Lenz's law, this induced emf is opposite to the direction of flow of current in the external circuit. This emf is known as the **back emf**. The current flow I in the circuit is hence given by

$$E - \varepsilon_b = IR, \tag{11.15}$$

where E is the emf of the battery, ε_b is the back emf, and R is the coil resistance.

11.6.2 The generator

Sometimes called the **Dynamo**, the electric generator is any device that converts mechanical energy into electrical energy. The source of the mechanical energy may be high speed steam jet from a nozzle, water falling from a height on a set of turbine blades, the pedalling stroke of a bicycle, an internal combustion engine, and many more. The generator converts mechanical energy into electricity using the principles of electromagnetic induction. A changing magnetic flux linkage induces an emf in a closed conducting circuit. The device works on principles that are almost the reverse of the electric motor.

There are two types of generators: the **direct current** (or d.c.) generator, and the **alternating current** (or a.c.) generator. Consider a coil of wire placed in a magnetic field – **Fig. 11.9**. As the coil rotates, the area of the coil exposed to the magnetic field varies from a minimum value of zero, to a maximum value equal to the area A of the coil. The minimum value occurs when the plane of the coil is parallel to the magnetic field, while the maximum value obtains when the plane of the coil is perpendicular to the magnetic field. At an instant t, the effective area A_{eff} threaded by the magnetic field is related to the area of the coil by

$$A_{eff} = A\cos\theta = A\cos\omega t, \tag{11.16}$$

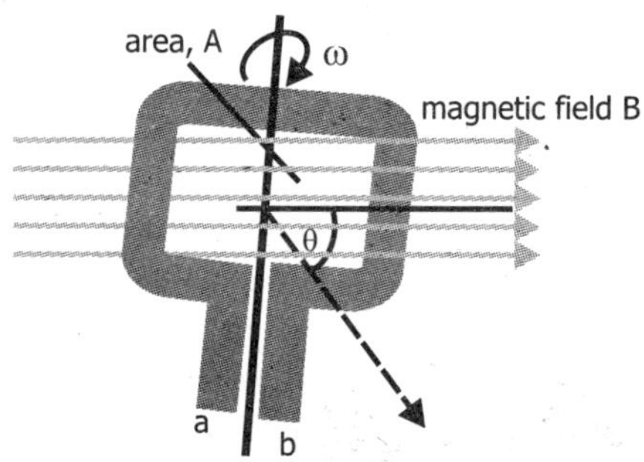

Fig. 11.9 The generator. A coil rotates about an axis in a magnetic field. The normal to the coil makes angle θ with the field.

where ω is the angular velocity of the coil, and θ is the angle between the normal to the plane of the coil and the direction of the magnetic field.

The induced emf is hence given by

$$\varepsilon = -\frac{d\Phi}{dt} = -NB\frac{dA_{eff}}{dt}.$$

Substituting for A_{eff} from (11.16), and performing the differentiation, we get

$$\varepsilon = NAB\omega \sin \omega t = \varepsilon_0 \sin \omega t. \tag{11.17}$$

The practical arrangement can either lead to a d.c. generator, or an a.c. generator.

(a) The d.c. generator

In the d.c. generator, a commutator is used to reverse the current so as to produce a voltage in the same direction at all times. **Fig. 11.10** (a) shows the practical arrangement, and (b) gives the voltage output waveforms for (i) a single-split commutator, and (ii) a commutator consisting of two pairs of metal plates. In both cases the output is not a pure d.c., but a ripple voltage. Special circuits containing combinations of capacitors and inductors are used to produce a pure d.c. voltage – a process known as "smoothing".

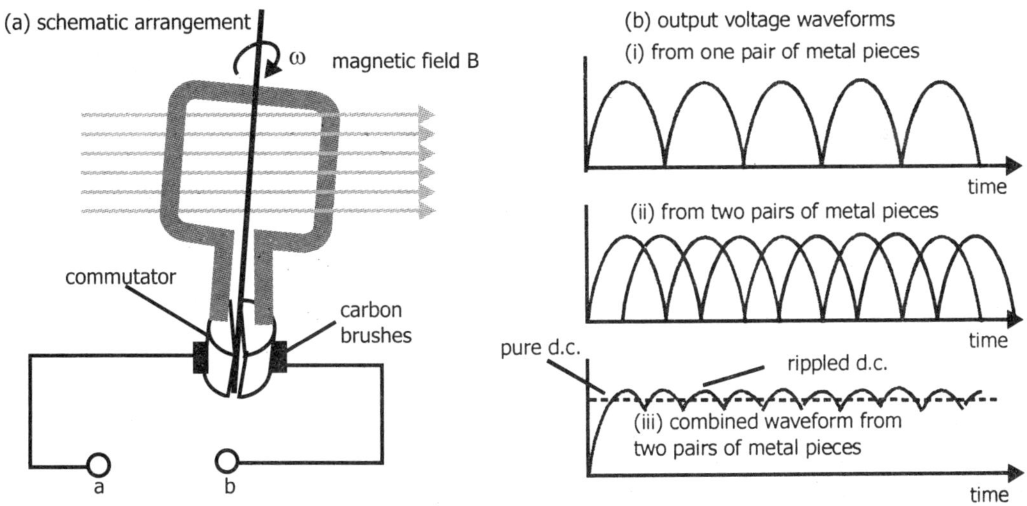

Fig. 11.10 The d.c. Generator, (a) schematic arrangement, and (b) output voltage waveforms.

(b) The a.c. generator

Fig. 11.11 is a schematic diagram of the a.c. generator. In this arrangement, the output

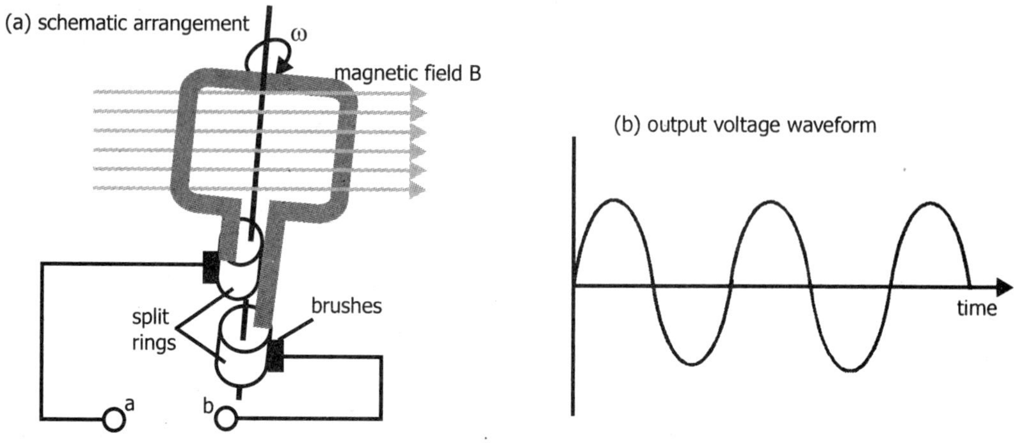

Fig. 11.11 The a.c. Generator, (a) schematic arrangement, and (b) output voltage.

voltage is obtained from a set of **split rings**. The output is varied from $-\varepsilon_0$ to $+\varepsilon_0$, in accordance with equation (11.17). Such a voltage is an **alternating voltage**. We study the behaviour of this voltage to greater detail in the next chapter.

11.6.3 The transformer

The alternating current generator is the major electric power source driving today's technology. Almost all countries in the world have invested huge amounts of money to establish and maintain power programs to drive industry and provide comfort for the home. The voltage delivered to homes and factories from power stations is either 115 V (in the Americas), or 220/240 V in most other parts of the world. This voltage is often referred to as the "mains" voltage.

The internal electronics of many appliances, e.g. cassette tape recorders, radio sets, need far less than the mains voltage to operate. By contrast, some industries may require more than this voltage. A device used to give out lower or higher voltage output from a given mains input is called the **transformer**. If the voltage output from the transformer is higher than the mains supply, the transformer is described as a **step-up** transformer. If the transformer output is lower than the mains voltage, the transformer is a **step-down** transformer.

The transformer represents the major application of mutual inductance between two coils. Basically a transformer's job is to transfer electrical energy from one circuit to another by means of electromagnetic induction. The winding connected to the energy source (or mains) is called the **primary**. The winding from which the energy is extracted to, for instance, operate the cassette tape recorder, is called the **secondary**.

Although electromagnetic coupling can be obtained by wrapping the two coils together, far greater degree of coupling is achieved when the coils are wrapped around a magnetic material. Thus, in all practical transformers we come across, the coils are wound around a soft iron core. In practice, the core is laminated, with the laminated pieces insulated from each other. This arrangement reduces losses due to eddy currents.

Fig. 11.12 (a) shows the schematics of two coils wound on a soft iron core to make a transformer, whilst (b) is the usual circuit symbol for the transformer. Transformers only work with alternating voltages. The alternating voltage through the primary induces an emf in the secondary, which in turn induces a back-emf in the primary.

If the induced emfs in primary and secondary coils by E_P, and E_S, respectively,

$$E_p = -N_p.\left(\frac{d\Phi}{dt}\right), \tag{11.18}$$

and

$$E_S = -N_S.\left(\frac{d\Phi}{dt}\right). \tag{11.19}$$

In (11.18) and (11.19), N_P and N_S are the number of turns of the primary and secondary coils, respectively, and $\left(\frac{d\Phi}{dt}\right)$ represents the commons flux change between the coils. It is assumed that there is total electromagnetic coupling between the primary and secondary coils. From these equations, we get

$$\frac{E_S}{E_p} = \frac{N_S}{N_p}. \tag{11.20}$$

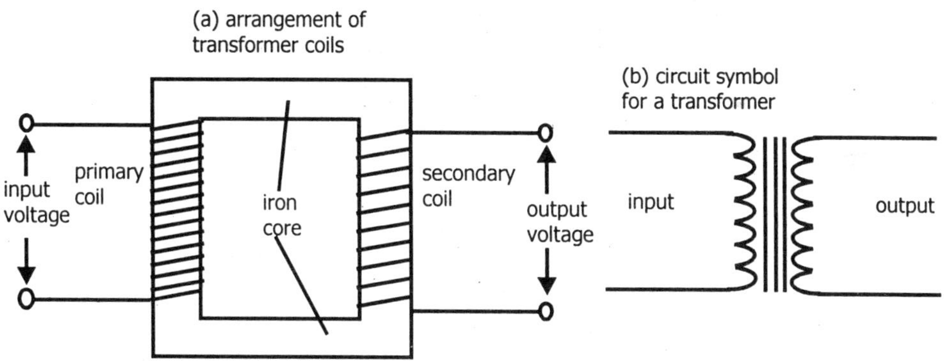

Fig. 11.12 The transformer (a) schematic diagram of the transformer. The primary and secondary coils are electromagnetically coupled together through the iron core, (b) transformer circuit symbol.

The quantity $\frac{N_S}{N_p}$ is known as the **turns ratio** of the transformer. When $\frac{N_S}{N_P} > 1$, the voltage at the secondary is greater than that at the primary coil. We therefore have a step-up transformer, in which the output voltage is greater than the input. When $\frac{N_S}{N_P} < 1$, a step-down transformer results, for which a smaller voltage appears at the secondary.

As stated above, it is assumed that there is total (i.e. 100%) electromagnetic coupling between the primary and secondary coils. Under these conditions, the power output from the secondary is equal to the power input at the primary coil. Writing these powers in terms of the respective currents and voltages, we get

$$I_S E_S = I_P E_P, \tag{11.21}$$

where I_P and I_S are the currents in the primary and secondary coils, respectively. Thus

$$\frac{I_S}{I_P} = \frac{1}{E_S/E_P}. \tag{11.22}$$

From (11.22), it follows that a step-up transformer produces less current output than the current input. Similarly, a step-down transformer produces more current output than the current input.

In the practical transformer, however, there is always some loss of electromagnetic coupling, leading to energy loss. This energy loss appears as heat, making the transformer warm during operation.

11.6.4 The induction coil

Many heavy machines require large voltages for normal operation. For some of these machines, it is desired to derive the required high voltage from a source of relatively low voltage. An everyday example of this situation is that of the motor car. To start the engine of a car into motion sometimes requires up to 4 kV to turn the flywheel. In this case the source of this high voltage is often a 12-volt battery, and an **induction coil**. The induction coil steps up the low voltage into the needed high voltage source.

Like in the transformer, the induction coil consists of a primary coil and a secondary coil. The primary coil consists of relatively few turns of heavy insulated wire, whilst the secondary coil consists of thousands of turns of insulated fine copper wire. Both the primary and secondary coils are wrapped around a cylindrical core of soft-iron wires. The primary is connected through an automatic interrupter, similar to the spring-loaded system we studied in connection with the electric bell. **Fig. 11.13** shows the circuitry for the induction coil.

On interrupting the current flow in the primary coil the current jumps across the interrupter even after the gap is opened slightly. The current thus behaves as if it possessed inertia. The result is a very high voltage across the output of the secondary coil. The higher the frequency of interruption, the higher the secondary voltage output voltage. Often a capacitor (i.e. condenser) is connected across the interrupter gap to minimize arcing during making and breaking.

As we saw in the transformer, a high output voltage means a low output current. The output current in the secondary coil is much lower than the primary current. Often in operation there is a loss of power in the windings of the coil. This appears as heat.

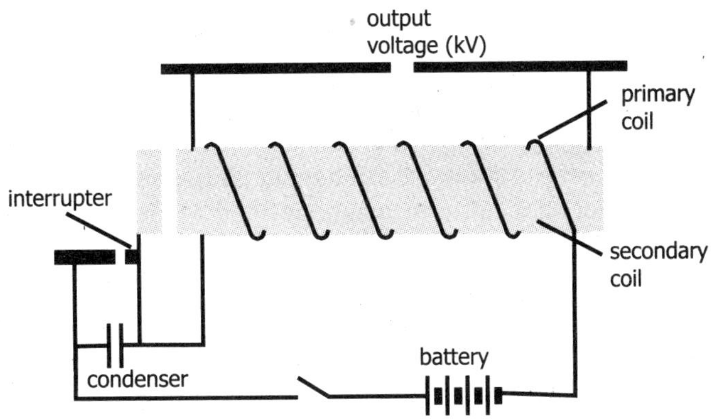

Fig. 11.13 Circuitry of the induction coil. The primary has far less turns than the secondary coil.

EXERCISES

1. A coil of area 10 cm^2 is placed with its plane perpendicular to a uniform magnetic field of 0.10 T. The coil is removed to a field-free region in a time of 1.0 ms. Calculate the magnitude of the induced emf in across the coil.

2. A rectangular coil of area 50 cm^2 has 100 turns of conducting wire of resistance 1.5 Ω, connected to a 10 V supply. The coil is free to rotate about an axis joining the midpoints of two opposite sides. When the coil is placed at right angles to a uniform magnetic field, a torque of magnitude 0.30 Nm is experienced. Calculate the magnetic field.

3. A flat coil of area 10 cm^2 with 500 turns rotates at 100 revolutions per minute about an axis at right angles to a uniform magnetic field of 0.5 T. Calculate
   (i) the maximum value of the emf induced in the coil,
   (ii) the instantaneous value of the emf when the plane of the coil is at an angle of 60° to the magnetic field.

4. A back emf of 20 mV is induced in the coil of an electric motor when the current in the coil is reduced from 2.5 A to 1.0 A in a time of 1.0 ms. Calculate the self induction of the coil.

5. A coil of area 20 cm^2 is placed in a uniform magnetic field of 0.5 T, with the field perpendicular to the plane of the coil. When the coil is moved to a field-free region, an emf of 2.0 V is induced in the coil. Determine the time taken to move the coil.

6. A horizontal metal bar of length 1.5 m falls from rest under gravity in the earth's

magnetic field of 1.70×10^{-7} T, in a direction perpendicular to the earth's magnetic field. Calculate the potential difference across the ends of the bar when it had fallen through a distance of 2.0 m.

7. A flat circular coil of mean radius 5.0 cm having 50 turns of conducting wire, is made to rotate about a diameter in a uniform magnetic field of flux density 0.4 T. The axis of rotation is set perpendicular to the magnetic field direction, and the coil rotates with an angular speed of 10 rad/s. Calculate
   (i) the peak value of the induced emf in the coil,
   (ii) the peak current flowing in the coil, given that the coil resistance is 0.5 Ω/cm.

8. A flat square coil of side 0.10 m and 20 turns, rotates in a uniform magnetic field of flux density 0.1 T. The coil is linked on to an external circuit with resistance 25 Ω. Calculate
   (i) the angular speed of the coil if the maximum induced emf is 30 mV,
   (ii) the maximum current in the external circuit.

9. A coil of 6 cm^2 cross-section, having 200 turns of wire, rotates in a uniform magnetic field of 0.5 T at a speed of 180 revolutions per minute. Determine the maximum emf induced in the coil.

10. A portable grinder operates with input voltages not exceeding 25 V. It is connected to the mains through a step-down transformer with 500 turns in the primary coil. Calculate the number of turns of the secondary coil if the input mains voltage is
    (i) 220 V,
    (ii) 115 V.

11. Explain the terms **self inductance**, and **mutual inductance**. Derive expressions for the effective inductance of a number of non-interacting coils connected
    (i) in series
    (ii) in parallel

12. Outline the principles of operation of the following electromagnetic devices
    (i) the transformer,
    (ii) the a.c. generator,
    (iii) the d.c. motor.

12

CHAPTER

Alternating Currents and Voltages

12

CHAPTER

Alternating Currents and Voltages

Commonly abbreviated A.C., an alternating current or alternating voltage (A.V.) is a current or voltage that reverses direction at regular intervals. The electricity that is supplied by power stations to our homes, factories, and laboratories is alternating voltage, as opposed to the direct current from storage batteries. We shall learn later why it is more convenient and technically better to produce and transmit alternating voltages, compared to direct voltages. For the moment, however, our study in this chapter would be confined to the behaviour of circuit components when alternating voltages are applied to them.

12.1 THE ALTERNATING CURRENT AND VOLTAGE

In our study of the generator in the last chapter, we saw how an alternating voltage is generated by rotating a coil in a magnetic field. The voltage output V from the generator varies with time according to the equation

$$V = V_0 \sin \omega t, \tag{12.1}$$

where ω is the angular frequency of rotation of the coil, and V_o is the amplitude of the voltage, known more commonly as the **peak voltage**. Similarly, the current I from the generator can be written as

$$I = I_0 \sin \omega t. \tag{12.2}$$

In (12.2) I_o is the **peak current**. A plot of the alternating voltage waveform is shown in **Fig. 12.1**

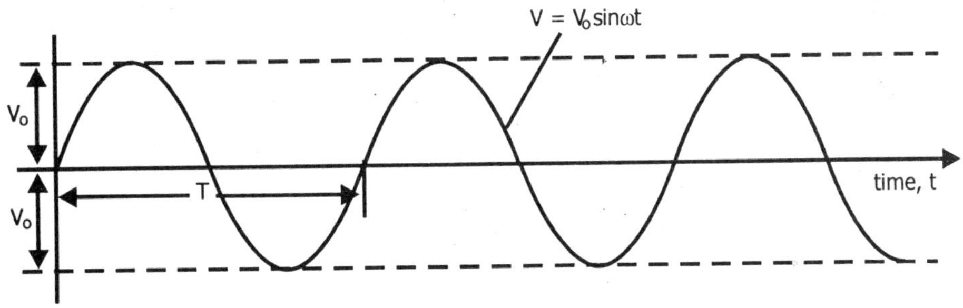

Fig. 12.1 Alternating voltage waveform, showing the period T, and the peak voltage V_o.

12.1.1 Effective values of voltage and current

From **Fig. 12.1**, in one-half cycle, an alternating voltage is positive, and negative in the next half-cycle. The voltage continuously varies with time over the period. The actual voltage (and current) measurement at any time gives its **effective value**. Also known as the **root-mean-square** value, the effective voltage V_{eff} is defined by the equation

$$V_{rms} = V_{eff} = \sqrt{\text{mean of the squares of the voltages over one cycle}}.$$

Using the equation for the alternating voltage, we get

$$V_{rms} = \sqrt{\overline{(V_o \sin \omega t)^2}} = V_o \sqrt{\overline{(\sin \omega t)^2}} = V_o \sqrt{\overline{(\sin^2 \omega t)}}.$$

The horizontal lines over the quantities under the square root indicate that we are finding the average value of the quantity over one cycle. Over one period (or cycle) of the alternating voltage, $\sin^2 \omega t$ varies continuously between 0 and 1, with an average of $\frac{1}{2}$. We hence have the effective voltage as

$$V_{rms} = V_{eff} = \frac{V_o}{\sqrt{2}}. \qquad (12.3)$$

Similarly, the effective value of the current is given by

$$I_{rms} = I_{eff} = \frac{I_o}{\sqrt{2}}. \qquad (12.4)$$

Equations (12.3) and (12.4) represent the voltage and current delivered from the supply. These are the values that would be given by a device measuring the voltage or current.

Example 12.1

An alternating current passes through a coil of resistance 3.0 Ω in a water container from which heat losses may be neglected. The thermal capacity of the whole arrangement is 840 J/K. If the temperature of the water rises from 5.2 °C to 27.3 °C in 5 minutes, calculate the r.m.s. value of the alternating current in the coil.

Solution

We equate the electrical energy from the coil to the thermal energy. In a time interval t, the electrical energy consumed = $I_{rms}^2 Rt$, where I_{rms} is the r.m.s. value of the current, and R is the coil resistance. The heat energy U is given by $U = h\Delta T$, where h is the thermal capacity of the arrangement, and ΔT is the rise in temperature. Thus,

$$I_{rms}^2 Rt = h\Delta T, \text{ giving } I_{rms} = \sqrt{\frac{h\Delta T}{Rt}} = \sqrt{\frac{840 \text{ J/K} \times (27.3 - 5.2)\text{K}}{3\,\Omega \times 300\text{s}}} = 4.54 \text{ A}.$$

12.1.2 Phase difference between current and voltage

From our study of direct current voltages, we know that the current through a circuit component depends on the resistance that component presents to current flow. For resistors, this current is proportional to the resistance of the element. Similarly, for alternating currents, the current depends on the resistance of the component. In addition, for alternating voltages, the current can be either in phase, or out of phase with the voltage. It therefore follows that if an alternating voltage $V = V_0 \sin \omega t$ is applied to circuit elements, the current I is given by the general relation

$$I = I_0 \sin(\omega t + \phi), \tag{12.5}$$

where ϕ is the phase angle between the current and the applied voltage. The phase difference depends on the elements in the circuit, and how these elements are connected to each other.

12.2 A.C. APPLIED TO RESISTOR, INDUCTOR, AND CAPACITOR

We begin our study of the behaviour of circuit elements to alternating voltages by considering the single elements. Later, we expand to the study of the behaviour of combinations of these components.

12.2.1 A.C. applied to resistor

Fig. 12.2 is a circuit showing a resistor R connected across an alternating voltage supply. We see from this diagram the usual circuit symbol for the alternating voltage. We apply Ohm's law to give the current I, as

$$I = \frac{V}{R} = \frac{V_0 \sin \omega t}{R}. \tag{12.6}$$

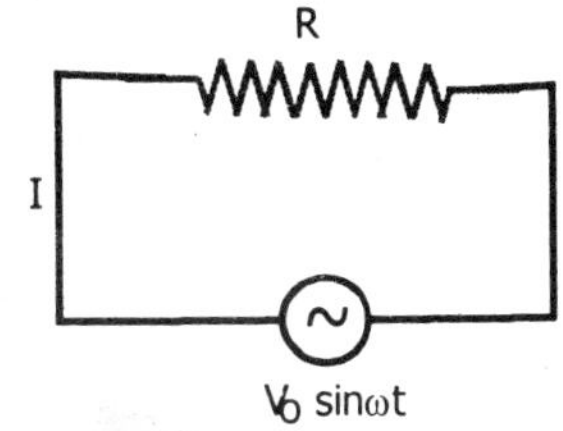

Fig. 12.2 Alternating voltage applied to a resistor.

From equation (12.6) we note the following about the current:

1. the current through the circuit is directly proportional to the resistance. This is the same as we saw in direct current circuits.
2. the current is **in phase** with the alternating voltage.

Example 12.2

The alternating current given by $I = 2.8 \sin 100\pi t$ amps is applied to an 80 Ω resistor. Write down

(i) the peak value of the current,
(ii) the r.m.s. value of the current,

(iii) the frequency of the input a.c.,
(iv) the waveform of the p.d. across the resistor.

Solution

We compare the given a.c. with the general alternating current $I = I_0 \sin \omega t$, and hence determine the parameters. From the current expression $I = 2.8 \sin 100\pi t$, we write

(i) peak value of the current I_o = 2.8 amps.

(ii) the r.m.s. value of the current $I_{rms} = \frac{I_0}{\sqrt{2}} = \frac{2.8}{1.414} = 1.98$ A.

(iii) the frequency f of the input a.c. is given by $2\pi f = 100\pi$, or f = 50 Hz.

(iv) the p.d. across the resistor is given by $V_R = IR = 224 \sin 100\pi t$.

12.2.2 A.C. applied to pure inductor

From our study in the last chapter, we found that an inductor is simply a coil of wire, having a resistance and an inductance when current flows through it. For our present purposes, we consider that the effect of the resistance is small compared to that of the inductance. This assumption appears reasonable, since most inductors have a length of wire short enough for the resistance to be considered negligible. **Fig. 12.3** shows an inductor connected across an a.c. supply voltage.

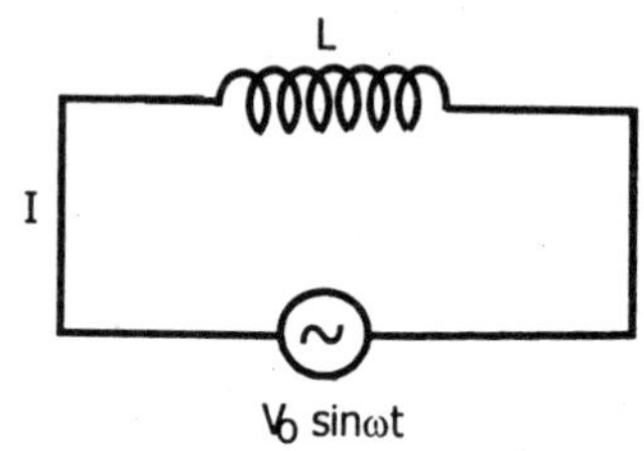

Fig. 12.3 Alternating voltage applied to a pure inductor.

Thus, we write

$$L\frac{dI}{dt} = V_0 \sin \omega t,$$

giving

$$I = -\frac{V_0}{\omega L}\cos \omega t = \frac{V_0}{\omega L}\sin\left(\omega t - \frac{\pi}{2}\right) \qquad (12.7)$$

From (12.7), it follows that the current through the inductor **lags behind** the alternating voltage by $\pi/2$. Alternatively, if an alternating current passes through an inductor, the p.d. developed across the inductor **leads** the current by $\pi/2$. Comparing equations (12.6) and (12.7), we see that the ωL plays a role similar to R in the resistive circuit. This quantity is known as the **inductive reactance**, and can be considered as a measure of the opposition to the flow of current by the inductor. The difference, however, is that unlike the resistor, this opposition depends on the frequency of the alternating voltage – the higher the frequency the larger the inductive reactance.

12.2.3 A.C. applied to capacitor

Fig. 12.4 shows a series circuit with a capacitor across the terminals of an a.c. supply. Proceeding in a manner similar to the previous two cases, we obtain an expression for the potential difference across the capacitor.

If Q is the charge stored in the capacitor,

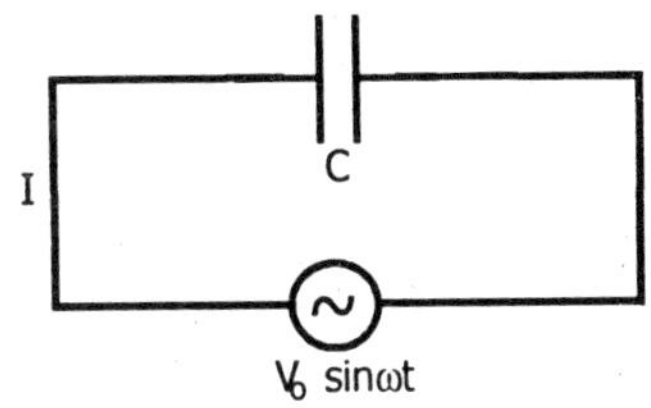

Fig. 12.4 A.C. applied to capacitor.

$$\frac{Q}{C} = V_0 \sin \omega t. \quad (12.8).$$

Differentiating both sides of (12.8), and bearing in mind that current is the rate of flow of charge, we get the current I as

$$I = \omega C V_0 \cos \omega t = \frac{V_0}{1/\omega C} \sin(\omega t + \pi/2) \quad (12.9)$$

From (12.9), it is seen that the current through the capacitor **leads** the alternating voltage by $\pi/2$. The quantity $1/\omega C$ is known as the **capacitive reactance**. Like the inductive reactance, the capacitive reactance depends on the frequency of the alternating voltage – the lower the frequency the higher the capacitive reactance.

Both the inductive and capacitive reactances perform roles similar to that of a resistor in a circuit. As a result, the units of reactance is the same as that of resistance – i.e. Ω (ohm). We study the relevance of the phase relations further in the next section. The normal symbol used to represent a reactance is X.

Example 12.3

The alternating current given in **Example 12.2** is applied to (a) an inductor of inductance 20 mH, and (b) a capacitor of capacitance 50 μF. Determine for each component

(i) the reactance of the circuit,
(ii) the p.d. across the component.

Solution

We compare the given a.c. with the general alternating current $I = I_0 \sin \omega t$, and hence determine the parameters. From the current expression $I = 2.8 \sin 100\pi t$, we write $\omega = 100\pi$. Thus, we have the following:

(a) for the inductor,
   (i) the reactance is given by $X_L = \omega L = 100\pi \times 0.020 = 6.3\,\Omega$
   (ii) the p.d. across the inductor is $V_L = I_{rms} X_L = 0.7071 \times 2.8 \times 6.3 = 12.47\ V$

(b) for the capacitor,

(i) the reactance is given by $X_C = \frac{1}{\omega C} = \frac{1}{100\pi \times 50 \times 10^{-6}} = 63.7\,\Omega$

(ii) the p.d. across the inductor is $V_C = I_{rms} X_C = 0.7071 \times 2.8 \times 63.7 = 126.1\,V$

12.3 VECTOR IMPEDANCE DIAGRAMS

The phase relationships obtained between current and voltage for the resistor, inductor, and capacitor enable us to represent the currents and voltages in these elements in terms of vectors. Such diagrams are called **vector impedance diagrams**. These are shown in **Fig. 12.5**.The word **impedance** is used for the overall opposition to current flow in a circuit containing combinations of resistors, capacitors, and inductors.

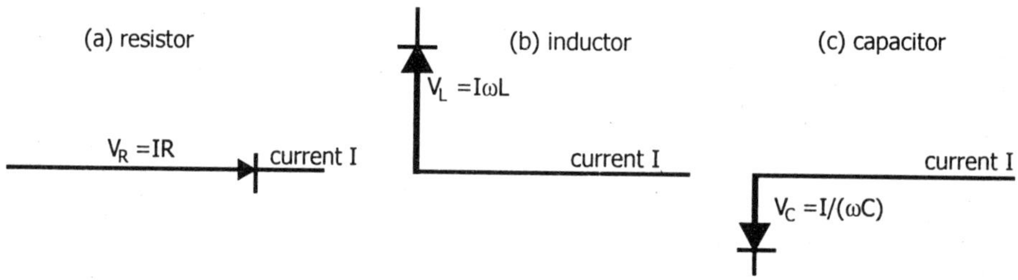

Fig. 12.5 Vectors representing the potential differences across (a) resistor, (b) inductor, and (c) capacitor. The current direction in each case is represented by the baseline.

In a vector impedance diagram, the baseline (or horizontal) is used for the current direction. The p.d. across the resistor is represented by a vector points in this direction, since the p.d. across the resistor is in phase with the alternating current. The p.d. across an inductor is represented by a vector in the positive y-direction, whilst the p.d. across a capacitor gives a vector in the negative y-direction.

12.4 LR, RC, AND RLC SERIES CIRCUITS

In this section we apply the concept of vector impedance diagrams to series combinations of resistors, inductors, and capacitors.

12.4.1 The RL circuit

Fig. 12.6 shows a circuit containing a resistor and a pure inductor connected in series across a source of alternating current. The sum of the potential differences across the resistor and the inductor is equal to the supply voltage.

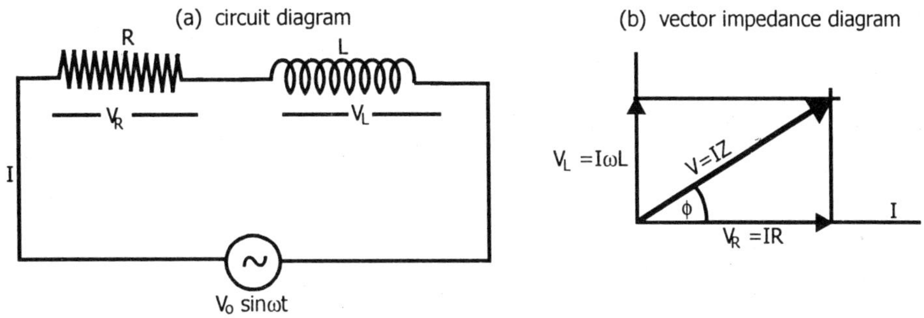

Fig. 12.6 Inductor and resistor connected in series across an alternating supply.

Thus, we have $V_L + V_R = V_0 \sin \omega t$. Writing the expressions for these potentials, we get

$$L\frac{dI}{dt} + IR = V_0 \sin \omega t, \tag{12.10}$$

where L and R are the inductance, and resistance, respectively.

Even though equation (12.10) can be solved analytically to get the current flowing through the circuit, it is easier to employ vector impedances. **Fig. 12.6** (b) shows the vector impedance diagram. The overall p.d. across the combination is

$$V = IZ = \sqrt{V_R^2 + V_L^2}.$$

Expressing V_R and V_L in terms of the current and the resistance (or reactance), we get

$$V = IZ = I\sqrt{R^2 + (\omega L)^2}. \tag{12.11}$$

Thus,

$$Z = \sqrt{R^2 + \omega^2 L^2}, \tag{12.12}$$

and the phase angle ϕ between the current and the voltage is given by

$$\tan\phi = \frac{\omega L}{R},$$

or

$$\phi = \tan^{-1}\left(\frac{\omega L}{R}\right). \tag{12.13}$$

In terms of the supply voltage, the current through the combination is given by

$$I = \frac{V_0}{Z}\sin(\omega t - \phi) = \frac{V_0}{\sqrt{R^2 + \omega^2 L^2}}\sin(\omega t - \phi). \tag{12.14}$$

Alternatively, for an alternating current supply $I = I_0 \sin \omega t$, the potential difference V across the RL circuit is given by

$$V = I_0 Z \sin(\omega t + \phi) = I_0 \sqrt{R^2 + \omega^2 L^2} . \sin(\omega t + \phi) \quad (12.15)$$

Example 12.4

The alternating voltage $310 \sin 100\pi t$ is applied to a series circuit consisting of a 100 Ω resistor, and an inductor of 200 mH. Determine

(i) the impedance of the combination,
(ii) the r.m.s. current flow in the circuit,
(iii) the power dissipated in the circuit.

Solution

From the data given, $\omega = 100\pi = 314.2\ s^{-1}$, $V_o = 310$ Volts, R = 100 Ω, L = 0.2 H.

(i) the impedance Z of the RL series combination is given by $Z = \sqrt{R^2 + \omega^2 L^2}$. Substituting the values of the quantities from the data we get the impedance Z = 118.5 Ω.

(ii) The current I in the circuit is given by $I = \frac{V_0}{Z} \sin(100\pi t - \phi)$, where the phase difference ϕ is given by $\phi = \tan^{-1}\left(\frac{\omega L}{R}\right) = \tan^{-1}\left(\frac{100\pi \times 0.2}{100}\right) = 0.56$ radians. The current has a peak value $I_0 = \frac{V_0}{Z}$, and r.m.s. value $I_{rms} = \frac{1}{\sqrt{2}} \frac{V_0}{Z}$. Substituting the data values, and calculating, we get $I_{rms} = 1.85$ A.

(iii) The power factor is given by $\cos\phi = 0.847$. The power dissipated in the circuit is obtained from $P = I_{rms}^2 Z \cos \phi = 1.85^2 \times 118.5 \times 0.847 = 343$ watts.

12.4.2 The RC circuit

In a manner similar to the RL circuit, we proceed to find the combined effect of a resistor and a capacitor connected in series across an alternating supply. **Fig. 12.7** (a) shows the circuit diagram, whilst (b) shows the vector impedance diagram.

If Q is the instantaneous charge in the capacitor, we write $RI + \frac{Q}{C} = V_0 \sin \omega t$, giving

$$R\frac{dI}{dt} + \frac{I}{C} = \omega V_0 \cos \omega t.$$

Again, we adopt the much easier method of vector impedances to solve this latter equation for the current. From the vector diagram,

$$Z = \sqrt{R^2 + \left(1/\omega C\right)^2}, \tag{12.16}$$

and

$$\phi = \tan^{-1}\left(\frac{1/\omega C}{R}\right). \tag{12.17}$$

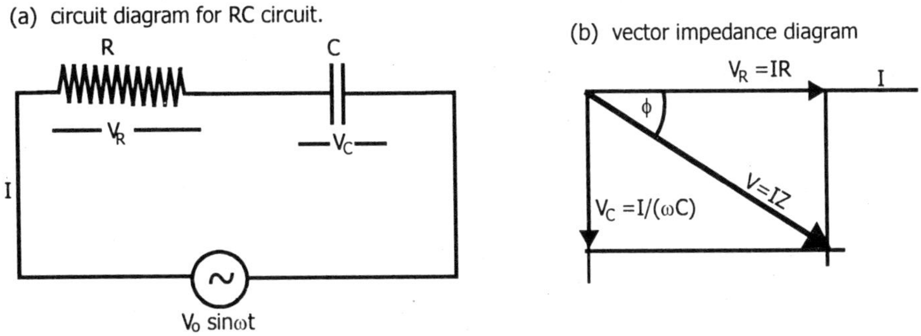

Fig. 12.7 Resistor and capacitor connected in series across an alternating supply.

For an alternating voltage supply $V = V_0 \sin \omega t$, the current flow is given by

$$I = \frac{V_0}{Z}\sin(\omega t + \phi) = \frac{V_0}{\sqrt{R^2 + \dfrac{1}{\omega^2 C^2}}}\sin(\omega t + \phi). \tag{12.18}$$

For an input alternating current $I = I_0 \sin \omega t$, the voltage across the combination is

$$V = I_0 Z \sin(\omega t - \phi) = I_0\sqrt{R^2 + \frac{1}{\omega^2 C^2}}.\sin(\omega t - \phi). \tag{12.19}$$

From equations (12.13) and (12.17), it follows that the phase angle ϕ is given by $\tan\phi = \dfrac{\text{reactance}}{\text{resistance}}$. For any combination of resistances and reactances in a circuit, the phase angle between the alternating current and voltage is given by

$$\phi = \tan^{-1}\frac{X}{R}, \tag{12.20}$$

and the impedance Z is given by

$$Z = \sqrt{R^2 + X^2}. \tag{12.21}$$

Example 12.5

The a.c. source of $2.8\sin 100\pi t$ is applied to a series combination consisting of a capacitor of capacitance 50.0 μF and a resistance of 80 Ω. Determine

(i) the capacitive reactance,
(ii) the impedance of the combination,
(iii) the potential difference measured across the series combination,
(iv) the phase angle between the current and voltage across the combination,
(v) the potential differences measured across the resistor, and the capacitor,
(vi) the voltage waveform across the combination.

Solution

In **Fig. 12.8**, we have the circuit diagram for the combination, and the diagram showing the vector impedances. From the given a.c. waveform, we set $\omega = 100\pi = 314.2\ \text{s}^{-1}$.

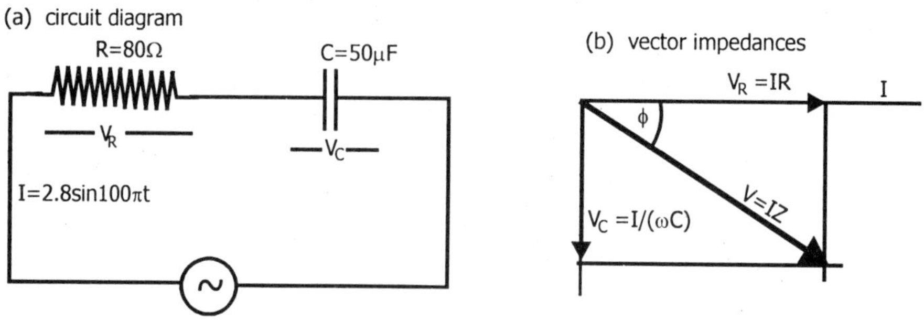

Fig. 12.8

(i) Capacitive reactance $X_C = \dfrac{1}{\omega C} = \dfrac{1}{314.2\ \text{s}^{-1} \times 50 \times 10^{-6}\ \text{F}} = 63.7\ \Omega.$

(ii) From the corresponding vector impedance diagram, the impedance Z of the combination is given by

$$Z = \sqrt{R^2 + X_C^2} = \sqrt{80^2 + 63.7^2} = 102.2\ \Omega.$$

(iii) the potential difference V measured across the combination is given by
$V_{rms} = I_{rms}Z = 1.98\ \text{A} \times 102.2\ \Omega = 202.35\ \text{V}.$

(iv) the phase angle ϕ between the alternating current and the voltage across the combination is given by $\phi = \tan^{-1}\left(\dfrac{X_C}{R}\right) = \tan^{-1}\dfrac{63.7}{80} = 0.796$ radians (or 45.6°).

(v) The potential difference measurements would indicate the r.m.s. values of the voltages across the components.

For the resistor, we have $V_R = I_{rms}R = 1.98\ A \times 80\ \Omega = 158.4\ V$.
For the capacitor, $V_C = I_{rms}X_C = 1.98\ A \times 63.7\ \Omega = 126.13\ V$.

We note here that the sum of the potential difference measurements across the components is greater than the potential difference across the combination, obtained in (iii). This results from the difference in phase between the overall voltage and the alternating current.

(vi) The general expression for the alternating voltage waveform across the combination is given by $V = I_0 Z \sin(\omega t - \phi)$, the symbols having their usual meanings. From the figures obtained for this circuit, we write the output waveform as follows:
$V = I_0 Z \sin(\omega t - \phi) = 2.8 \times 102.2 \sin(100\pi t - 0.796) = 286.16 \sin(100\pi t - 0.796)$ Volts.

12.4.3 The series RLC circuit

The analysis of the series RLC is slightly more complicated than the previous two connections. The method of solution, however, is the same. In **Fig. 12.9**, we show the circuit in (a), whilst the vector impedance diagram is presented in (b).

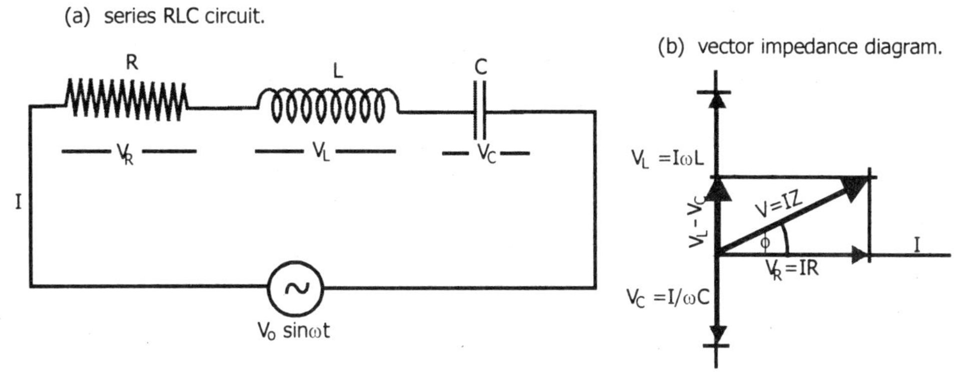

Fig. 12.9 Series circuit consisting of resistor, inductor and capacitor connected in series across an alternating voltage supply. The vector impedance diagram is given in (b).

Since the p.d. across the inductor is opposite in direction to that across the capacitor, we subtract the vectors representing the potential differences across the two reactances, and combine the result with the vector representing the potential across the resistor. We thus have

$$Z = \sqrt{R^2 + (X_L - X_C)^2} = \sqrt{R^2 + \left(\omega L - \frac{1}{\omega C}\right)^2}, \tag{12.22}$$

where X_L and X_C are the inductive and capacitive reactances, respectively, and R is the total resistance in the circuit. The phase angle ϕ is given by

$$\phi = \tan^{-1}\left(\frac{X_L - X_C}{R}\right) = \tan^{-1}\left(\frac{\omega L - 1/\omega C}{R}\right). \qquad (12.23)$$

The current I through the circuit for an input voltage $V = V_0 \sin \omega t$, is given by

$$I = \frac{V_0}{Z}\sin(\omega t + \phi) = \frac{V_0}{\sqrt{R^2 + \left(\omega L - \frac{1}{\omega C}\right)^2}}\sin(\omega t + \phi). \qquad (12.24)$$

For an alternating current $I = I_0 \sin \omega t$, the overall potential difference V across the combination is given by

$$V = I_0 Z \sin(\omega t - \phi) = I_0\sqrt{R^2 + \left(\omega L - \frac{1}{\omega C}\right)^2}\sin(\omega t - \phi). \qquad (12.25)$$

In this analysis, we have taken the inductive reactance to be greater than the capacitive reactance. We would still have the same impedance if the reverse were true.

Example 12.6

The alternating voltage $160\sin 120\pi t$ is applied to a series circuit consisting of a 100 Ω resistor, a 50 μF capacitor, and an inductor of 200 mH. Determine

(i) the impedance of the combination,
(ii) the r.m.s. current flow in the circuit,
(iii) the expression for the current flow through the circuit.

Solution

Using the information, $\omega = 120\pi = 377.04\ s^{-1}$; $C = 50\ \mu F$; $R = 100\ \Omega$; $L = 0.20$ H, and $V_o = 160$ Volts.

(i) We first calculate the reactance. $X_C = \frac{1}{\omega C} = \frac{1}{377.04\ s^{-1} \times 50 \times 10^{-6} F} = 53\ \Omega.$,

$X_L = \omega L = 377.04\ s^{-1} \times 0.2\ H = 75.4\ \Omega$. Thus, $X = X_L - X_C = 22.4\ \Omega.$.

The impedance Z is hence given by $Z = \sqrt{R^2 + X^2}$. Substituting the numerical values obtained above, we get $Z = 102.5\ \Omega$.

(ii) The r.m.s. current through the circuit is given by $I_{rms} = \frac{V_{rms}}{Z} = \frac{V_0}{Z\sqrt{2}}$.

Substituting the numerical values, we get $I_{rms} = \frac{160\ \text{Volts}}{102.5\ \Omega \times 1.414} = 1.1$ A.

(iii) The phase angle ϕ is given by $\phi = \tan^{-1}\left(\frac{X}{R}\right) = \tan^{-1}\left(\frac{22.4}{100}\right) = 0.22$ radians.

Thus the current waveform is given by $I = \frac{V_0}{Z}\sin(\omega t - \phi) = 1.56\sin(120\pi t - 0.22)\,A$.

12.4.4 Power in a.c. circuits

An alternating voltage $V_0 \sin\omega t$ applied to a circuit produces a current given by $I = I_0 \sin(\omega t + \phi)$, where I_0 is the peak value of the current, and ϕ is the phase angle between the current and the alternating voltage. The power p dissipated in the circuit at any instant is expressed in terms of the current and the voltage by

$$p = VI = V_0 \sin\omega t . I_0 \sin(\omega t + \phi) = I_0 V_0 \sin\omega t . \sin(\omega t + \phi).$$

Using trigonometric identities (see Appendix 3), we write the above expression as

$$p = I_0 V_0 \left[\sin^2 \omega t . \cos\phi + \frac{1}{2}\sin 2\omega t . \sin\phi\right]. \quad (12.26)$$

Over one cycle of the alternating voltage, the average power dissipated is given by

$$P = \bar{p} = I_0 V_0 \left[\cos\phi . \overline{\sin^2 \omega t} + \frac{1}{2}\sin\phi . \overline{\sin 2\omega t}\right]. \quad (12.27)$$

In (12.27), it is seen that because ϕ is constant, the trigonometry functions of ϕ have been taken out of the averaging process. Over one cycle, $\sin 2\omega t$ varies continuously from -1 to $+1$, with an average of 0. On the other hand, $\sin^2 \omega t$ varies continuously from 0 to $+1$, with an average of $\frac{1}{2}$. Putting these average values into (12.27), we get the power dissipated in one cycle in an a.c. circuit as

$$P = \frac{1}{2} I_0 V_0 \cos\phi. \quad (12.28)$$

In terms of the r.m.s. values of the current and voltage, equation (12.28) becomes

$$P = I_{rms} V_{rms} \cos\phi. \quad (12.29)$$

The quantity $\cos\phi$ is known as the **power factor** of the circuit. For a given current and voltage, maximum power is derived from a circuit when $\phi = 0$ (i.e. the current is in phase with the voltage).

Equation (12.29) can be written in alternative forms using the connection between the current, the voltage, and the impedance. Thus we have

$$P = I_{rms}^2 Z \cos\phi = \frac{V_{rms}^2}{Z} \cos\phi. \tag{12.30}$$

12.4.5 Series resonance

In **Fig. 12.10** we reconstruct the series circuit with resistor, inductor, and capacitor, this time across a supply voltage of variable frequency. In this illustration, we measure the current flowing through the series combination as the frequency of the supply voltage is

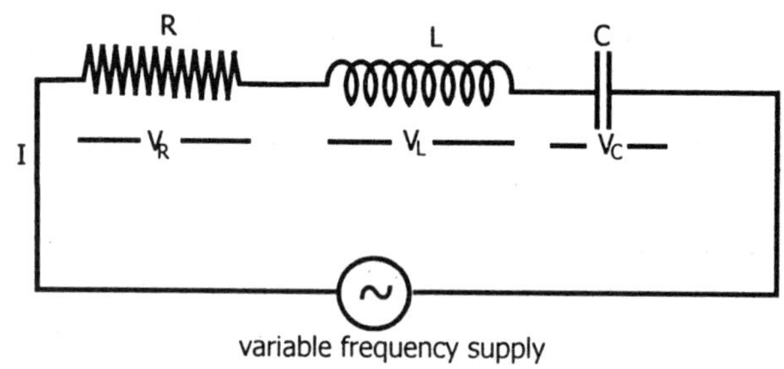

Fig. 12.10 The series RLC circuit across a variable frequency supply.

slowly increased. Alternatively, using fixed component values, we could measure the p.d. across each component as the frequency of the supply is varied. In the latter case, it is seen that the p.d. across the inductor is small at low frequencies, and increases as the frequency of the supply increases. By contrast, the p.d. across the capacitor is large at low frequencies, and drops off as the frequency is increased. The impedance of the combination at any frequency $f\ \left(= 2\pi/\omega\right)$ of the supply is given by

$$Z = \sqrt{R^2 + \left(\omega L - \frac{1}{\omega C}\right)^2}. \tag{12.31}$$

At low frequencies $1/\omega C > \omega L$; the reactance of the inductor-capacitor combination is mainly due to the capacitor. At high frequencies, however, $\omega L > 1/\omega C$, making the overall reactance mainly inductive. In between these two extremes, there is a frequency at which the inductive reactance is equal to the capacitive reactance, for the given values of the capacitance and the inductance. At this frequency, the impedance in the circuit is a minimum, making the current a maximum. This condition is described as **resonance**, and the frequency of the alternating voltage at which it occurs is called the **resonance frequency**. **Fig. 12.11** (a), (b), and (c) show the variations with frequency of the r.m.s. current, the inductive and capacitive reactances, and the impedance.

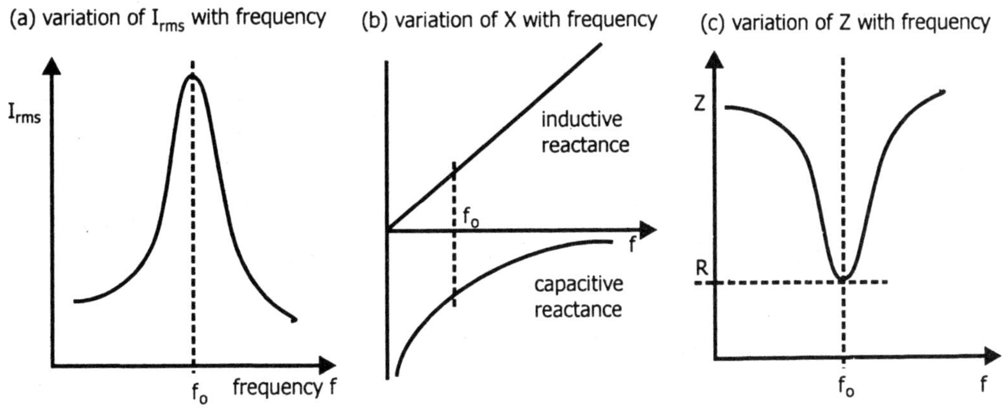

Fig. 12.11 Behaviour of series resonance circuit. Frequency variations of (a) current, (b) reactance, and (c) impedance. At resonance, (i) the current is a maximum, (ii) the inductive reactance is equal and opposite to capacitive reactance, and (iii) the impedance is a minimum.

The angular frequency ω_o ($=2\pi f_o$) for resonance is obtained by setting the reactance in (12.31) to zero. Thus, we have $\omega_0 L = \frac{1}{\omega_0 C}$, giving

$$\omega_0 = \frac{1}{\sqrt{LC}}. \qquad (12.32)$$

The resonance frequency is hence given by

$$f_0 = \frac{1}{2\pi\sqrt{LC}}. \qquad (12.33)$$

The following conditions apply at resonance:

1. The impedance of the circuit is a minimum. This impedance is purely resistive. The circuit behaves as though there is no inductance or capacitance present.
2. The current in the circuit is a maximum.
3. The power factor is unity. This results from the fact that, with the impedance being purely resistive, the current flow in the combination is in phase with the supply voltage.

Example 12.7

At 1.0 kHz, the r.m.s. voltages across a series combination consisting of a 100 Ω resistor, a capacitor, and an inductor are respectively 10.0 V, 5.0 V, and 15.0 V. Calculate

(i) the capacitance and inductance in the combination,

(ii) the resonance frequency f_0.

Solution

(i) the current through the series combination can be obtained from the p.d. across the resistor V_R and the resistance. Thus $I = \frac{V_R}{R} = \frac{10.0\ V}{100\ \Omega} = 0.10$ A. Using this current, we find the capacitance C of the capacitor, and the inductance L of the inductor.

For the capacitor, $V_C = \frac{I}{\omega C}$, or $C = \frac{I}{2\pi f V_C} = \frac{0.10}{2 \times 3.142 \times 1000 \times 5.0} = 3.18\ \mu F$.

For the inductor, $V_L = I\omega L$, or $L = \frac{V_L}{2\pi f I} = \frac{15.0}{2 \times 3.142 \times 1000 \times 0.10} = 23.8$ mH.

(ii) Maximum current flows at the resonance frequency f_o, given by $f_o = \frac{1}{2\pi\sqrt{LC}}$.

With the values of L and C from above, $f_o = \frac{1}{2\pi\sqrt{0.0238 \times 0.00000318}} = 578.4$ Hz.

12.4.6 Tuning circuits

The main use of the series RLC circuit is in tuning. In this, the capacitor or inductor in the series circuit is varied so as to obtain resonance with an incoming alternating voltage of a given frequency. This alternating voltage is usually made up of signals from a radio transmitting station. At resonance, the current in the series circuit is a maximum, thereby producing maximum p.d. across the resistor. In a radio receiver, this p.d. is then amplified and fed to the loudspeaker.

Tuning may be achieved by adjusting the value of the capacitor, or of the inductor. These modes of tuning are accordingly known as capacitive tuning, and inductive tuning. Inductive tuning is commonly done by altering the position of a piece of soft iron or other ferromagnetic material relative to the coil in the circuit.

When a series circuit is tuned to some resonance frequency, there may often be large potential differences across the capacitor and the inductor. These two potential differences cancel each other out, leaving only the that across the resistor. The ratio of the voltage V_L dropped across L (or C) to that dropped across the resistor at resonance is

$$\left(\frac{V_L}{V_R}\right)_{res} = \frac{\omega_0 L I}{R I} = \frac{\omega_0 L}{R} = \frac{1}{R}\sqrt{\frac{L}{C}}. \qquad (12.34)$$

This ratio is known as the **quality factor** (or Q-factor) Q of the circuit, and defines the sharpness of resonance. The smaller the Q-factor, the sharper the tuning. Thus,

$$Q = \frac{1}{R}\sqrt{\frac{L}{C}}. \qquad (12.35)$$

Since it is the ratio of two similar quantities the quality factor is a pure number.

Example 12.8

A capacitive tuning circuit is to receive a radio signal of frequency 1.205 MHz. The circuit contains a series RLC combination with a 22 kΩ resistor, an inductor of 0.05 mH, and the variable capacitor. Calculate

(i) the capacitance required to receive the signal,
(ii) the quality factor of the tuned circuit.

Solution

(i) Tuning the circuit to receive the radio signal is achieved by adjusting the value of the capacitor such that resonance is obtained at the frequency of 1.205 MHz. From equation (12.33), the frequency f_o is given by $f_o = \frac{1}{2\pi\sqrt{LC}}$, from which the required capacitance $C = \frac{1}{4\pi^2 f_o^2 L}$. Substituting values of the quantities, C = 1.4 nF.

(ii) From (12.35), the quality factor is $Q = \frac{1}{R}\sqrt{\frac{L}{C}} = \frac{1}{100}\sqrt{\frac{5.0\times10^{-5}}{1.4\times10^{-9}}} = 1.89$.

12.5 ELECTRIC POWER

We conclude our study of alternating currents and voltages in this chapter with a brief account of the production, and handling of the biggest source of alternating voltage that drives today's technology. Usually referred to as the electricity generating power station, in any country, these stations are responsible for the generation and distribution of electrical power to homes, factories, government establishments, and other institutions.

In Volume 1, we studied energy conversions that eventually lead to the creation of electrical power. In coal-fired stations, for example, the primary source of energy is chemical. The heat produced by the burning coal is used to produce steam in a boiler. Superheated steam from the boiler is sent through a nozzle from which the steam exits at high pressure to enter into a turbine. A drive belt between the turbine and the generator turns the generator coil as the turbine blades rotate under the action of the high pressure steam. **Fig. 12.12** is a layout of the final stages in the generation of electric power. Nuclear power stations operate on similar principles, with the difference that steam is produced from the energy harnessed in the nuclear fissions occurring in the reactor core.

The hydroelectric power stations and windmills are different in this regard. In the hydroelectric power station, water at high velocity from a waterfall or dam is fed directly to the nozzle through a valve. High-speed water from the nozzle turns the turbine blades in much the same way as the high pressure steam in an oil-fired station. In wind-powered stations, fan blades installed in the wind course turn as the wind blows over them. The blades are attached to coils that lie in a magnetic field, thus generating electricity. Since the power that can be harnessed from one windmill is usually very small, power stations using wind power have many windmills distributed over a wide area. The electrical energies of these mills are added together to produce the lighting and heating needed for the homes. As the cost of coal and oil increase, both hydro-power and wind-power stations are becoming more common nowadays especially in developing countries with natural features that can be far less expensively tapped.

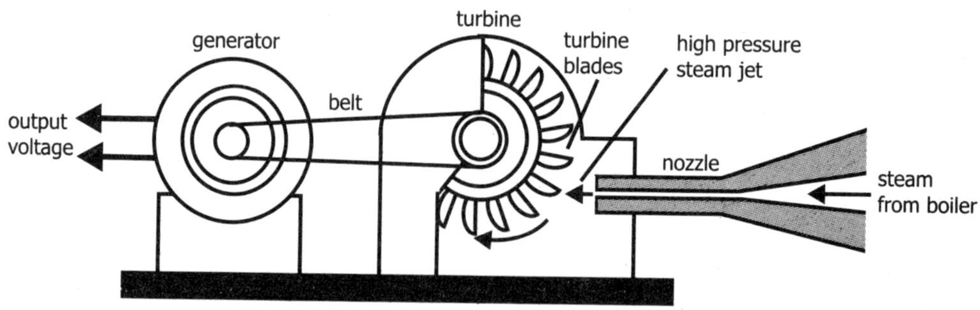

Fig. 12.12 Layout of the final stages in the generation of electricity

The output voltage from the generating station is of the order of thousands of volts. This voltage is transmitted along overhead cables to the homes and factories. The electricity fed into our homes is typically 220 Volts (in Britain, Europe and Australia) or 115 Volts in the Unites States and the Americas. Transformers at distribution points step down the voltage to the required level for consumption.

EXERCISES

1. An alternating current $3\sin 100\pi t$ A passes through a heating coil of resistance 30 Ω. Determine
   (i) the r.m.s. value of the current flow,
   (ii) the voltage waveform across the resistor
   (iii) the r.m.s. voltage across the resistor.

2. The coil in the above question is used to heat 500 g of water at 20 °C, in a calorimeter with thermal capacity 400 J/K. Neglecting any heat losses, determine how long it will take to bring the water to the boil. Specific heat capacity of water = 4183 J/kg/K.

3. An induction coil with negligible resistance, has an inductance of 2.5 mH. The inductor is connected across a 12-volt, 60 Hz supply. Calculate the current through the coil.

4. An alternating voltage $310\sin 100\pi t$ is applied to a 30 Ω resistor and a 50 mH inductor in series. Determine
(i) the impedance of the circuit,
(ii) the r.m.s voltages across the resistor, and across the inductor,
(iii) the phase angle between the current and the applied voltage.

5. An alternating current with r.m.s. value 4.5 A passes through a coil of resistance 3.0 Ω in an ice-cold water container from which heat losses may be neglected. The thermal capacity of the whole arrangement is 840 J/K. Calculate how long it will take for the temperature of the water to rise to 20 °C. (Specific heat capacity of water = 4183 J/kg/K).

6. Radio waves from the transmitter of a radio station are broadcast at frequency 1.205 MHz. A capacitive tuning circuit in a receiver uses a coil of inductance 0.004 mH. Calculate the capacitance needed for this tuning.

7. The alternating current given by $I = 2.8\sin 100\pi t$ A, is applied to an 80 Ω resistor. Determine
(i) the peak value of the current,
(ii) the r.m.s. value of the current,
(iii) the power dissipated in the resistor.

8. An inductive tuning circuit is to receive a radio signal of 30 mV at frequency 3.336 MHz. The circuit contains a series RLC combination with a 30 kΩ resistor, a 100 nF capacitor, and a variable inductor. Calculate
(i) the inductance required to receive the signal,
(ii) the current through the combination at resonance,
(iii) the p.d. across the capacitor at resonance,
(iv) the quality factor of the circuit.

9. At 1.0 kHz, the r.m.s. voltages across a series combination consisting of a 100 Ω resistor, a capacitor, and an inductor are respectively 10.0 V, 5.0 V, and 5.0 V. Calculate
(i) the current through the circuit
(ii) the impedance of the combination at this frequency,
(iii) the resonance frequency of the circuit,

10. The alternating current $3\sin 100\pi t$ A is applied to a series circuit consisting of a 100 Ω resistor, and an inductor of 200 mH. Determine
(i) the impedance of the combination,
(ii) the r.m.s. voltages across and resistor, and the inductor,
(iii) phase angle between the current and the voltage,
(iv) the power dissipated in the circuit.

11. A 50 Ω resistor is connected in series with a capacitor, and an inductor, across an alternating voltage source of frequency 50.0 Hz. The r.m.s. voltages across the inductor

and the capacitor, measured separately are 120 V and 50 V, respectively. If the total voltage across the series combination is 130 V, calculate
(i) the current flow in the circuit,
(ii) the capacitance of the capacitor,
(iii) the inductance of the inductor,
(iv) the power dissipated in the whole circuit

12. A series circuit contains a resistance of 25 Ω, an inductor of reactance 40 Ω, and a capacitor of reactance 50 Ω, across a supply voltage of frequency 50 Hz. Calculate
(i) the supply voltage required to send a current of 2.5 A through the circuit,
(ii) the power dissipated in the circuit.

13. The alternating current $1.6\sin 120\pi t$ A flows through a series circuit consisting of a 100 Ω resistor, a 100 μF capacitor, and an inductor of 500 mH. Determine
(i) the impedance of the combination,
(ii) the r.m.s. potential difference across the resistor, the capacitor, the inductor, and the whole combination,

14. For the circuit in Exercise 13 above, determine
(i) the phase difference between the current and the alternating voltage,
(ii) the expression for the voltage waveform,
(iii) the power consumed by the whole circuit over one cycle.

15. A series RLC circuit uses a resistor of 24 Ω, an inductance of 30 mH, and a capacitor of capacitance 50 μF, across a 120 V, 60 Hz supply. Calculate
(i) the inductive and capacitive reactances in the circuit,
(ii) the impedance in the circuit,
(iii) the phase angle between the voltage and current in the circuit,
(iv) the current flow in the circuit,
(v) the power dissipated in the circuit.

16. The circuit shown in **Fig. 12.13** is set to resonate at the frequency of the supply indicated. Determine

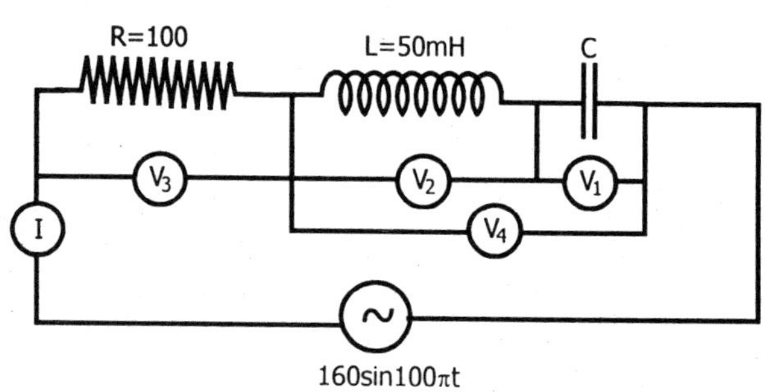

Fig. 12.13

(i) the capacitance in the circuit,
(ii) the readings on the voltmeters,
(iii) the reading on the ammeter,
(iv) the power dissipated by the whole circuit,
(v) the quality factor of the circuit.

17. A series RLC circuit uses a resistor of 100 Ω, an inductance of 50 mH, and a capacitor of capacitance 50 μF, across a 220 V, 50 Hz supply. Calculate

(i) the inductive and capacitive reactances in the circuit,
(ii) the impedance in the circuit,
(iii) the phase angle between the voltage and current in the circuit,
(iv) the current flow in the circuit,
(v) the power dissipated in the circuit.

13

CHAPTER

Transient Currents

13
CHAPTER

Transient Currents

In electrostatics, we studied the role of a capacitor in a circuit with an electrolytic cell as the source of power supply. Various capacitor combinations were studied. The study of direct current electricity enabled us to expand our knowledge of circuit behaviour to study the effects of resistor combinations. Our study of electromagnetic induction revealed further that there is yet another circuit component, the inductor, which can be connected in combinations similar to those of the capacitor and the resistor. In the previous chapter on alternating currents, we saw how these elements are connected together across an alternating voltage supply. In this chapter, we study the effects of the combinations of these components across a direct current supply voltage.

When a capacitor and a resistor are connected in series across a d.c. supply, the current flow in the circuit is only for a very short time, as the capacitor charges. A series combination of an inductor and a resistor shows current flow when the circuit is opened, after being closed for a long time. Currents of this nature – flowing for a short time only – are known as **transient currents**. Combinations of resistors, capacitors, and inductors in d.c. circuits lead to the flow of such currents. We study the production, behaviour, and uses of these currents in this chapter. For simplicity, we here consider only the RC circuit, and the RL circuit.

13.1 THE RC CIRCUIT

An RC circuit is a series combination of a resistor and a capacitor connected across a direct current supply. The circuit diagram shown in **Fig. 13.1** can be used to study the behaviour of this circuit. We consider the case where the switch S_1 is closed, with S_2 open. At an instantaneous time t after the closure of S_1, the potential differences across the resistor and the capacitor are related through the equation

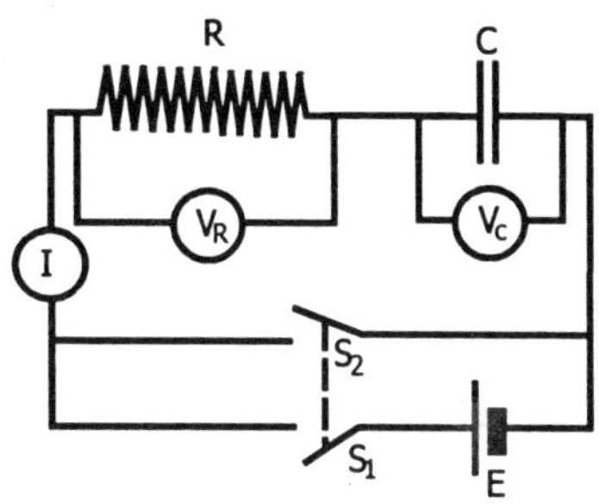

Fig. 13.1 The RC circuit

$$V_R + V_C = E,$$

where E is the emf of the supply. Writing these voltages in terms of the current, the resistance R, and the charge Q stored in the capacitor, we get

$$RI + \frac{Q}{C} = E, \tag{13.1}$$

or

$$R\frac{dI}{dt} + \frac{I}{C} = 0, \tag{13.2}$$

where C is the capacitance of the capacitor. Equation (13.2) is obtained from (13.1) by differentiating both sides of (13.1), bearing in mind that by definition, the rate of flow of charge is the current flow in the circuit. From (13.1), we get

$$I = I_0 e^{-t/RC} = \frac{E}{R} e^{-t/RC}, \tag{13.3}$$

where e is the exponential function, and I_o is the current at the instant the switch is closed. From (13.3), it follows that the current flows only momentarily through the circuit.

Writing (13.1) alternatively as $R\frac{dQ}{dt} + \frac{Q}{C} = E$, we get the charge flow as

$$Q = Q_0\left(1 - e^{-t/RC}\right) = CE\left(1 - e^{-t/RC}\right). \tag{13.4}$$

From (13.3) and (13.4), is it seen that the current in the circuit decreases from the moment the switch is closed, to zero over a some time. The charge in the capacitor, on the other hand, rises exponentially to a maximum value Q_o. Thus, as the switch is closed, the capacitor charges up until such time when it becomes fully charged. In this situation, $V_C = E$, and $V_R = 0$. These changes of current in the circuit, and charge in the capacitor are shown in **Fig. 13.2**.

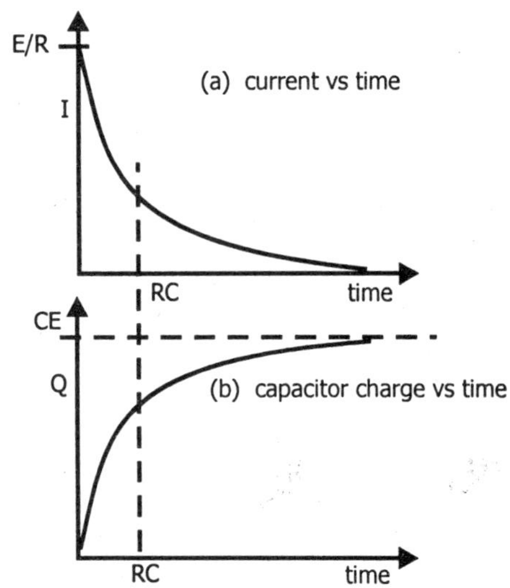

Fig. 13.2 Plots of (a) of current vs time, and (b) capacitor charge vs time, for an RC circuit. After the time RC, the current flow is reduced to 36.8% of its initial value, while the capacitor charge rises to 63.2% of the saturation value.

About the most important property of the RC circuit across a d.c. supply is the **time constant.** The time constant is a measure of how fast the current drops, or how fast the capacitor charges. The time constant τ is defined as

the time it takes for the current in the circuit to drop to 1/e (or 36.8%) of its maximum initial value.

From (13.3), it follows that

$$\tau = RC. \tag{13.5}$$

In one time constant, the current falls off to about 36.8% of its maximum initial value, whilst the capacitor would have attained a charge of 63.2% of the maximum charge.

The discharge of the capacitor is studied by opening S_1 after being closed for a long time, and closing S_2. The capacitor discharges through the resistor. The charge in the capacitor diminishes to zero at a rate depending on the time constant. The charge reduces to 36.8% of the maximum value within one time constant. **Fig. 13.3** shows the variations of the charge with time for both charging and discharging. The discharge is governed by the equation

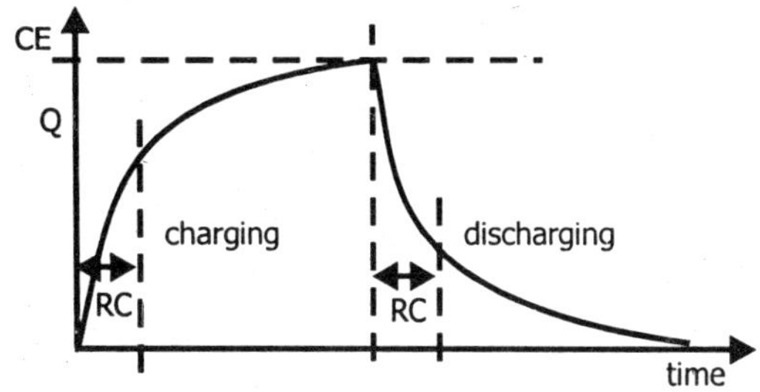

Fig. 13.3 Charging and discharging of the capacitor through the resistor.

$$Q = CEe^{-t/RC}, \qquad (13.6)$$

where Q is the charge in the capacitor at time t after S_1 is opened and S_2 closed. In the section following, we explore some of the practical uses of the RC combination.

Example 13.1

A series RC circuit is constructed using a 100 kΩ resistor and a 100 μF capacitor across a d.c. supply of 200 Volts. Calculate

(i) the time constant of the circuit,
(ii) the p.d. across the resistor after 10 s,
(iii) the charge stored in the capacitor after 10 s,

Solution

(i) From (13.5), the time constant $\tau = RC = 10^5\,\Omega \times 100 \times 10^{-6}\,F = 10\,s$.

(ii) When the circuit is just switched on, the current I_o flowing is given by $I_o = \frac{E}{R} = \frac{200\,V}{10^5\,\Omega} = 2.0\,mA$. Since the time constant is 10.0 s, the current flowing in the resistor after 10.0 s is I = 0.368 x 2 mA = 736 μA. The p.d across the resistor is therefore given by $V_R = IR = 73.6 \times 10^{-5}\,A \times 10^5\,\Omega = 73.6\,V$.

(iii) The maximum charge in the capacitor $Q_o = CE = 100 \times 10^{-6}\,F \times 200V = 0.02\,C$. After one time constant (or 10 s), the charge stored is 0.632 x 0.02 C = 0.013 C.

Example 13.2

A 500 μF capacitor is fully charged using a 12-volt battery, and connected across a resistor of resistance 20 kΩ. Calculate

(i) the charge in the capacitor before the resistor is connected,
(ii) the time constant of the circuit,
(iii) the charge remaining in the capacitor after 10.0 s.

Solution

(i) the full charge of the capacitor is $Q = CE = 500\ \mu F \times 12\ V = 0.006\ C$.

(ii) the time constant of the circuit is $\tau = RC = 20\ k\Omega \times 500\ \mu F = 10.0$ s.

(iii) After 10 s the charge remaining in the capacitor is $0.368 \times 0.006\ C = 0.0022\ C$.

13.2 SOME APPLICATIONS OF THE RC CIRCUIT

The RC circuit has a number of standard applications. These range from highly specialised electronic applications to its use in everyday simple devices. We here study a couple of the simple everyday applications.

13.2.1 Camera flashgun

Fig. 13.4 shows a typical RC circuit arrangement for the flashgun in the camera. In its simplest form, the circuit has a switch S, which can be switched to positions A and B of the RC. With the main power switch closed, connecting S to A charges the capacitor. A warning light comes on when the capacitor becomes fully charged, which normally takes only a few seconds. When S is connected to B, the capacitor discharges through the resistance filament of the flash bulb. Using a lamp with very low resistance, this discharge takes only a fraction of a second, thus turning the flash light on.

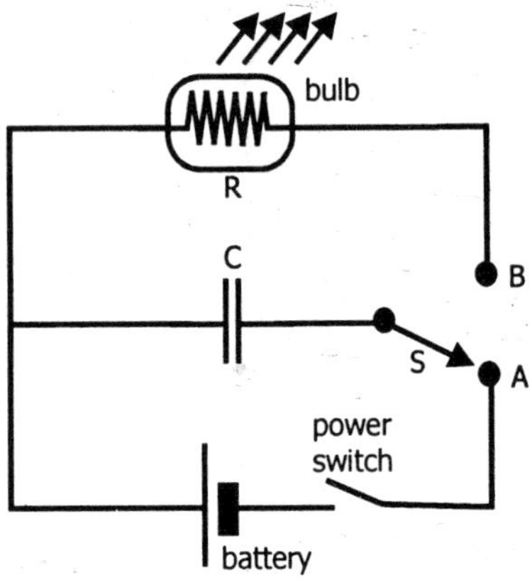

Fig. 13.4 Capacitor charging and discharging in the camera flash gun.

13.2.2 Separation of a.c. and d.c. voltages

Apart from storage batteries, fuel cells, and photocells, the biggest source of direct current supply is through **rectification**, the process by which an alternating voltage is converted to direct voltage. The general principles of rectification involve sending the alternating voltage through **diodes**. The details of this rectification process is covered in chapter 15.

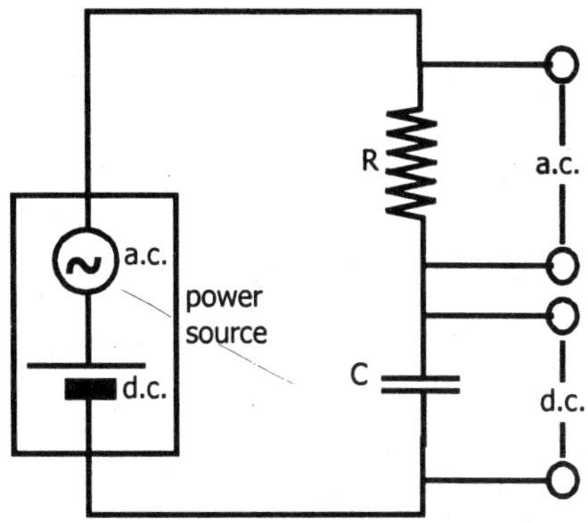

Fig. 13.5 RC circuit for separating a.c. and d.c. If $\omega CR >> 1$, almost all the a.c. voltage appears across the resistor, while the d.c. appears across the capacitor.

Most often, the output voltage from the rectifying circuit is not pure d.c., but contains small components of the original alternating voltage. Certain sensitive devices of today's technology are designed to operate only with d.c. voltages, and may malfunction if fed with a.c. voltage. It therefore becomes necessary to separate both the a.c. and d.c. components from the source. This can be done using an RC combination. **Fig. 13.5** is the diagram of a circuit designed to achieve this separation. Such a circuit is referred to as a **separating circuit**.

Bearing in mind that the current in an RC circuit with a d.c. supply diminishes to zero over a long time, it is seen that over a long period of time most of the d.c. voltage in the circuit appears across the capacitor, whilst the a.c. is dropped across the resistor. To achieve near perfect separation, it is arranged that at the frequency of the alternating component, the reactance of the capacitor is much less than the resistance in the circuit. This leads to the condition that $\frac{1}{\omega C} << R$, or $\omega CR >> 1$.

Using the same principle, an RC circuit is used in many ways to block out the d.c. component of a combined voltage. Used in this way, the capacitor is said to be a **blocking** capacitor.

There are many other specialised uses of the RC circuit – in pulse shaping, in high resistance measurements, and in filters. The details of these lie outside our current scope.

THE LR CIRCUIT

Fig. 13.6 (a) shows an RL circuit connected across a direct current supply. When the switch is closed, the current in the circuit rises from zero. While the current is rising, there is electromagnetic opposition due to the inductor.

The current thus builds up to its maximum steady state value. Before this steady value is attained, however, the current goes through a transient state. At any time t following the closure of the switch, the voltage in the circuit is the sum of the potential differences across the inductor and the resistor. Thus, we have

$$V_L + V_R = E,$$

giving

$$L\frac{dI}{dt} + RI = E. \tag{13.7}$$

From (13.7), the current in the circuit at time t is given by

$$I = \frac{E}{R}\left(1 - e^{-\frac{R}{L}t}\right), \tag{13.8}$$

where e is the exponential function. The variation of this current with time is shown in the plot of **Fig. 13.6** (b). The time constant τ for this circuit is defined in much the same manner as for the RC circuit. In terms of the inductor and the resistance, this time constant is given by

$$\tau = \frac{L}{R}. \tag{13.9}$$

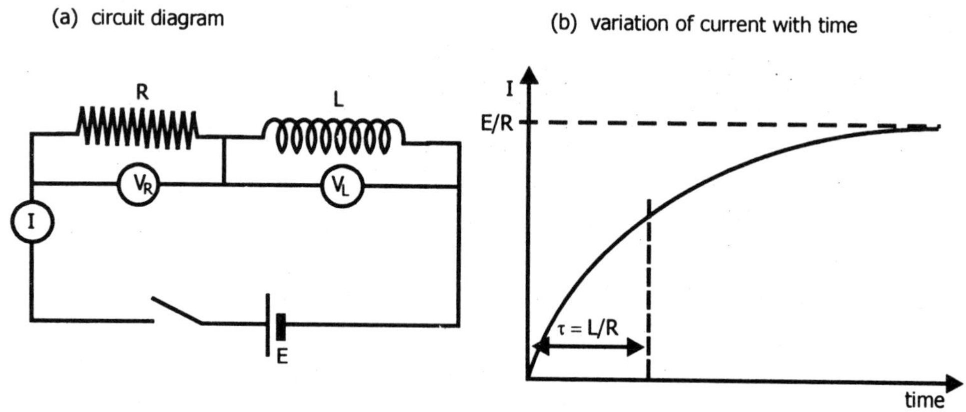

Fig. 13.6 The RL circuit across d.c. supply; (a) shows the circuit diagram, whilst (b) shows the variation of current with time.

When the switch is opened after being closed for a long time, the time constant for current decay becomes much smaller than the rise time when the circuit is closed. This is because of the near infinite resistance of the open circuit. The current therefore drops to zero a lot faster than the rise when the circuit is closed.

There are many uses of the RL circuit. About the most common is in the application of the high voltage produced when the RL circuit is opened. This is employed in the ignition system of the motor car. The RL circuit here uses an induction coil of high inductance in series with the 12-volt battery. The circuit is opened by the breaker points in the distributor to produce the high voltage needed for each spark plug to fire. Voltages in excess of 6,000 volts are generated in this way.

Of far lesser importance in d.c. applications are circuits containing capacitor and inductor (i.e. the LC), and circuits containing all three components (i.e. the RLC), across a direct current supply. We shall leave the study of these to more advanced texts.

EXERCISES

1. Calculate the time constants for the following corresponding pairs of circuit components when connected in series across a direct current supply.

(a)	R = 500 Ω, C = 50 μF,	(b)	R = 10 Ω, L = 200 mH,
(c)	R = 50 kΩ, C = 50 μF,	(d)	R = 1.0 MΩ, C = 2.0 μF,
(e)	R = 20 Ω, L = 12 mH,	(f)	R = 20 Ω, L = 200 mH,
(g)	R = 5.0 MΩ, C = 50 pF,	(h)	R = 10 Ω, L = 20 mH,

2. A series circuit consisting of a 100 kΩ resistor and a 2.0 μF capacitor is connected across a d.c. supply of 100 V. Calculate
   (i) the maximum current obtained when the circuit is closed,
   (ii) the time taken for the capacitor voltage to rise to 63.2 V.

3. Using the circuit in **Exercise 2**, the power source is disconnected and the capacitor discharges through a 100-Ω resistor. Determine
   (i) the time constant for the discharging circuit,
   (ii) the p.d. across the resistor after one time constant.

4. When a 1.0 nF capacitor is connected in series with an unknown resistor across a d.c. supply voltage, the time constant of the circuit is measured to be 75 μs. Calculate the value of the unknown resistor.

5. A series RC circuit is constructed with a resistor of 1.0 MΩ resistor and a 10 nF capacitor across a d.c. supply of 200 Volts. Calculate
   (i) the time constant of the circuit,
   (ii) the p.d. across the resistor after 0.01 s,

14

CHAPTER

Electrical Measurements

14
CHAPTER

Electrical Measurements

In this chapter we study the makeup and role of electrical measuring instruments. Without exception, these measurements use principles involving the nature of electricity and magnetism, and the interactions between electric and magnetic fields. Our studies of electricity and magnetism so far have enabled us to understand the principles of these interactions. Hence we can now appreciate how the principles are employed in practical measurements of electrical quantities.

We begin our study of measurements with the measurement of direct currents and voltages. We study the theory behind the operation of these instruments, and how we convert one instrument, say, the milliammeter, into an ammeter and voltmeter. Measurements of alternating currents fall into a different category. We study how measurements of these voltages are made. We end the chapter on combination instruments for measuring both alternating and direct currents and voltages.

14.1 MOVING COIL INSTRUMENTS

14.1.1 The milliammeter

In our study of magnetic field interaction, in chapter 10, we saw how the interaction between a current wire and a magnetic field is used in an instrument to measure a direct current. Known as the moving coil milliammeter, or galvanometer, **Fig. 14.1** shows the important parts.

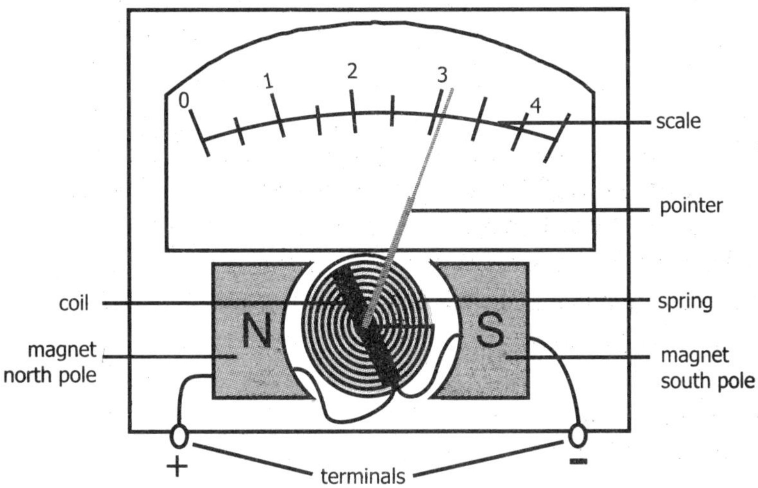

Fig. 14.1 Parts of the moving coil galvanometer (or milliammeter).

A radial magnetic field ensures that the magnetic lines of force are at right angles to the coil. The angular deflection ϕ of the pointer for current I through the coil is

$$BANI = k\phi. \tag{14.1}$$

giving

$$I = \frac{k}{BAN}\phi, \tag{14.2}$$

where B is the magnetic flux density, threading through a coil of area A, with N turns of wire, and k is the torsion constant for the spring.

The coil can withstand currents of the order of milliamps. As a result, the meter is also called a milliammeter. The meter can be used for the measurement of direct current and voltages. Because of this property, the meter is sometimes referred to as a universal meter. Addition of suitable resistances in parallel and in series with the coil enables conversion of the instrument into an ammeter (for measuring currents of the order of amps), and voltmeter – for measuring voltages.

14.1.2 Conversion of milliammeter to ammeter

A milliammeter is converted into an ammeter by connecting a resistor in parallel with the coil. This resistor is known as a **shunt** resistor. The value of the shunt resistor is such that most of the desired current flow through the shunt, leaving only that which the coil can withstand. In **Fig. 14.2** (a) shows the shunt connection, whilst the current flow is depicted in (b). The current i (in mA) flowing through the coil corresponds to the full-scale deflection reading of the milliammeter. To convert the milliammeter into an ammeter reading current up to I (amps), the value of the shunt should be such that current $(I - 10^{-3}i)$ flows through the shunt. The value of the shunt resistor is calculated in terms of the coil resistance, the full-scale deflection of the meter, and the desired current to be read by the ammeter. **Example 14.1** illustrates the principles of this calculation.

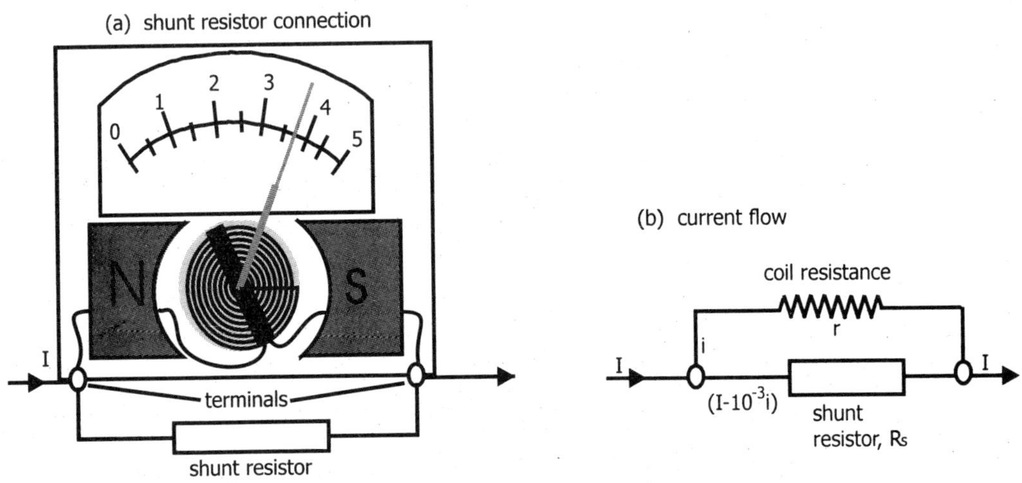

Fig. 14.2 Conversion of milliammeter to ammeter

Example 14.1

A milliammeter with coil resistance of 10.0 Ω gives full-scale deflection for a current of 5 mA. Calculate the shunt resistor that should be placed in parallel with the coil in order to convert this meter into an ammeter capable of reading currents up to 1.0 A.

Solution

The current through the shunt resistor R_s is given by $\left(I - 10^{-3} i\right) = 1.0 - 0.005 = 0.995$ A. Since the shunt is in parallel with the coil, the same potential appears across the two. Equating these potential differences, we get $\left(I - 10^{-3} i\right) R_S = 0.005\, r$.

Substituting the numerical values, we get $R_S = \dfrac{0.005 \times 10}{0.995} = 0.05025\ \Omega$.

14.1.3 Conversion of milliammeter to voltmeter

Connecting a **bobbing** resistor in series with the galvanometer coil does this conversion. The same current flows through the bobbing and the coil. **Fig. 14.3** (a) and (b) show the connection, and the current flow, respectively. **Example 14.2** illustrates how the value of the bobbing resistor is calculated.

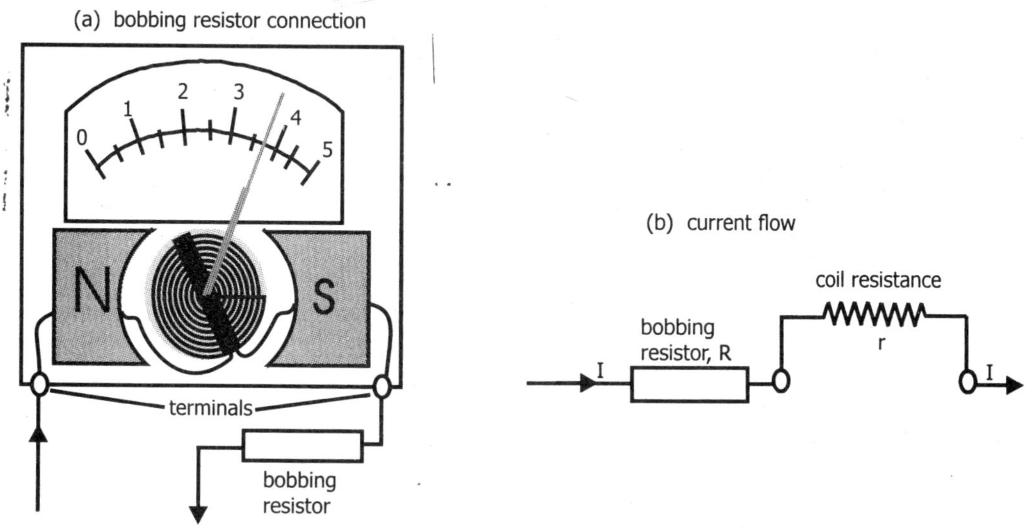

Fig. 14.3 Conversion of milliammeter to voltmeter

Example 14.2

For the meter in **Example 14.1** above, calculate the bobbing resistor that would convert the milliammeter into a voltmeter for measuring up to 5.0 V.

Solution

The current flow through the series resistor and the galvanometer coil is the full-scale deflection current of the galvanometer – i.e. 0.005 A. The potential difference across the coil is thus $ir = 0.005 \times 10 = 0.05$ V. The potential difference across the bobbing resistor is hence equal to 4.95 V. The bobbing resistance R is given by $R = \frac{V}{i} = \frac{4.95 \text{ V}}{0.005 \text{ A}} = 990\ \Omega$.

From these examples, it is seen that to convert a milliammeter into an ammeter, we add a small shunt resistor in parallel with the coil. To convert a milliammeter into a voltmeter requires a large resistor in series with the coil.

14.1.4 Measurement of resistance

Measurement of resistance is done with an instrument called **ohmmeter**. This meter is also a moving coil instrument, using the same principles as in the above instruments. The difference between the ohmmeter and the other meters we have just studied is that the ohmmeter has in it a battery or power source. **Fig. 14.4** is a schematic diagram of the ohmmeter in its simplest form. Contact with the external resistor is made at the terminals, thereby closing the circuit. Current flows through the coil and the external resistor in series with the battery. The deflection produced is proportional to the current flowing through the coil when the circuit is closed.

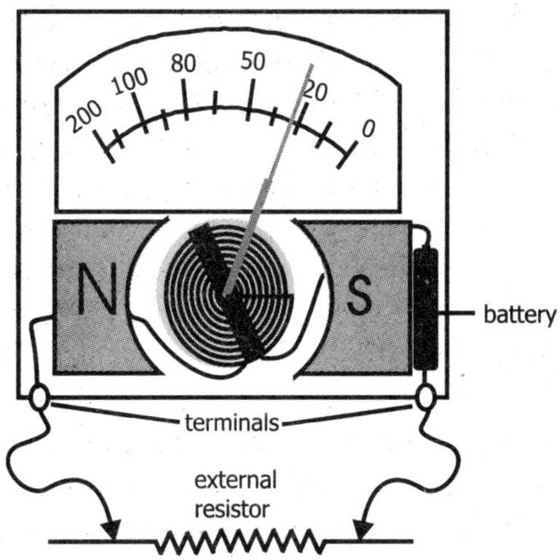

Fig. 14.4 The Ohmmeter

The resistance R corresponding to current I through the coil is given by

$$R = \frac{V}{I}, \tag{14.3}$$

where V is the p.d. across the external resistor, assumed constant. From (14.3), it follows that current measured is greater for a smaller resistor than for a larger resistor. Thus, the smaller the current the greater the resistance. The reciprocal relationship between the current and the resistance makes the deflection scale nonlinear. The calibration points on the scale are hence not equally spaced, being closer together for large resistances, as shown in the figure.

In measurements with the ohmmeter, it is imperative that coil resistance is small compared with the value of the external resistance to be measured. Errors are introduced when the resistance of the coil is comparable to the external resistance.

About the most frequent use of the ohmmeter is in testing circuit continuity. In this use, the resistance between two points in the circuit is measured. A near zero value indicates that the circuit is continuous. Most meters measuring resistance are equipped with a buzzer that gives an audible sound when the resistance very small, or nearly zero.

14.2 MEASUREMENT OF ALTERNATING CURRENTS

As we studied in chapter 12, an alternating current changes polarity many times every second. The moving-coil galvanometer is designed to measure currents flowing in one direction. The instrument is therefore not designed to measure alternating currents. It can, however be modified to enable it to measure alternating currents.

One of these modifications has given rise to an early instrument known as the **hot-wire ammeter**. The alternating current passes through a segment of resistance wire, making the latter hot. The expansion of the hot wire is communicated via a small spring to the pointer which moves over a scale. The deflection produced is proportional to the r.m.s. current flowing through the wire. In modern versions of the instrument, the hot wire heats one junction of a thermocouple, thereby producing a thermoelectric emf. This emf drives current through the galvanometer coil, producing deflection that can be related to the magnitude of the current in the usual way.

14.3 THE AVO METER AND MULTIMETER

The AVO meter is so called because it measures currents (A), voltages (V), and Ohms (O), all combined into one instrument. The instrument is capable of measuring both direct and alternating currents and voltages. A selection switch allows the user to select the range of measurement as well as the type of measurement. Another switch is used to select between A.C. and D.C. inputs. **Fig. 14.5** shows the layout of the meter. The scales are highly simplified for purposes of clarity. The diagram shows the range switches and the scales of measurement. The A.C./D.C. selector switch is also shown.

Test probes allow measurement of currents, voltages and resistances. Using the range select switch it is possible to measure currents and voltages over a wide range.

Today, the AVO meter has been developed into the more versatile instrument called the **multimeter**. In addition to currents, voltages and resistances, the multimeter

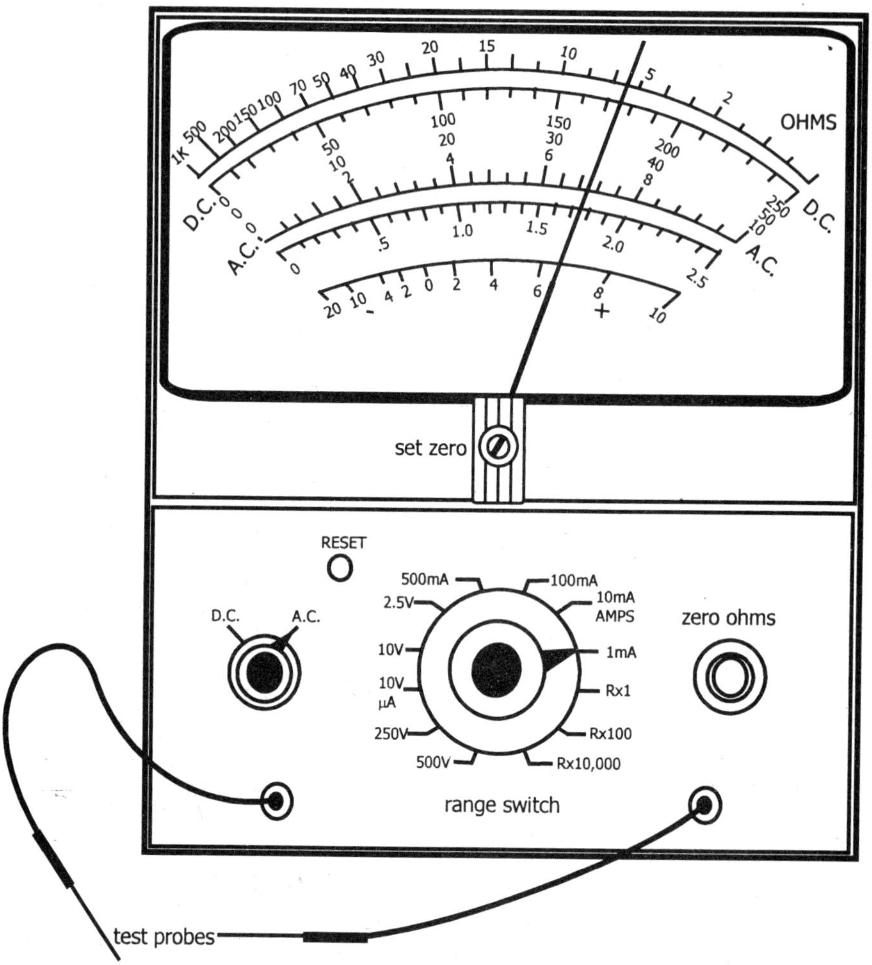

Fig. 14.5 The AVO meter

can be used to test the performance of capacitors, diodes, and transistors. In many multimeters, the output is digital, rather than the analog scale shown above. Some multimeters are equipped with miniature computer interfaces that enable them to store the measured data, and finally transfer these to a computer.

14.4 THE CATHODE RAY OSCILLOSCOPE

The cathode ray oscilloscope (CRO) is another electrical measuring and testing instrument that is in popular use throughout industry and in laboratories. Most times it is simply referred to as "scope". The scope works very much like a voltmeter. It is, however, a more versatile voltmeter in the sense that it can measure voltage changes over a fraction of a second. With this property, the instrument can be used to measure the durations of voltage pulses, and the periods of alternating voltages. Like the AVO meter, the front panel of the CRO is fairly complicated, with a number of switches to enable the measurements to be made.

The cathode ray oscilloscope is used to display visible patterns that are graphical representations of electrical signals. The graphs plot the relationships between two or more variables, with the horizontal axis normally being a function of time, and the vertical axis usually a function of the voltage generated by the input signal to the scope. Because almost every physical phenomenon can be used to generate electricity, the CRO is today the most versatile tool in all forms of physical investigation. Examples of the use of the oscilloscope range from detecting the small electrical pulses generated by nerve activities, to fault finding in complicated electronic circuits.

At the heart of the oscilloscope is a cathode ray tube that bears close similarity to our household television picture tube. The inside face, or screen of the tube is coated with a fluorescent substance, allowing the screen to glow at the spots where the electrons strike it. **Fig. 14.6** is a schematic diagram of the interior of the oscilloscope.

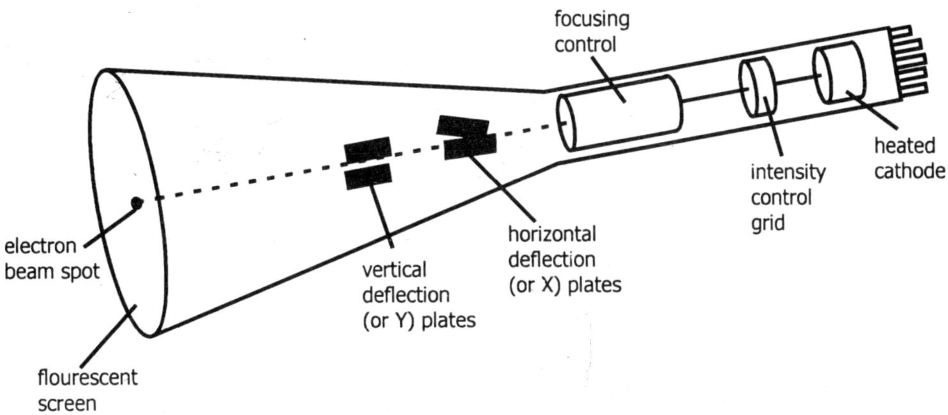

Fig. 14.6 Schematic diagram of the cathode ray tube, showing the screen, the deflection plates, the focusing and intensity controls, the source of electrons.

Electrons are emitted from the heated cathode, and travel to the screen under the action of a steady potential difference between the two ends of the tube. Control circuits are used to vary the intensity and sharpness of the beam spot. The intensity control works by having a negative potential applied to the electron beam. The negative potential applies a force of repulsion on the electrons leaving the cathode. The larger the potential applied the smaller the intensity. Focusing works in a manner similar to that of focusing a beam of light using a lens. The smaller the size of the spot, the better the focus. It is normally advisable not the leave the scope in a highly focused state for an extended period of time, with the electron beam falling on only one spot. This will burn that part of the screen. Next we see a set of deflecting plates. There are two sets of these – the horizontal deflecting plates, and the vertical deflecting plates. Normally referred to as the X-plates, the horizontal deflecting plates will move the beam sideways across the screen, while the vertical deflecting (or Y) plates allow us to move the beam up and down on the screen. In its usual operation, the unknown voltage is applied to the Y-plates, while a linear time-varying voltage is applied to the horizontal plates. In this way, unknown voltages and frequencies of alternating voltages, pulse sizes and their corresponding durations, can be determined.

Fig. 14.7 is a simplified diagram of the front panel of the cathode ray oscilloscope. The front panels of most oscilloscopes are far more complicated than is depicted in this diagram. The panel contains buttons and knobs for varying intensity, focusing, calibration and trigger levels. The figure shows a dual trace oscilloscope – one that can be used to study two separate inputs. Combinations of the calibration settings and the height and width of the voltage waveform on the display screen are used to measure the voltage and determine the frequency.

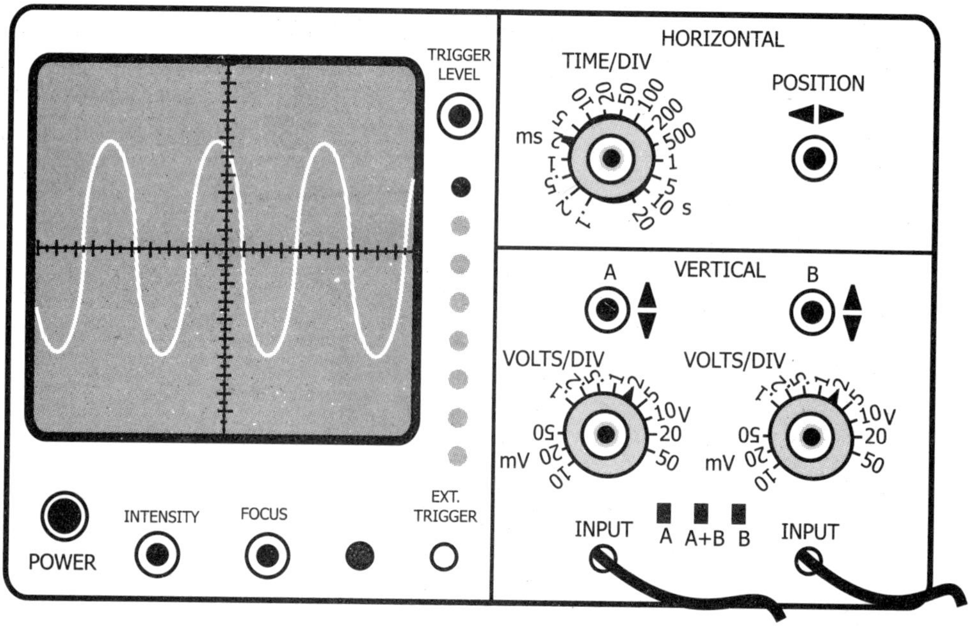

Fig. 14.7 Simplified front view of a dual input cathode ray oscilloscope. Only one input is displayed.

14.4.1 Measurement of voltage

Measurement of voltage uses the voltage calibration knob, and a reading of the height of the trace from the baseline. Using the alternating voltage waveform in the figure as example, we see that the voltage calibration, or voltage sensitivity, is 2.0 V/division. From the trace, it is seen that the height of the waveform from the baseline is 6.0 divisions. Thus, the amplitude V_o of the alternating voltage is given by

$$V_0 = \text{voltage sensitivity x height}$$
$$= 2.0 \text{ V x } 6.0 \text{ div} = 12.0 \text{ Volts.}$$

In the same manner, we can calculate the peak-to-peak voltage.

14.4.2 Measurement of frequency

For these measurements, we use the time base sensitivity. The frequency of the input voltage is determined in two stages. We determine the period of the waveform, and hence calculate the frequency. From our study of waves, the period of a wave is the time it takes the wave to travel one wavelength. For the trace shown in the diagram, the period T is given by

$$T = \text{time base sensitivity x time interval for one wavelength}$$
$$= 2 \text{ ms x } 6.0 \text{ div} = 12 \text{ ms.}$$

The frequency is hence given by $f = \frac{1}{T} = \frac{1}{12\text{ ms}} = 83.3\text{ Hz}.$

EXERCISES

1. A galvanometer with a coil of resistance of 5 Ω has a full-scale deflection of 10 mA. Calculate the values of the shunt and series resistances for converting this instrument into
   (i) an ammeter capable of reading up to 5 A,
   (ii) an ammeter reading up to 500 mA
   (iii) a voltmeter reading to 20 V,
   (iv) a voltmeter that reads up to 5 V.

2. **Fig. 14.8** shows the display on a two-channel input cathode ray oscilloscope screen, with the following voltage sensitivity settings:
   Channel A: 0.20 mV/division,
   Channel B: 20 mV/division,
   the time base sensitivity setting for both channels is 2 ms/div. Use the above information to estimate
   (i) the peak-to-peak voltage of input A,
   (ii) the amplitude of input B,
   (iii) the frequency of input A,
   (iv) the frequency of input B,
   (v) the difference in periods of the two inputs.

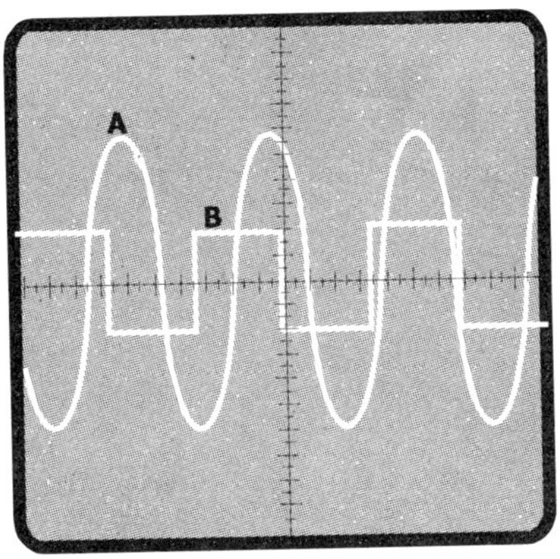

Fig. 14.8

15
CHAPTER

Solid State Electronics

15
CHAPTER

Solid State Electronics

15.1 THE ORIGINS OF ELECTRONICS

In its simplest form, electronics can be defined as the scientific study of phenomena dealing with the motions of electrons, and how to control this motion. From our study of direct current electricity, we note that conduction of electricity through metals is due to the motion of electrons. How does this electron motion differ from the study of electronics? The major difference between the two has to do with magnitude. In electronics, the currents resulting from the electron motions are very small – typically of the order of milliamps at best. In normal electricity, however, small as well as large currents are encountered. The small size of the currents in electronics makes it easy to control circuits based on such devices. Various scientific and technological disciplines, such as music, photography, physics, are concerned with different aspects of electronics.

The history of the development of electronics put the early beginnings of the subject to experiments in the late 1880s on electron currents in discharge tubes – evacuated glass tubes with electrodes at either end. Construction of the first so-called thermionic valve is attributed to the work of the English engineer John Fleming. Fleming showed that with this device, electron current could be restricted to flow in one direction only. This process, called rectification, produced an output current, which was used to operate such things as telephone receivers or recording devices.

Improvements to the thermionic valve saw the development of the triode in 1907 by the American engineer, De Forest. De Forest added a third element – the grid – to the tube. This latter electrode he placed between the anode and the cathode. By varying the negative potential on this third electrode, the overall electron current in the tube could be controlled. Further work using the triode revealed how the tube could be used to amplify input voltages. This discovery ushered in the age of voltage and current amplification – a means whereby a large output voltage can be produced from a small input. These amplifiers were then at the heart of the early electronic devices, such as sound systems, musical organs, and telephone communication systems. From the triode, the electron tube industry grew to produce, pentodes, photomultiplier tubes, klystrons, and so on.

15.2 MODERN ELECTRONICS

The vacuum tube was indeed a big technological step. Within the first half of the last century many consumer electronic devices have appeared on the market. As demands for various electronic devices grew, it was soon realised that electronic technology based on thermionic tubes had certain inherent disadvantages that would limit their widespread adoption. The most prominent of these disadvantages are the following:

1. thermionic tubes needed to be operated at fairly high voltages between the electrodes. These high potential differences lead to heat generation within the equipment. The equipment hence gets hot, and needs to be cooled if the device is to be on for an extended length of time. Thermal deterioration meant the tubes needed to be replaced at regular intervals.

2. thermionic tubes are bulky. Accordingly, equipment based on these tubes was bulky and at times presented handling problems, especially in circumstances where space is a premium.

Because of the above disadvantages, the industry soon realised the need to develop electronic devices that were small, and do not require high operating voltages. The breakthrough came in 1947 when three scientists – John Bardeen, William B. Shockley, and Walter M. Brattain - working at Bell Laboratories introduced the transistor. The transistor is made up of a solid block of semiconducting material into which are fused three electrodes. This simple arrangement was shown to perform the functions that were performed by thermionic tubes. The transistor was small, often less than one-tenth the size of the thermionic valve. Moreover, the transistor requires far less voltages for operation compared to vacuum tube. Because the transistor is made up of a solid semiconductor material, devices based on the transistor are described as **solid state** devices. Accordingly, the branch of electronics based on these devices has come to be described as **solid state electronics**.

The development of the transistor was the first step in the development of electronic devices with very small sizes. By the late 1980s – just about one hundred years since the invention of the first thermionic valve – the electronics industry had achieved the capability to manufacture solid state devices containing as many as three million transistors on a single semiconductor chip. These chips, known as **microprocessors**, lie at the heart of the modern computer used today, and in computer-based robotics. The availability of inexpensive microprocessors has also brought about the computerization of an enormous array of consumer products, ranging from the control of washing machines, automatic focusing of cameras, the remote control of hi-fi and television sets, etc.

In this chapter of the book, we broadly outline the field of solid state electronics. Starting with the electron structure of the semiconductor, we follow through with a brief survey of some practical applications.

15.3 THE INTRINSIC SEMICONDUCTOR

In our study of direct current circuits, we skipped mention of the third class of materials in addition of conductors and insulators. This third class is the **semiconductor**. As the name implies, semiconducting materials have electrical conducting properties between those of the conductor and the insulator. Conductors are normally found in groups 1, 2 and 3 of the Periodic Table of Elements. Insulators are in the higher groups – above group 4. Semiconductors are found in Group 4. **Table 15.1** shows the portion of the Periodic Table related to semiconductors. The complete Periodic Table of Elements can be found in standard chemistry texts.

Table 15.1 Portion of the Periodic Table of Elements related to semiconductors.

	Column (Groups)				
Period	II	III	IV	V	VI
2		Boron B	Carbon C	Nitrogen N	
3	Magnesium Mg	Aluminium Al	Silicon Si	Phosphorus P	Sulphur S
4	Zinc Zn	Gallium Ga	Germanium Ge	Arsenic As	Selenium Se
5	Cadmium Cd	Indium In	Tin Sn	Antimony Sb	Tellurium Te
6	Mercury Hg		Lead Pb		

On our simplified pictorial view, the semiconductor atom has four valence electrons. Examples of semiconductors are carbon (chemical symbol C), silicon (symbol Si), and germanium (Ge). Silicon is the most widely used of these elements. Its pictorial electronic configuration is shown in **Fig. 15.1** (a). The atom is shown containing a nucleus with a charge of 14 protons, and 14 electrons orbiting the nucleus. The outermost orbit has 4 electrons – called valence electrons. These are the electrons that take part in bonding with different atoms to form compounds. Germanium has a similar but more complex electronic configuration as it contains many more electrons and orbitals than silicon.

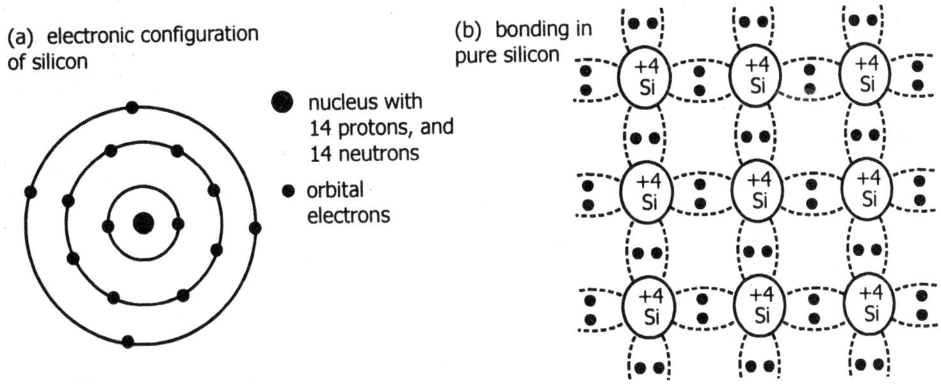

Fig. 15.1 Electron configurations of (a) the silicon atom, and (b) pure silicon crystal. In (b), the nuclear core is shown with a charge of +4, corresponding to the four valence electrons. Each silicon atom in the crystal shares its valence electrons with four surrounding atoms in an arrangement referred to as covalent bonding.

When many silicon atoms come together to make up the silicon crystal, each atom shares its four valence electrons with four neighbouring atoms, in what is known as covalent bonding. In this bonding, the valence electrons are shared between atoms in such a manner that each atom ends up with filled electronic shells – the so-called octet configuration. Because it involves the pairing of electrons, this bonding is also called **electron-pair** bonding. Details of how exactly this bonding takes place lie outside our current scope. **Fig. 15.1** (b) depicts this bonding in the silicon crystal.

At very low temperatures all the valence electrons are part of electron-pair bonds. At normal room temperatures, however, there is enough thermal energy to set some electrons of the pair bonds free. When this happens, a **hole** is created. The hole can be considered as the absence of an electron in an electron-pair bond. The hole carries a positive charge equal in magnitude to that on the electron. The positive charge on the hole acts as an attractive force on an electron coming close. This way, the unpaired electron is paired, and the hole moves to a new position. The hole thus moves in the crystal in much the same way as the free electron.

In the absence of an external field, the holes and the freed electrons move haphazardly throughout the semiconductor crystal. When an external field is applied to the crystal, the electrons and holes move in opposite directions, the holes moving in the direction of the field. This movement of charges in opposite directions in the crystal is responsible for the electrical conduction in the semiconductor. While the holes and electrons move in the semiconductor, the crystal as a whole remains electrically neutral, the overall positive charge of the holes balancing the overall negative charge of the free electrons.

When heat is applied to a semiconductor, the thermal energy is used to produce more free electrons and holes, thereby increasing the electrical conductivity of the crystal. Thus, for a given potential difference applied to the crystal, more current results at the higher temperature. The resistance of the semiconductor therefore **decreases** as the temperature of the semiconductor is increased. The temperature coefficient of resistance of a semiconductor is therefore **negative**. It can be recalled from chapter 9 that the temperature coefficient of resistance of metals is positive. Notwithstanding this lower resistance at a higher temperature, the currents measured for the semiconductor are much smaller – sometimes by as much as a thousand times – than those for the metal.

15.4 DOPED SEMICONDUCTORS

Although in principle we can increase the conduction properties of the semiconductor by heating, in practice, this method is never used, as it would damage the semiconductor. The conduction properties are improved immensely by **doping**. In doping, atoms of an element with different valency are added as impurities to the crystal structure of the semiconductor. Semiconductors produced by doping are called **extrinsic** semiconductors – as distinct from pure or intrinsic semiconductors. There are two types of doped semiconductors: the **n-type**, and the **p-type**. The techniques used in the manufacture of doped semiconductors are highly complex. Standard methods include substitution doping, and interstitial doping. In substitution doping, a semiconductor atom is substituted by the impurity atom, whilst in interstitial doping the impurity atom is inserted into the internal structure of the semiconductor crystal. Doping leads to the production of large numbers of charge carriers of one type or the other, thereby increasing the conduction properties. Using substitution doping as example, we depict the electronic configuration of the n-type and the p-type semiconductors in **Fig. 15.2**.

15.4.1 The n-type semiconductor

The n-type semiconductor is produced by doping the pure semiconductor with a pentavalent element – i.e. an element with five valence electrons. Examples of elements used for this purpose are arsenic and phosphorus. Four of the five valence electrons of the impurity atom form covalent bonds with four neighbouring atoms of silicon, leaving one valence electron free. This extra free electron adds to the free electron population in the crystal, thereby increasing its conductivity.

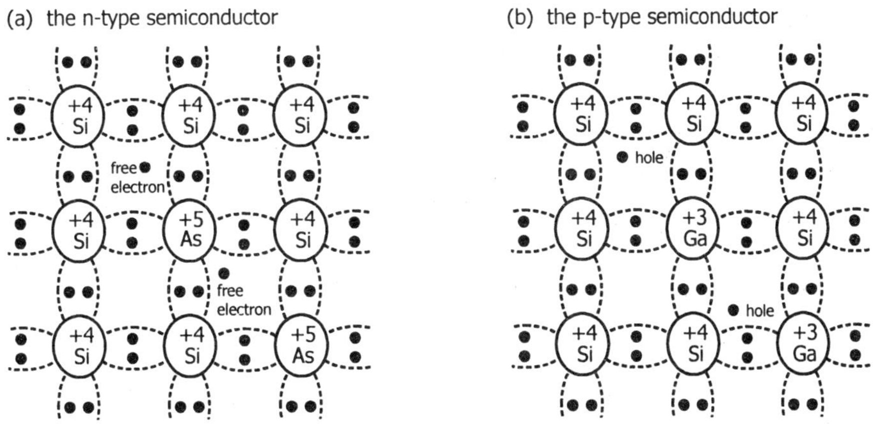

Fig. 15.2 Representations of (a) the n-type semiconductor, and (b) the p-type semiconductor. Majority carries for the n-type are free electrons (n), whilst holes (p) are the majority carriers for the p-type.

The presence of the extra free electron per impurity atom added increases the free electron population of the crystal in excess of the population of holes, the latter arising from thermal excitations of the pure crystal. Because of their relative numbers, the electrons in the n-type semiconductor are referred to as the **majority** carriers, whilst the holes are the **minority** carriers.

15.4.2 The p-type semiconductor

In the p-type semiconductor the crystal is doped with a trivalent element – i.e. an element with three valence electrons. Elements commonly used are aluminium, gallium and indium. The three valence electrons of the semiconductor atom form covalent bonds with three neighbouring impurity atoms, leaving one valence electron of the semiconductor atom unpaired. The absence of the electron in the electron-pair bond is a hole. Thus a p-type semiconductor has extra holes which contribute to the electrical conduction properties.

The presence of the extra hole per impurity atom added increases the hole population on the crystal in excess of the population of free electrons, the latter arising from thermal excitations of the pure crystal. In the p-type semiconductor, holes are the majority carriers, whilst free electrons are the minority carriers.

The doped semiconductor crystal as a whole remains electrically neutral. In the case of the n-type semiconductor, for every extra free electron introduced in the crystal, there is an extra positive charge on the impurity core. This makes the n-type crystal as a whole electrically neutral. Similarly, a p-type semiconductor as a whole remains electrically neutral. Here, for every extra hole introduced in the crystal, there is the extra negative charge of the unpaired electron of the semiconductor atom. Often, schematic representations of the n-type semiconductor would have a preponderance of negative charges, whilst a preponderance of positive is used for illustrations of the p-type semiconductor. These negative and positive signs are meant only to indicate the charge of the majority carriers, and not the overall charge on the crystal.

15.5 THE SEMICONDUCTOR DIODE

The more useful properties of the semiconductor arise from what happens when two or more oppositely doped semiconductor crystals are joined together. The simplest combination results from joining a p-type semiconductor to an n-type semiconductor. When this is done, we get what is known as the **diode**, and the interface between the two semiconductors is called the **PN junction**.

Let us now investigate the processes taking place when a p-type semiconductor is joined to an n-type semiconductor. Remember that the majority carriers in the p-type are holes, and those in the n-type are free electrons; but each crystal as a whole is electrically neutral. When a p-type semiconductor is joined to an n-type semiconductor, some initial recombination of holes and free electrons occurs at the junction of the two crystals. This recombination process occurs only for a brief moment, and stops. Each hole in the p-type semiconductor recombining leaves behind a net negative charge on the p-type semiconductor. Similarly, each recombining electron from the n-type crystal leaves behind a net positive charge on the n-type semiconductor. The excess negative charge on the p-type semiconductor repels further electrons from crossing the junction to recombine with holes in the p-type. In a similar way, the net positive charge on the n-type crystal prevents further holes from the p-type from recombining with electrons in the n-type. In this way, a potential barrier is created at the junction of the two crystals. This barrier potential is called **junction barrier**. This potential barrier would have to be overcome if further recombination is to take place. Controlling this potential across the junction holds the key to controlling the current in the semiconductor. For silicon, this junction barrier is 0.7 V, and for germanium it is 0.3 V.

Within a short distance on either side of the junction, there is a high concentration of positive charges in the n-type semiconductor, and negative charges in the p-type semiconductor. Due to this high concentration of immobile charges, this region is commonly called the **space charge region**. This region is also known as the **depletion region**, since it contains no mobile charges – i.e. it is depleted of free electrons and holes. The typical thickness of the space charge region for silicon is around 10^{-4} cm. **Fig. 15.3** depicts the PN junction.

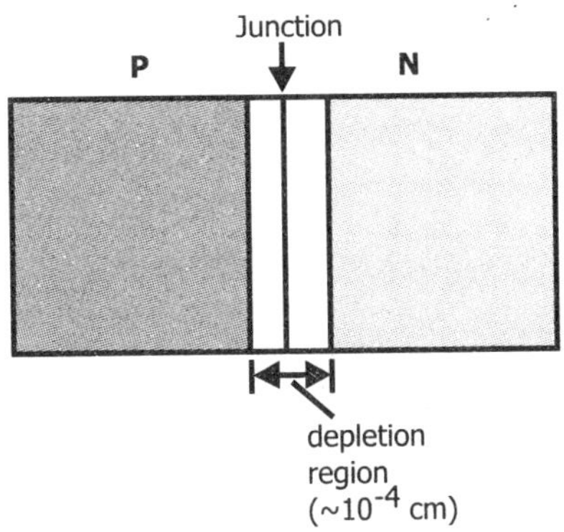

Fig. 15.3 The PN junction

15.5.1 Biasing the diode

Biasing is the term used to describe the connection of the diode to a d.c. voltage supply. There are two methods of biasing: **forward bias** and **reverse bias**. **Fig. 15.4** shows circuit symbol for the diode, and diagrams for both biasing connections.

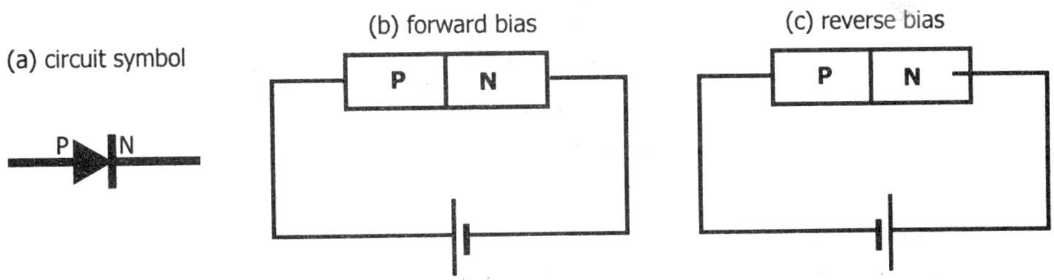

Fig. 15.4 The PN junction; (a) circuit symbol, (b) forward bias, (c) reverse bias.

In forward bias, the p-type semiconductor is connected to the positive terminal of the supply, while the n-type semiconductor is connected to the negative terminal. The holes in the p-type are repelled from the positive terminal and attracted to the negative terminal. Similarly, the free electrons of the n-type are repelled from the negative terminal and attracted to the positive terminal. This cross movement of charges across the depletion region constitutes the current that is obtained. Thus current flow across the depletion region is due to the movement of majority carries in both semiconductors.

In reverse bias, the p-type semiconductor is connected to the negative terminal of the supply, while the n-type is connected to the positive terminal. The holes in the p-type are attracted to the negative terminal, while the free electrons of the n-type are attracted to the positive terminal. This further increases the depletion region, thereby greatly limiting the current flow. There is, however, some small current flow in this reverse bias connection. This current flow is due to the movement of the minority carriers in both semiconductors. For these minority carriers, the PN junction would be forward biased. Because of the relatively small number of minority carriers compared to the majority carriers, currents in the reverse bias connection are much smaller than those in the forward bias connection. Reverse bias currents are usually a thousand times less than the corresponding forward bias currents.

15.5.2 I-V characteristics of the PN junction

The I-V characteristic of a PN junction is the variation of the current through the junction with the potential difference applied across the junction. **Fig. 15.6** shows the typical current-voltage characteristics for the PN junction.

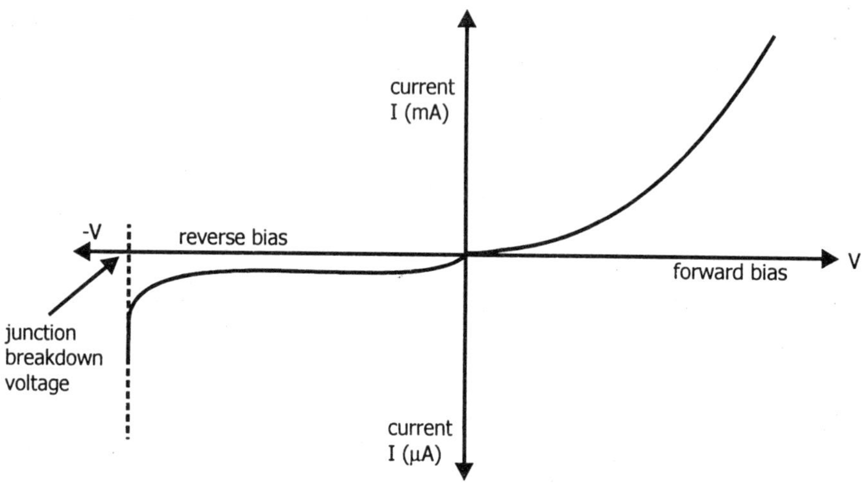

Fig. 15.6 I-V characteristics for the PN junction

Due to the small current measured in both the forward bias and reverse bias modes, the PN junction is considered to have high resistance; the resistance being much higher (μA currents) in the reverse bias mode than in forward bias. In the forward bias mode, the current through the junction increases nonlinearly. Thus, the current-voltage relationship does not obey Ohm's law. The diode is an example of a non-ohmic device.

In the reverse bias mode, the current is small but constant, until the reverse bias voltage reaches the **junction breakdown voltage**. Thereafter, the current increases immensely. For silicon, this reverse voltage has magnitude around 25 V.

15.6 USES OF THE DIODE

In general, solid state devices play one of the following basic roles in any circuit:

1. To stop the flow of electron current,
2. To start the flow of electron current, and
3. To control the amount of current flow.

When connected in a circuit, the diode allows electron current flow in one direction whilst blocking the flow in the opposite direction. One use of the diode is in **rectification** – i.e. converting an alternating voltage into a direct voltage. The other common use of the diode is in switching circuits. We treat these uses below.

15.6.1 As a rectifier

The word 'rectification' is used in electronics to describe the process of converting an alternating current input into a direct current output. Any device that achieves this is called a **rectifier**.

In the section on biasing the diode, we found that the reverse-biased diode carries very little current (of the order of microamps), whilst in the forward bias mode the device can conduct a thousand times more current. Thus if a diode is connected across an alternating supply voltage, the device will conduct in one half cycle when it is forward biased, and conducts little or no current in the other half cycle when it is reverse-biased. The output therefore would be a direct voltage in only one direction. With only one diode, we have **half-wave** rectification, whilst **full-wave** rectification is obtained with two or four diodes. These circuits are accordingly referred to as half-wave and full-wave rectifier circuits. The more common arrangement is the full-wave rectifier. The usual circuit diagram for this is shown in **Fig. 15.7** (a), whilst (b) shows the waveforms of the input and output voltages. In practice, combinations of capacitors and resistors are used to produce a constant output.

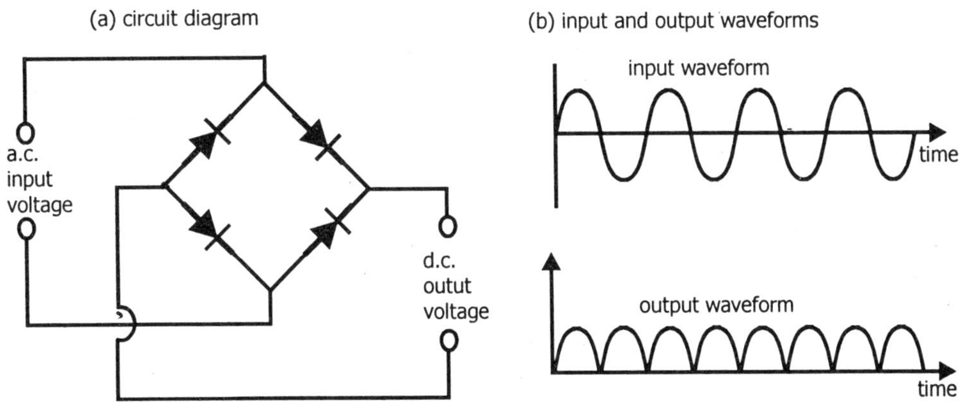

Fig. 15.7 Full wave rectification using the semiconductor diode; (a) basic circuit diagram, (b) input and output voltage waveforms.

15.6.2 As a switch

This is most common in logic gates in digital circuits. In this application, the gate opens when the diode conducts, and is closed when the diode is in the reverse bias mode – i.e. not conducting. We shall study more on logic gates later on in the chapter.

15.7 THE TRANSISTOR

From our study of the PN junction, the following can be said:

A PN junction biased in the forward direction acts as a low resistance element (on account of the high current),

and

A PN junction biased in the reverse direction acts as a high resistance element (on account of the low current).

In 1947, workers at Bell Laboratories experimented a new device formed by joining together two PN junctions, one biased in the forward direction, and the other in the reverse direction. In these experiments, a current was sent through the forward biased junction into the reverse biased junction. This device was named the **transistor**, a word representing a short form of trans-resistor. The device **TRANS**fers current from a low res**ISTOR** into a high res**ISTOR**.

The transistor can be made by having an n-type semiconductor sandwiched between two p-type semiconductors, to make up what is known as a PNP transistor. Alternatively, a p-type semiconductor can be sandwiched between two n-type semiconductors. This makes an NPN transistor. These are shown in **Fig. 15.8**. The biasing of the device is shown in the figure. As can be seen, one junction is forward-biased, whilst the other is reverse-biased. The middle element is called the **base**, on either side of which are the **emitter**, and the **collector**. The emitter-base junction is forward biased, while the base-collector junction is reverse biased. The circuit symbol of the transistor is drawn with the three legs, and an arrow between the base and the emitter. By convention, the arrow is set to point in the direction of the hole current – i.e. the flow of current due to the holes. For the PNP transistor, holes flow from the emitter to the base; for the NPN transistor, holes flow from the base to the emitter.

There are three main manufacturing requirements for a transistor to operate. The first is that the emitter must be heavily doped. Heavy doping means there is a large supply of free charges that can be made to flow through the base into the collector. The second requirement is that the base must be very thin and lightly doped. Thirdly, the collector voltage must be relatively high. As the emitter-base junction is forward biased, small emitter-base voltages are required for operation. Reverse biasing the collector-base

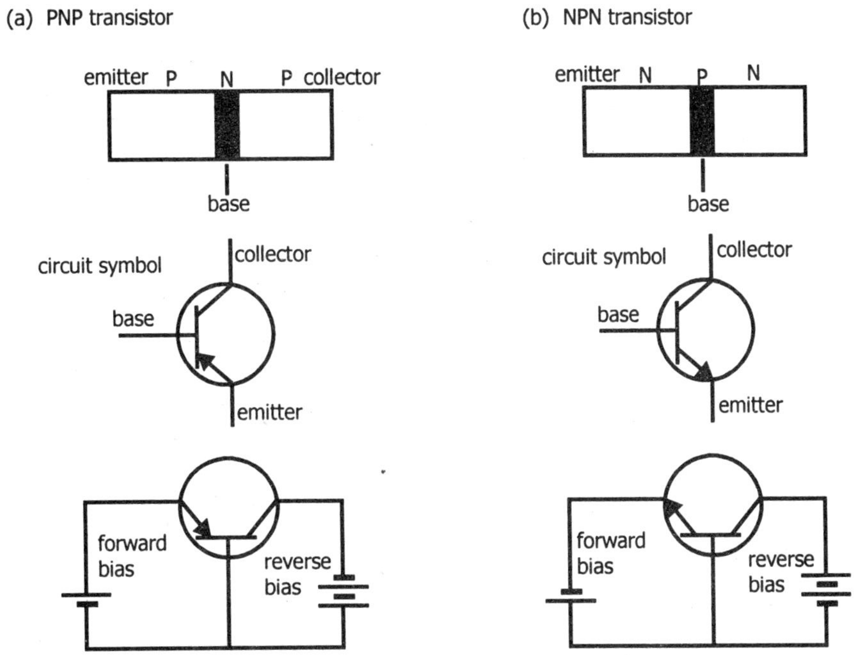

Fig. 15.8 PNP and NPN transistors and their connections

junction implies that large collector-base voltages are required to transfer all the mobile charges from the emitter into the collector. Typical values of emitter-base and base-collector voltages are of the order of 0.6 V, and 21 V, respectively.

The PNP transistor behaves in exactly as the NPN transistor. The main difference in connection lies in the polarities of the bias voltages. For the PNP transistor, a negative voltage is placed on the base, a positive voltage on the emitter, and a positive voltage on the collector. In the NPN transistor, a positive voltage is placed on the base, a negative voltage on the emitter, and a negative voltage on the collector. Because most application voltages are positive, the NPN transistor is more widely used than the PNP transistor.

15.8 TRANSISTOR APPLICATIONS

No matter the type, a transistor is used to do one of two things: to amplify a given input voltage, and to switch power on/off a given device.

15.8.1 Amplifier configurations

In general, the purpose of an amplifier is to produce an output voltage/current larger than the input. The word amplifier is, however, also used in circumstances where the output is less than the input. We refer to the degree of amplification by a parameter called the **gain** of the amplifier. The gain of the amplifier is the ratio of the output voltage/current to the input voltage/current. The voltage gain A_v is given by

$$A_V = \frac{\text{voltage output}}{\text{voltage input}} = \frac{V_{out}}{V_{in}} \tag{15.1}$$

Similarly, the current gain A_I is given by

$$A_I = \frac{\text{current output}}{\text{current input}} = \frac{I_{out}}{I_{in}}. \tag{15.2}$$

There are three types of transistor amplifier. These are the **common-emitter** amplifier, the **common-base** amplifier, and the **common-collector** amplifier. We here briefly review these.

(a) Common-emitter amplifier

The most common class of transistor amplifier is the common-emitter amplifier. This amplifier has the emitter common to both input and the output circuits. The input voltage to the transistor is applied between the base and the emitter, and the output obtained between the emitter and the collector.

(b) Common-base amplifier

The common-base amplifier has the base element common between the input and output circuits. The input is fed between the base and the emitter, and the output is obtained between the collector and the base. The gain of this amplifier is lower than that of the common-emitter amplifier.

(c) Common-collector amplifier

As the name suggests, the common-collector has the collector element of the transistor common to the input and output circuits. The input is between the collector and the base, whilst the output is between the collector and the emitter. Calculations and measurements show that the output voltage from this amplifier is less than the input voltage – i.e. a gain of less than unity.

Using the NPN transistor, this amplifier configuration is shown in **Fig. 15.9**. If the PNP transistor were to be used, the polarities of the bias voltage would be reversed.

15.8.2 Switching applications

In addition to amplifying, the transistor is also used as a switch – i.e. turning power to a given device on or off. From our study so far, current flows in a transistor when emitter-base junction is forward-biased, and the current flow stops when the junction becomes reverse-biased. If the output of a transistor is connected to a device, the device comes on

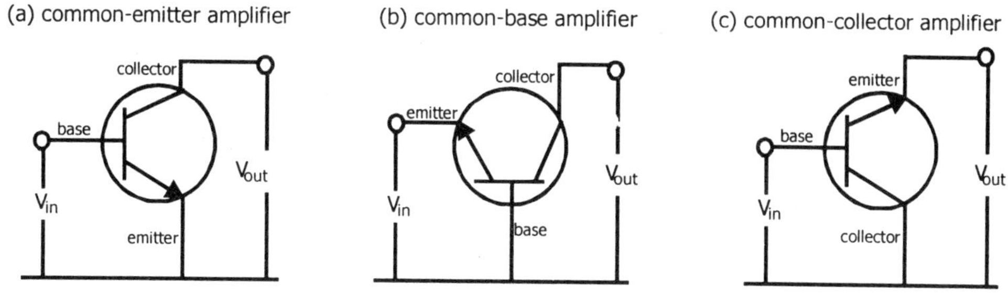

Fig. 15.9 Transistor amplifier configurations; (a) common-emitter amplifier, (b) common-base amplifier, and (c) common-collector amplifier.

when the emitter-base junction becomes forward-biased, and off when the junction is reverse-biased. This way the transistor functions as a switch.

Switching transistors are used in circuits that are either turned "on" or "off". Many of these transistors are found in automated circuits, the most complex being the computer. These transistors control circuits in which the signal is either present or not present – as, for example, in applications in which a light beam comes on automatically, either at a particular time, or at regular intervals.

15.9 INTEGRATED CIRCUITS

A typical NPN or PNP transistor operating as an amplifier produces a voltage gain of around 200. Thus an output of around 3.0 V can be obtained from an input voltage of around 15 mV. While this can be considered good for some applications, much smaller voltages need to be amplified in many other situations. An amplifier with much higher gain is necessary for these applications. A common approach toward increasing the gain is to use more than one transistor, and link them up in such a way that the output from one serves as input for the next, and so on. Such **multistage** amplifiers, as they are called, are rather expensive to put together, especially when very large gains are desired.

As a switch, a single transistor can be very effective to control a simple operation – e.g. turning on a lamp. Where many similar switching operations are to be undertaken in tandem, many transistor switches are required. Again, putting together a large bank of single transistors to achieve these is rather expensive and time consuming.

Modern technology has addressed the above problems with the manufacture of the **integrated circuit**, simply referred to as **IC**. As silicon is more common than germanium in solid state devices, the IC is also commonly referred to as the **silicon chip**. Even in one of its simplest forms, the integrated circuit can behave in a manner equivalent to a large number of transistors and diodes put together. Some ICs produce amplification factors of the order of 3000 or more. These are known as **operational amplifiers**. Other ICs used for switching do also perform the switching action of a large number of transistors and diodes put together. In many cases, ICs are manufactured to perform these functions as desired. As technology has developed, today's electronic component market has an increasing range of these devices, designed to play the many switching and amplifying roles in circuits. We find some of these control circuits in the household washing machine, Hi-Fi system, automated cameras, and many other devices.

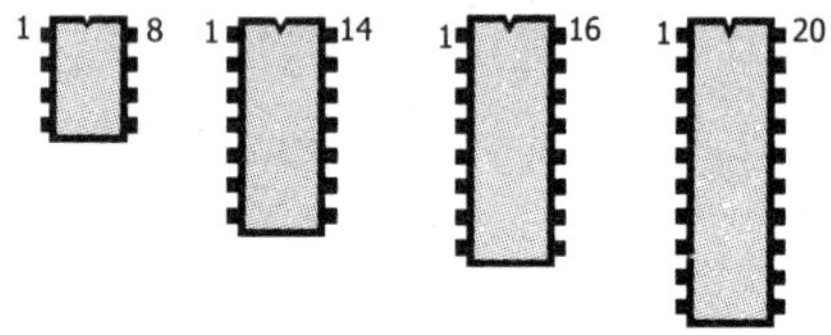

Fig. 15.10 Some integrated circuits.

The silicon chip is normally in the shape of a rectangular block with metal connecting legs (usually called 'pins') along each of the longer sides. These legs provide connections to the various elements of the circuitry inside the chip. The number of legs of ICs are usually 8, 14, 16, 20, and more. **Fig. 15.10** is a top view of some of the common chips.

Each integrated circuit comes with a set of notes indicating the kind of chip it is, the pin identifications, showing the connections of the input, the bias, and output voltages. The booklet on the device would also show the recommended operating conditions – temperature, voltage limits, gain, etc.

15.10 DIGITAL ELECTRONICS

Integrated circuits fall into a number of categories. These include operational amplifiers, logic gates, timers, and microprocessors. Operational amplifiers employ a large batch of miniature multistage transistor amplifiers connected such that overall gains in the order of thousands are obtained. With these devices, millivolt inputs can be amplified to produce outputs of the order of volts.

Whilst operational amplifiers give amplified outputs, logic gates produce the same output – typically 5 volts – when triggered. The electronic gate is either open or closed, switching a device on or off. The electronics industry today has a wide range of these gates, some having one, two, or more inputs and outputs. Some of these chips are used as timers, switching a device in some state for a prescribed period (or at specific times). This latter application underpins the behaviour of many automated gadgets we take for granted today.

Logic gates are at the heart of the modern branch of electronics known as **digital electronics**. The electronics of valves and amplifiers give out **analog** outputs – i.e. outputs of volts, or millivolts, and amps, or milliamps. Logic gates are either open or closed, corresponding to the switching off or on of an electrical device. The number 0 is returned when the gate is closed, (or device switched off) and 1 when the gate is open (or device turned on).

The behaviour of logic gates and their combinations is analysed using a special branch of mathematics developed for the purpose. The branch is called **Boolean Algebra**, so named in honour of its inventor, Arthur Boole. This algebra works with the numbers 0 and 1, corresponding to a circuit component being off or on, respectively. Whilst the complete analysis of the laws of Boolean algebra lie outside our current scope, we can gain an insight into some of the basic operations through the switching action of logic gates. Below are the basic gates.

15.10.1 The AND gate

This is also called "all or nothing" gate. The gate accepts two or more inputs. An output of zero is returned if any of the inputs is zero. An output of 1 is returned when **all** of the inputs are 1. The simplest AND gate has two inputs. **Fig. 15.11** (a) shows the circuit symbol for the AND gate, whilst (b) illustrates the switching mode of the gate. In this illustration, the lamp L comes on only when both switches A and B are conducting. The behaviour of any logic gate is presented in a **truth table**. The truth table for the AND gate is also given in the figure.

(a) symbol for AND gate

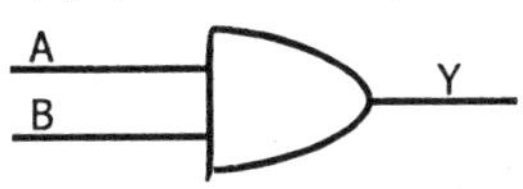

Truth table

Inputs		Output
A	B	Y
0	0	0
0	1	0
1	0	0
1	1	1

(b) AND circuit using switches

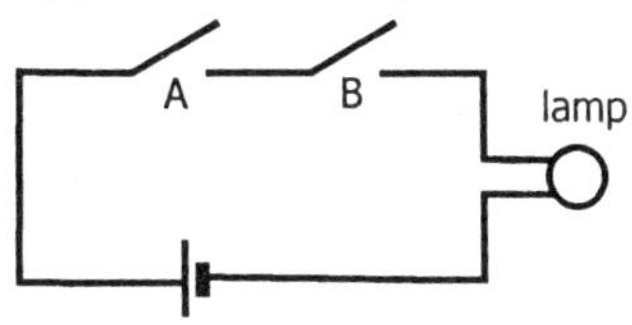

Fig. 15.11 The two-input AND gate; (a) circuit symbol, (b) AND switching

The three-input and the four-input AND gates work on the same all-or-nothing principle, giving output only when there is a voltage on all of the inputs.

15.10.2 The OR gate

The OR is also called the "any or all" gate. The gate accepts two or more inputs. An output of zero is returned only if there is aero input on all gates. An output of 1 is returned if any of the inputs is 1. **Fig. 15.12** (a) shows the circuit symbol for the OR gate, whilst (b) illustrates the switching mode of the gate. The lamp comes on when any or both switches A and B are conducting. The behaviour of this gate is shown in the adjoining truth table.

The three-input and the four-input OR gates work on the same any-or-all principle, giving output when any input is on.

(a) symbol for OR gate

Truth table

Inputs		Output
A	B	Y
0	0	0
0	1	1
1	0	1
1	1	1

(b) OR circuit using switches

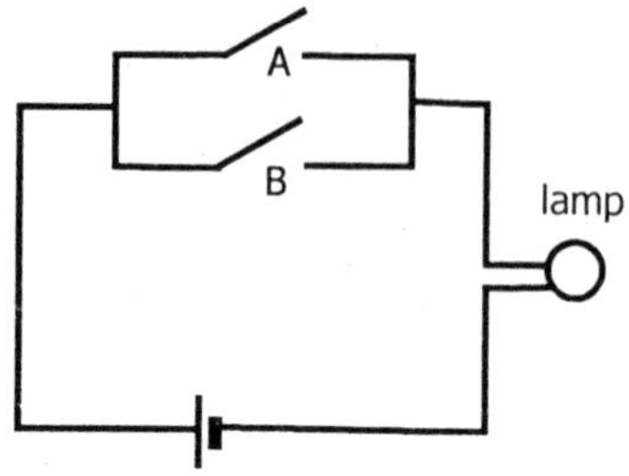

Fig. 15.12 The two-input OR gate; (a) circuit symbol, (b) OR switching

15.10.3 The NOT gate

This gate is also called the **inverter**, or inverting gate. A NOT gate has only one input, and one output. An output of 1 is returned for zero input, and vice versa. **Fig. 15.13** (a) shows the circuit symbol for the NOT gate, and its truth table. Two NOT gates connected in series would produce an output of 1 for input of 1. This is shown in diagram (b). The behaviour is shown in the truth table.

(a) symbol for NOT gate

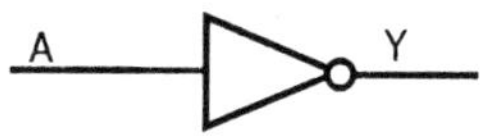

Truth table

Input	Output
A	Y
0	1
1	0

(b) two inverters in series

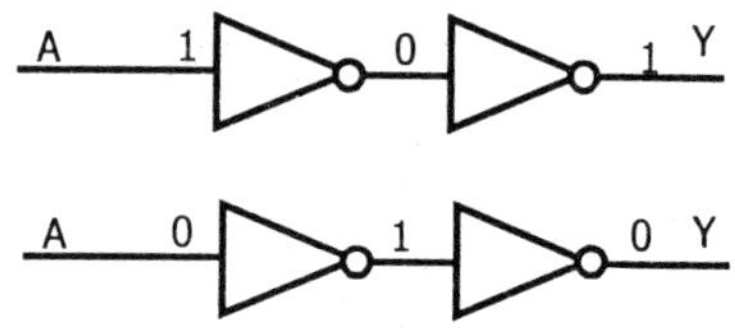

Fig. 15.13 The NOT gate; (a) circuit symbol, (b) two inverters in series.

In the analysis of the performance of the NOT gate, a common term used is **complement**. In normal circuitry, the opposite of a device being "on" is being "off". In digital electronics, "0" corresponds to the "off" state, and "1" corresponds to the "on" state. The complement of "0" is hence "1", and the complement of A is written as $\overline{A}$.

15.10.4 Combining logic gates

Fig. 15.13 (b) gives a simple case of combining logic gates. Logic gates of any type can be combined to give any desired output when the inputs vary. This has provided modern electronics with an almost endless set of combinations for the control of complicated circuits. These circuits are used in many devices – from household gadgets to laboratory instruments. Special useful gates have resulted from combining two or more basic gates. These are the NOR gate and the NAND gate.

The NOR gate

The NOR gate performs the inverse of the OR gate. A simple equation representing the NOR gate is

$$\text{NOR} = \text{OR} \bullet \text{NOT},$$

and **Fig. 15.14** (a) is a schematic of the combination that gives the NOR function, whilst (b) is the circuit symbol for the NOR gate.

(a) OR and NOT gates combined

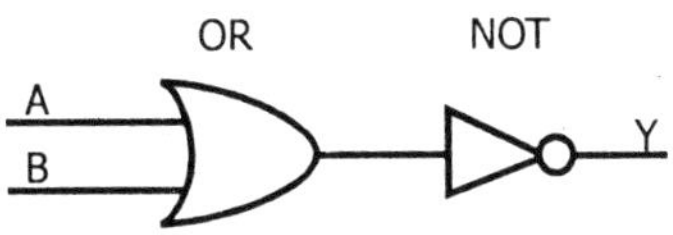

truth table

Inputs		Output
A	B	Y
0	0	1
0	1	0
1	0	0
1	1	0

(b) symbol for NOR gate

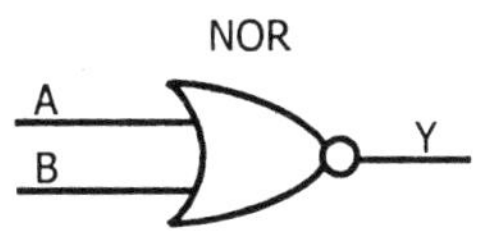

Fig. 15.14 The NOR gate and its truth table

From its truth table, it is seen that this combination conducts only when there is no input on either A or B; alternatively, there is an output voltage when neither A **nor** B has an input. In practice, many electronic logic devices perform the NOR function easily, thereby making it possible to achieve the nor function without the need for two gates, as this is often more expensive.

The NAND gate

NAND is short for "NOT AND". The expression for NAND is

$$\text{NAND} = \text{AND} \bullet \text{NOT}$$

This gate is formed by putting an AND gate and a NOT gate together. **Fig. 15.15** shows the combination, the truth table, and the usual circuit symbol. It is clear from the truth table that this combination does not conduct when both A and B have inputs.

(a) AND and NOT gates combined

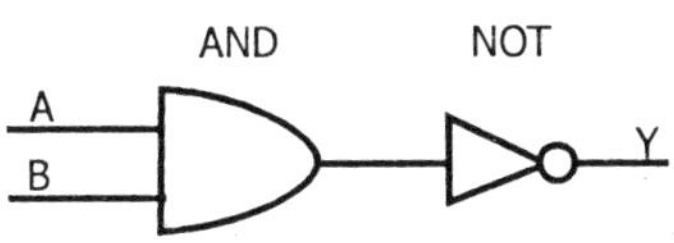

truth table

Inputs		Output
A	B	Y
0	0	1
0	1	1
1	0	1
1	1	0

(b) symbol for NAND gate

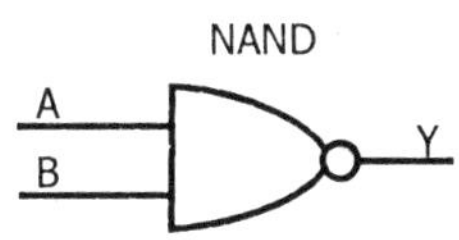

Fig. 15.15 The NAND gate and its truth table

The EXCLUSIVE-OR (or XOR) gate

The exclusive-or (abbreviated as XOR) gate is referred to as the "any but not all" gate.

There is an output for any input, but no output for all "0" or all "1" inputs. **Fig. 15.16** shows the XOR combination, the truth table, and the usual circuit symbol.

(a) NOT, AND and OR gates combined

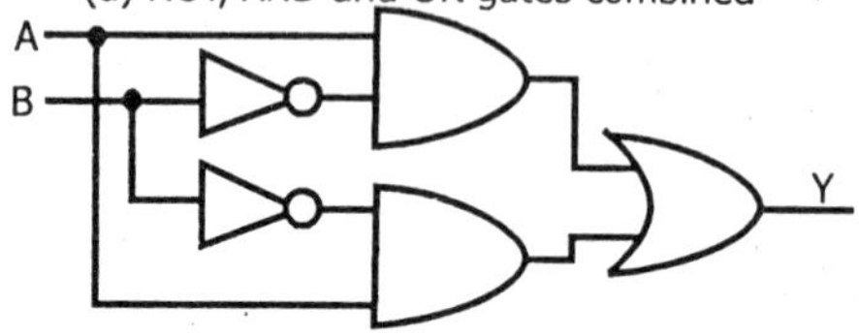

truth table

Inputs		Output
A	B	Y
0	0	0
0	1	1
1	0	1
1	1	0

(b) symbol for XOR gate

Fig. 15.16 The exclusive-OR (XOR) gate and its truth table

Other combinations

In addition to the above, other combinations of gats are used in order to obtain the performance required. **Fig. 15.17** gives an example of one such combination that is common in many intruder alarm circuits.

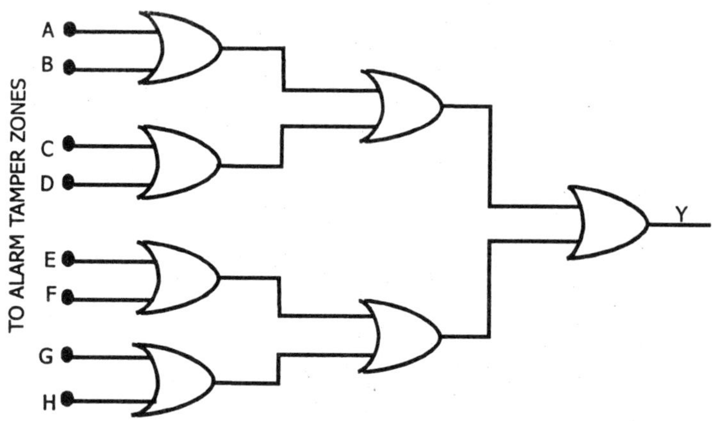

Fig. 15.17 An example of gate combination in an intruder alarm circuit.

The inputs A, B, ..., are obtained from tamper circuits which come on when their respective switches are opened. There is always an output when any or all of the tamper switches are opened. In operation, this output is used to trigger a buzzer or siren. It is left as an exercise for the student to construct the truth table for this arrangement.

In addition to operating switches, logic gates can be combined in circuits to perform arithmetic operations such as addition, subtraction, multiplication, division, powers, logarithmic, exponential, and trigonometric functions. These advancements has given rise to the pocket calculators, advanced teller machines used at supermarket outlets, and in complicated scientific and industrial applications. At the very advanced end of digital electronics is the **microprocessor**. This is a single chip fabricated in such a manner that it does the functions of a large number of chips put together. This single chip can sometimes have as many as 40 connecting points or more. The microprocessor can be programmed to perform many complicated switching operations, perform arithmetic operations such as addition, subtraction, multiplication and division. It also has the ability to store the results of these operations in certain memory locations, for use when required. The microprocessor allows the operator to **program** the operations desired. e and the operator in the form of an **input**. The microprocessor then carries out the instructions at the input and gives an **output** information.

The microprocessor is today at the heart of any complicated electronic devices. Examples are in toys, remote control units for certain household appliances, washing machines, household intruder alarm circuits, dishwashers, microwave ovens, and many more. The most intricate use is in the microcomputer, discussed in the next chapter.

In the technology today, we take the operations of switches and buttons so much for granted. It is nice to note that what brings about the desired results have been possible with the revolutionary advances in the field of electronics. Bearing in mind that the beginning of it all was the diode, we come to appreciate the huge developments in the field of electronics – a technology that has had immense impact on our lives today.

QUESTIONS

1. Using examples where necessary, explain the following terms used in electronics:
- (i) intrinsic semiconductor,
- (ii) extrinsic semiconductor,
- (iii) depletion region,
- (iv) barrier potential,
- (v) rectification,
- (vi) transistor,
- (vii) logic gates,
- (viii) truth tables,
- (ix) integrated circuits,
- (x) microprocessor.

2. (a) List the three basic gates used digital electronics, describing the function of each.
(b) Give two examples of how basic gates can be combined.

3. A two-input AND gate has two trains of pulses on its inputs A and B. The levels of pulses at different times intervals are shown, the duration of each time interval being 100 ms.
(a) Complete **Table 15.1** by filling in the output level for each time interval
(b) Sketch a graph of the output level as a function of time.

Table 15.1

Time (ms)	Level of inputs		Output level
	A	B	Y
100	0	1	
200	1	1	
300	1	0	
400	1	1	
500	1	0	
600	0	0	
700	0	1	
800	0	0	

4. Complete Exercise 3 of for a two-input OR gate, with the same inputs.

5. (a) Draw a sketch of the four-input OR gate
   (b) Draw the truth table for the four-input OR gate.

6. Prepare the truth table for each of the logic combinations depicted in **Fig. 15.18**.

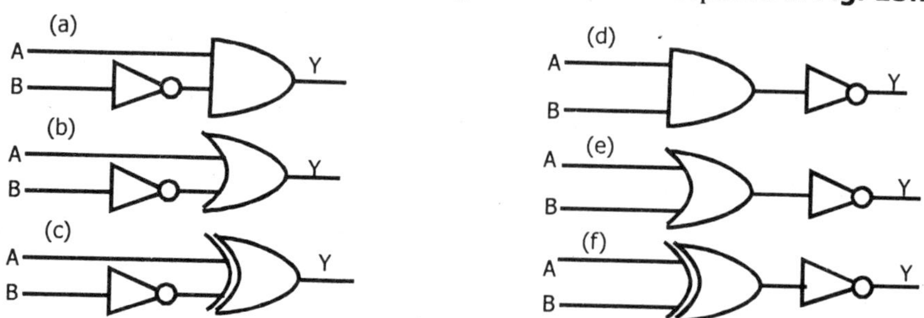

Fig. 15.18

7. In the circuit segments shown in **Fig. 15.19**, the output of each logic combination is connected to a lamp. For each combination state the input conditions that would switch the lamp on. Verify using a truth table. Assume all gates are correctly biased.

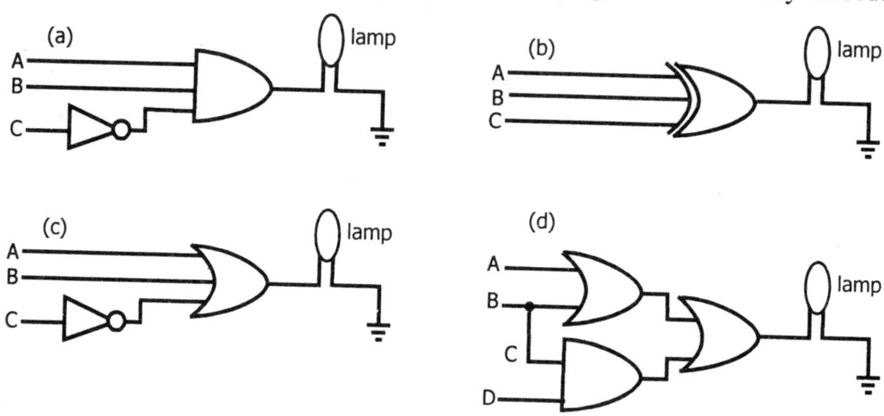

Fig. 15.19

8. Prepare the truth table for each of the logic combinations depicted in **Fig. 15.20**.

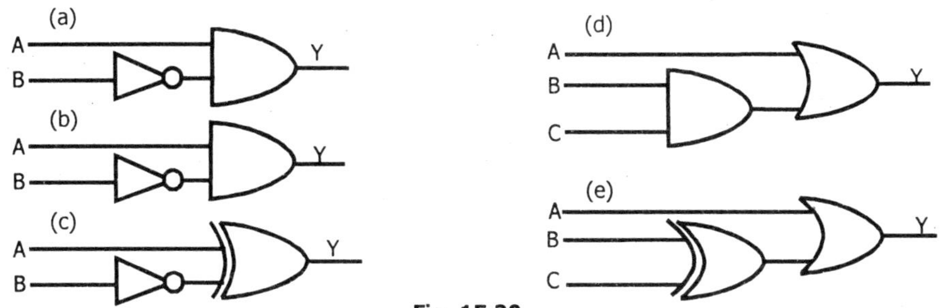

Fig. 15.20

16

CHAPTER

An Introduction to Co puters

16
CHAPTER

An Introduction to Computers

The word 'computer' has become a household name in our technology world. Computers have opened up lots of possibilities into areas of advancement that would otherwise not be achievable. As our societies grow larger and larger, there is need to process the ever increasing amount of information that is put out everyday. In the last twenty years, we have witnessed what amounts to an explosion in computer use in a manner very similar to the impact of the industrial revolution in the middle of the nineteenth century. With the enormous expansion in computer technology, this age is often described by many as the **age of the information revolution**.

With the kind of impact they have had on society and technology, it is clear that any book of this nature would not be complete without an account, albeit brief, of what computers are, and how they manage to play the roles we all have come to take so much for granted. This last chapter is devoted to providing an introduction to the amazing world of computers. A lot of details of computer makeup are left out for obvious reasons, leaving the reader with an account that puts emphasis only on its broad knowledge and uses.

16.1 THE INFORMATION AGE

Our societies have become information based. Much of this information is on paper; these papers are found moving everyday in files in offices, in filing cabinets, and in storage rooms. Generating and handling these huge amounts of information requires an army of workers. These workers can be classified as '**knowledge workers**'. They are concerned with producing information either through manual of automated processes. To handle all the information that society finds necessary requires the use of computers. Consequently, computers are commonplace and, with the kind of impact they have had on our lives, everyone should be familiar with computers and their capabilities and limitations. In fact, you may already own some computers without being aware of it.

As computer use has increased, we have unwittingly entered into a new world. Most businesses, industries, institutions, and even homes have felt the influence of computers. Computers calculate our telephone and electricity consumption bills, make it easier for governments to manage natural resources (e.g. fish stocks), make weather stations process the large information necessary to predict weather patterns, etc.

Computers are excellent in performing repetitive calculations. They are better at such tasks than humans. Humans suffer fatigue and outside influences, making the task of repetitive calculations very boring and prone to errors. Often, because of our extreme reliance on computers for fast routine calculations, we get the impression that the computer is infallible. As we seek to understand how computers function, it is important to appreciate the fact that computers do fail. Computers do make mistakes, and their operations rely on information fed into them by their human counterpart.

Computers are so much a part of our lives today that computer terminology has entered into our everyday language expressions. For example, we often see job advertisements where the salary is quoted as '25K' to '35K' (meaning 25,000 to 35,000). Also, we talk of someone's 'input' (meaning contribution) on a discussion topic. We 'interface' (meaning relate) daily with our peers and people in the workplace. There are many similar terminologies we use without knowing exactly where they come from.

There is nothing wrong with this, as long as we use these terms in their right contexts. Common computer-related terms do give us a good start in learning about computers.

In its application, the computer takes in **data** fed to it by the user. Data are raw facts and figures. Examples are ages of children in a school, temperature measurements at certain times of the day, etc. By themselves, these data have no meaning. They only acquire meaning when they are processed to give, for example, the average age of students in a given class, and variation of temperature in a given season. In other words, the data has been processed. Processed data is called **information**. The information age therefore relies on processing data into useful information that the society needs.

16.2 WHAT IS A COMPUTER?

Computers come in many forms. Some perform only one task routinely all the time; others perform many tasks either simultaneously or in sequence. For these reasons, it is not always easy to give a succinct definition of a computer. We here adopt what is considered a 'working definition' of the computer, based on its properties and what can be achieved with it. We proceed as follows:

1. A computer is an **electronic device**, using the movement of electrons in solid-state devices, rather than the physical movement of internal parts to achieve its results. This aspect excludes mechanical calculators – e.g. the abacus.
2. A computer has **internal storage**. It can store instructions and data very fast. With this property, the computer can execute program statements in sequence at high speed.
3. A computer is capable of using a **stored program** to direct activities. It internally stores a series of instructions that detail the methods of handling specific operations, and the order in which the operations are to be performed. This requires us to be able to 'communicate' with the computer. The languages used in this communication are known as **programming languages**. Detailed knowledge of these languages is a specialised field, and people with this task are called **programmers**. Only a tiny fraction (less than one percent) of computer users today are programmers.
4. A computer allows **program-execution modification**. It can make decisions that change the course of a program's execution, depending on the value of the data in storage or on the results of arithmetic or logical operations.

In summary, therefore, we can define a computer as

an electronic machine possessing internal storage capabilities, a stored program of instructions, and the capabilities to modify the course of executing the instructions during processing.

As stated earlier, the above serves only as a working definition. No doubt you will find other similar definitions in other books. Some definitions equate the computer to the microprocessor upon which it is built. We shall learn in this chapter that the computer is a lot more than just the microprocessor.

Computers come in many models, makes and sizes, varying in complexity from the so-called single task versions (e.g. scanning machine at the supermarket checkout) to the more sophisticated multi-task versions designed to handle jobs of high complexity. These latter machines often come in a range of sizes, contained in a few boxes. To cover the varying versions of computers, the term **computer system** is considered a more fitting description. A computer system is composed of machines, and programs for operating the computer. In the language of the computer industry, these are called the **hardware** and the **software**.

Hardware is the tangible computing equipment. It is the physical machinery. It is a complicated web of electronic connections all interrelated and working in tandem to produce the computing results. In addition to this machinery, there are other components that are also classified as hardware. Printing equipment, scanners, pointing devices, etc. fall into this category. These latter are called **peripheral devices**, or simply **peripherals**. They are also called **input/output devices**, or just **I/O** devices. The hardware is the most evident part of the computer system. It is readily visible, and appears to be the most expensive.

Software is the general name given to the sets of instructions that describe to the computer system what to do. Software sometimes are called **computer programs**. Specific software is designed to instruct the computer on the method of solving a specific problem. Examples of software are **word processing** and **spreadsheet** programs. A given software outlines a systematic course of action for the computer to follow. Programming languages enable the computer to communicate with the programmer. Raw data is fed into the computer; the computer then follows the software directions for processing the data into information.

16.3 USE OF COMPUTERS

With explosion of computer technology, computers are used increasingly in all places of work in our societies. The machines are used in stores, in schools, in manufacturing plants, in medicine, etc. There is an ever-expanding list of these uses; all we do here is to give a few of the common areas of application.

16.3.1 Computers in society

People use computers regularly to make their jobs of data handling easier. The business world generates lots of data. Examples are data on sale of items from a store, data on tax returns, and data on employee wages. Constant monitoring of sales data over the years helps the business house to project on future activities and market trends. If a company is to remain competitive, performance analyses of various kinds are bound to be part and parcel of the company's operations. These analyses and projections are done using computers.

Banks constantly keep records of deposits and withdrawals by their clients. This accounting procedure makes it possible for the bank to make decisions on their clients' status when it comes to assessing the suitability of clients for certain loans. The bank also keeps data of its own investments and dealings with other banks and other financial institutions. Like any business, this information allows the banks to make projections.

Foreign trade between countries is based on realistic market assessment of one country in relation to another. Currency valuations depend on certain economic performance indicators. These indicators result from data on the economic activities – e.g. mining activities, agricultural produce, manpower valuations, natural resources, etc. These data are used to work out the rate of exchange with other currencies, and the price of commodities on the world market.

Modern methods of weather forecasting have been possible thanks to the application of computers. Weather patterns from wide regions are analysed using computers. Information obtained is very useful to the transportation industries. The flight programs of airplanes, and shipping timetables depend crucially on these weather forecasts.

16.3.2 Computers in schools and colleges

In the school or college, computers are used for purposes of data management, as an aid to teaching and learning, and in laboratory and other technical exercises.

(a) Data management.

A typical school or college of medium size in most places would have in excess of 600 students, about 30 teachers, a Principal's Office with administrative staff consisting of a bursar, secretarial staff, and other support staff. Add to these personnel in special school programs like agriculture, mechanic workshop, and you easily have the equivalent of a small village of almost a thousand people. The staff are in different categories, with different managerial needs. The school management keeps records of the activities and jobs of all these people on a routine basis.

The school needs to manage the have records on all students. These records would include details on the parental background of each student, payment of fees, subjects taken by the student, test and examination performance in these subjects, performance in sporting and other activities. These performance indicators are used to generate terminal testimonials for the students.

Information on teachers also forms part of the school data. Academic background of each teacher, subjects taught, classes taught, and assessment records.

Extracurricular activities need to be planned and executed in such a manner as to cause minimal disruption to the school's general program. These activities would include open days, special days in the school calendar, visits to the school by dignitaries, and sporting events.

To successfully administer all the above records needs good data management facilities and personnel. Data from all the activities are processed into useful information by computers. Decisions based on such information can then be made into sound policy to benefit the school as a whole.

(b) As an aid to teaching and learning

In the teaching tasks, computers are used increasingly to help get the teacher prepared. Many institutions today are turning to computers to put the teaching timetables together. In such cases the tedium in this job had been removed or minimised.

Our societies still rely heavily on the use of paper. Teachers fall very squarely into this category of paper users. Schools have seen an increase in the use of the computer to prepare teaching aids. These aids include, the class notes, complicated diagrams, handouts, and test questions. These materials are prepared only once and saved on the computer. Thereafter, modifications to the original document can be made quite easily. An added advantage of this method is that teachers are today better able to compare and exchange teaching information more than before.

A growing area of computer use in education is in what is commonly known as **computer-aided instructions**. Often abbreviated as **CAI**, these computer programs are designed for self-tuition. The programs are widely available in almost every subject in the school or college curriculum. In their use, the student goes through a set of instructions on a particular topic, and on many occasions is presented exercises to test understanding and reinforce confidence. The word **'interactive'** is often used to describe this one-to-one relationship between student and the computer. Although these programs do not entirely replace the teacher, they make the teacher's job a lot easier; concentrating on topics the student has difficulty with, rather than teaching every topic. These programs have also become very useful in helping the teacher prepare for classes.

(c) In laboratory and technical exercises

The advent of the computer has made it possible for some complicated procedures to be simulated. In simulation, we are able to carry out a procedure using special programs designed for the purpose. It is, for instance, possible to simulate the addition of one chemical to another in a test tube and study the result. Approaching complicated procedures in this way allows the student to learn quite a bit about what to do when carrying out the procedure in practice. This sometimes helps to minimise accidents, especially in the chemistry laboratory. Computers also help demonstrate some aspects of theory (e.g. motion of electrons in atoms, the behaviour of an electronic circuit under certain conditions, or rocket propulsion) that can be difficult to demonstrate practically. In many ways computers have reduced much of the chore from teaching and learning to mere fun.

16.3.3 Computers in industry

Computers have revolutionised practically all operations in today's industry. Graphics design is a lot easier now using specialised computer programs that enable the design engineer to produce complicated scale diagrams of plants and models in a time much shorter than before when things were done manually.

Manufacturing process control involves the use of computers to enable automation of continuous industrial operations. Examples of these operations include the control of electricity generating plants, processes involved in petroleum refineries, and automobile assembly plants. Whilst automation has been in existence in industries before the computer, the computer has made it a lot easier and more reliable.

Computers have become the main tool in the control of air traffic in the aviation industry. Many of the busy airports around the world use computer systems to safely handle the high volume of traffic – planes landing and taking off in quick succession.

The computer is also used to handle personnel information in the offices that control industry, in manner similar to the use in any other office. These are just a few of the many and ever growing list of industrial applications of the computer.

16.3.4 Computers in medicine

No area has received more attention in computer use than the health industry. Computers have become a vital aid to all of hospital work. These uses range from records of patient conditions, to monitoring patient conditions during the all-important surgical operations in the theatre. Many of these computers arc dedicated to single or at most a few dedicated tasks. Their use has impacted a lot on modern medical practices.

16.4 TYPES OF COMPUTERS

Computers come in many sizes and capabilities, depending on the dedicated uses. There are four types of computers, namely, the **supercomputer**, the **mainframe**, the **minicomputer**, and the **microcomputer**. Whilst you probably are going to use the microcomputer, it is good to know a little bit about the other computers.

16.4.1 Supercomputers

Supercomputers are very powerful machines. They are used to handle the most complex data and scientific calculations. Their costs make them prohibitive to be owned by an individual. Supercomputers are usually owned by government establishments, or by large organisations. Two examples of the supercomputer in use are the **CRAY** computer, and the **CYBER**. Both of these are very large heavy machines that require to be operated under very special conditions by well-qualified personnel. A typical CRAY computer can perform billions of calculations per second. The electronic circuitry needs to be cooled by some special liquid. All these requirements translate to a huge cost. A typical price for such a computer is in excess of thirty million US Dollars (i.e. $30,000,000.00). This price might include the salary of its maintenance engineer for the life of the machine. Using the CRAY computer, the United States National Weather Service can give the weather forecast for the entire globe in just a few minutes.

16.4.2 Mainframes

The mainframe computer is smaller than the supercomputer. Mainframes are large and powerful, and are often housed in big computer centres. They require special environmental controls, and must have special power supplies. Operating the mainframe also requires specially trained personnel and assistants. They usually cost in the range of $500,000.00 or more.

People who use mainframes normally work from remote terminals. They sit at a unit with a keyboard and a screen at a location that may be kilometres away from the computer. The terminal is hooked on to the mainframe via cables. A single mainframe can simultaneously have many users connected to it. Some users may not even be in the same city or town where the computer is located. In the latter case, such remote users are normally linked to the computer via telephone cables – a process known as **digital telecommunication**. Typical users of mainframes are banks, government agencies, and large commercial and industrial corporations.

16.4.3 Minicomputers

Minicomputers are smaller than mainframes and supercomputers, but are dedicated to handling the computer needs of many people. Minicomputers are smaller, slower, and less expensive than mainframes. Minicomputers are not as big as mainframes, and do not normally require special environmental conditions. They cost about one-tenth of a typical mainframe.

Being smaller in capacity, minicomputers support leas than the number of users that mainframes can support. Like mainframes, minicomputers also allow telecommunication. Although smaller, minicomputers offer a viable alternative to the more expensive mainframes for small to medium-sized businesses.

16.4.4 Microcomputers

This is the most common type of computer. It is small, and intended for the single user. For this reason, the microcomputer is also referred to as the **Personal Computer** – abbreviated as **PC**. Microcomputers vary greatly in size, ranging from the desktop version to the palm computer. The desktop version comes in two sizes. There is the 'tower' type, where the main computer box stands like a tower beside the rest of the system. There is also the version in which the display unit sits on top of the main electronic box on the tabletop. Both these versions suffer from the disadvantage that they are cumbersome to carry around. Carry flexibility is provided by the 'laptop' version. As the name implies, this computer is small enough to sit on the user's lap. A specially designed carry bag makes it possible for the user to carry the machine around. The laptop computer is becoming increasingly popular on the market because of this portability advantage. Like the desktop version the laptop fully satisfies the computer needs of a single user. It is also commonly called the 'notebook' computer, a name deriving from the manner in which the computer is opened for operation – very much like a notebook. Still smaller, and of recent arrival on the market, is the 'palmtop' computer.

This, too, is a version of the microcomputer. Unlike the laptop and the desktop, the palmtop is limited in its application. It is designed to satisfy only a limited range of individual computer needs.

For as little as $1,000.00, or less, one can purchase a microcomputer. Due to the advantage of portability, the laptop computer costs more than the desktop. **Fig. 16.1** shows the minicomputer, and versions of the microcomputer.

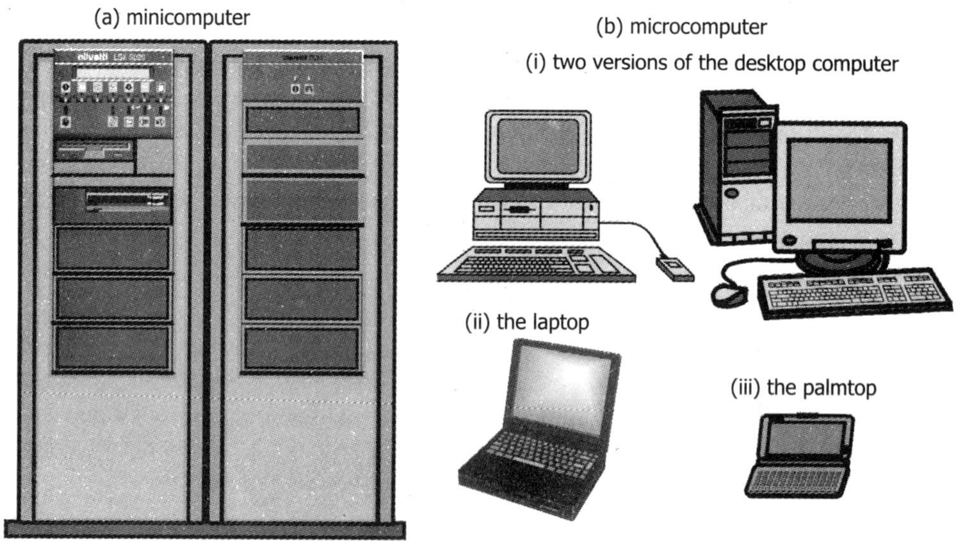

Fig. 16.1 Computers come in different sizes and shapes. Here shown are the mainframe, and different types of the microcomputer. Still smaller than the laptop is the palmtop - shown in (iii).

The microcomputer derives its name from the microprocessor, the electronic device at the heart of the machine. The microcomputer, however, is a lot more than the processor. Over the past twenty years, applications of the microcomputer have grown so much that practically every business office or school even in the developing world has a microcomputer, and almost every home in the developed countries has at least one microcomputer. Whilst the majority of users employ the desktop, the laptop is becoming increasingly popular with the business community. In the following sections, we study details of microcomputer hardware and software, and consider applications in networking and communications. In the rest of this chapter, we study the microcomputer exclusively.

16.4.5 Computer data

No matter its complexity, a computer manipulates data as a bank of switches. The computer combines a very large number of logic gates, producing 1 when data is present, and 0 otherwise. Thus computer data comes as a string of zeros and ones. This type of data representation is called the **binary system**, or **binary code**. Figures we use everyday are in the **decimal** code. Data presented using the binary code is called **bit**. It is possible to represent any figure in the binary code as a combination of zeros and ones.

Thus

the decimal code number 10 = 1010 in binary code, written as $10_{10} = 1010_2$,

and

decimal code number 100 = 1100100 in binary code, or $100_{10} = 1100100_2$.

As the computer uses switches to store information, it is clear that every number is first converted into binary. From the two conversions examples above, it is seen that representing numbers we use everyday into binary form can produce long rows of zeros and ones, some too long for easy manipulation. Other codes have been developed to make this number crunching easier. The hexadecimal code gives numbers to base 16. Details of these number systems can be found in standard mathematics textbooks.

Just as representing numbers in binary can lead to very long strings, so also expressing computer data in the corresponding bits is not very practical in many instances. Computer data can either be in bits or in **bytes**, where 1 byte = 8 bits. One byte of data can be thought of as equivalent to a bank of four switches, each of which can be turned on or off. With the huge amounts of data encountered, multiples of the byte are commonly used. The usual multiples are as follows:

1 kilobyte (or kB) = 1,000 bytes = 10^3 bytes
1 Megabyte (or MB) = 1,000,000 bytes = 10^6 bytes
1 Gigabyte (or GB) = 1,000,000,000 bytes = 10^9 bytes

We note that, no matter the representation, the basic data manipulation and storage ultimately works in bits. Representing the data in bytes offers an easy way of working with the data; it is also less prone to error in misreading.

16.5 MICROCOMPUTER HARDWARE

As we studied earlier, the computer system is made up of the hardware and software. The hardware comprises the electronic gadgets that make up the machine, and other peripheral devices the user may have connected. A functional approach to the study of computer hardware is to consider the system as a collection of **the system unit case, input devices**, **output devices**, **data storage devices**, and **other peripheral devices**. We adopt this approach in this chapter.

16.5.1 The system unit case

Fig. 16.2 shows examples of the **system unit case** in the desktop computer. The system unit case is the main piece of hardware of the computer system. All device connections to the computer are made to the system unit case. Communications between devices go via the system unit case. For this reason, the unit case is at the heart of the computer.

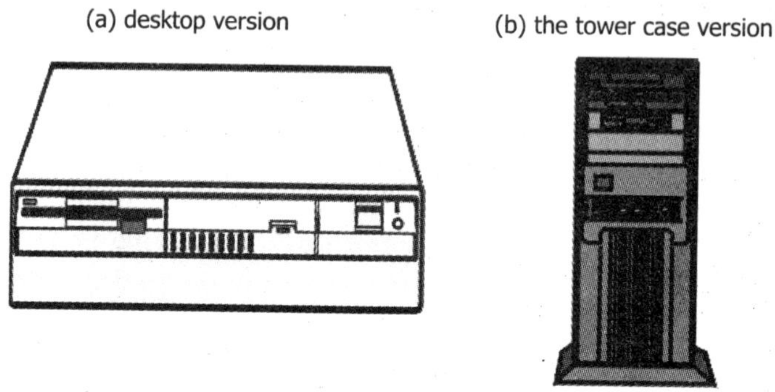

Fig. 16.2 The system unit case; (a) desktop version, and (b) tower case version.

At the back of the system unit case are the **ports** for connecting devices to the computer. There are two types of ports: the **serial port**, the **parallel port**. The parallel port is normally used for connecting printers, whilst the serial port is used mainly for communication devices – e.g. fax modem. A third class of port is becoming very popular. This is the **universal serial bus port**, or **USB port**. Computers equipped with this latter port are generally faster at data communication using this port than the normal serial or parallel port. This USB port is fast replacing the serial port in computers today.

Inside the unit case are the **power supply unit** (or **PSU**), the main circuit board called the **motherboard**, and internal **memory** media. A look into the system unit case would reveal the motherboard as a very complex electronic component. Indeed it is, and it controls all the functions of the computer. All device connections go to the motherboard. A glance at the motherboard would reveal the following

1. The **central processing unit**, usually abbreviated as **CPU**. This is the microprocessor. The CPU is like the brain in the human body.
2. The display chip. This controls the functions of the visual display unit.
3. The memory area, consisting of the **RAM** – short for **random access memory,** and the **ROM** – short for **read-only memory**.
4. Connectors for status indication.
5. Connections for the keyboard, the mouse main power, any peripheral components and the disk drives.

The modern motherboard is a lot more complex than this short account given here. For instance, sound card, modem and network card may also be part of a motherboard. Being very complex, the manufacturers not surprisingly have it concealed; the user would hardly have to open the system unit case. As a first rule in computer use, opening the case must be referred to qualified computer technician.

The CPU is a small but very complicated electrical device. From the outside, it looks very much like a square or rectangular box measuring about 7 cm on a side. Its tiny connectors stick out of the sides like legs, allowing the CPU to communicate with other components of the computer. **Fig. 16.3** (a) is a block diagram of the motherboard in a typical desktop computer, showing the positions of the components, whilst (b) shows the microprocessor. In technical terms, the processor is an integrated circuit that contains the **arithmetic and logic unit**, (or **ALU**), and the **control unit**. The CPU consists of thousands or millions of transistors, organized into groups. Each group carries out a specific class of functions.

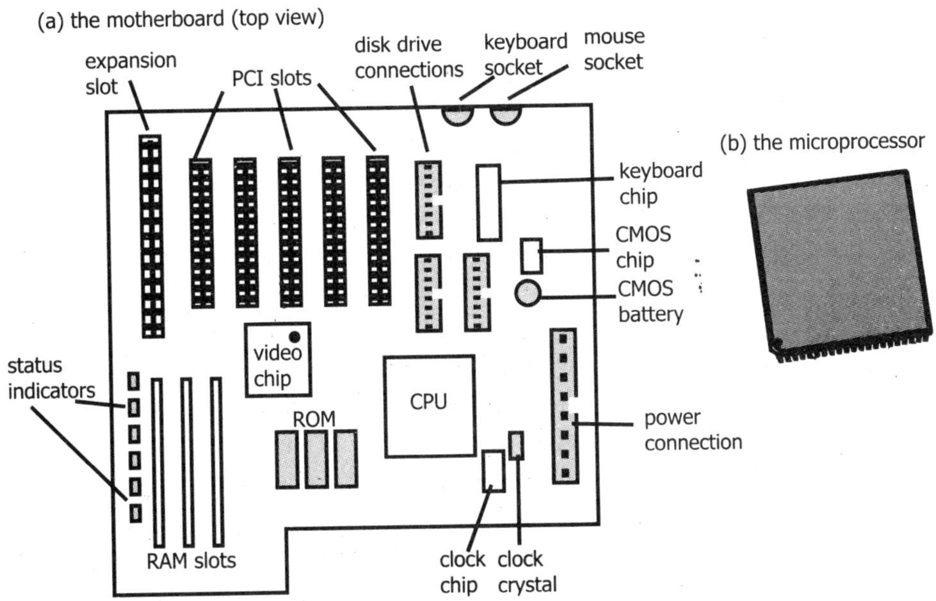

Fig. 16.3 (a) Simplified block diagram of the motherboard, (b) side view of the microprocessor.

The functions of the groups can be summarised as follows:

1. The **program counter**. This group keeps track of the instructions that the CPU must carry out.
2. The instruction **register** and **decoder**. This class of transistors stores the instructions given to the computer, and also interprets the instructions.
3. The **data register** and **accumulator**, which store information to be processed.
4. The **arithmetic and logic unit** that performs the mathematical and logical operations the computer undertakes.
5. **Address decoder** and **address bus**. The processed information is sent to this bus prior to being sent out as output.
6. **Cache**, which holds data and commands that are about to be executed.

The CPU works by comparing pieces of information with each other and making decisions about the information, and storing the results of the comparison. The comparisons do not have to be real thinking, although in some cases the computer is said to have 'artificial intelligence'. The comparisons simply involve determining whether a switch is on or turned off. This way, the microprocessor carries out all the complex data manipulations and outputs the results. Because of its complexity, the CPU might sometimes account for nearly half the price of the basic computer hardware.

One of the major ways of classifying the modern computer is to talk of its operating speed. This speed indicates the rate at which the machine processes information. Typical operating speeds for the modern computer are in the region of nanoseconds. Slower processing speeds often translate into slightly less expensive machines.

16.5.2 Input devices

Computers can process data into information at very high speeds. First, though, the data must get into the main memory of the computer. Only the can the central processing unit gain access to the data. Getting data into the computer is called **input**, and the mechanism used to get the data into the computer is called an **input device**.

An input device enables the user to 'talk' to the computer, and give instructions to the machine. These instructions are d in the computer r the machine to decipher. Computer programmers and hardware manufacturers give the details of this instruction code. The overwhelming majority of computer users simply learn how to use the device. **Fig. 16.4** is an array of some common input devices. Shown here are the **keyboard**, the **mouse**, the **trackball**, and the **joystick**.

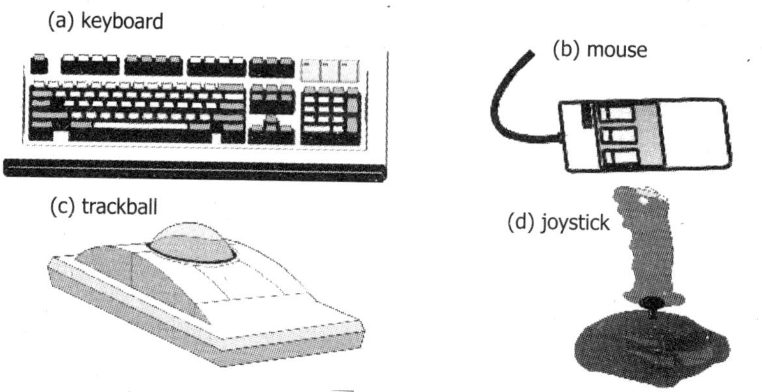

Fig. 16.4 Some input devices; (a) keyboard, (b) mouse,(c) trackball, (d) joystick.

The keyboard is perhaps the oldest of the input devices. It is modeled around the much older device – the typewriter. The keyboard, however, is e complicated than the typewriter, and has many more keys. A glance at the computer keyboard would reveal the similarity with the typewriter, but in addition has many other keys. At the top row are the **function keys**, labelled **F1** to **F12**. These keys play special roles under the command of the software. The numerical row of keys, and some special signs follow below the function keys. There is also the **numeric keypad**, which allows the user the flexibility of inputting numbers and mathematical signs. The other keys allow easy navigation and modification around documents. Like the function keys, these are also under the command of the software.

The computer industry employs two major keyboard technologies. One is **capacitive** technology, and the other is the **hard contact** technology. A capacitive keyboard consists of a printed circuit board that completely fills the space under the keys. Above the circuit board, and below each key is a capacitor. The current flow in the circuit containing the capacitor is altered each time a key is depressed and released. This information is sent by the keyboard chip (also called the **keyboard controller**) to the CPU. The information is then stored in RAM and displayed on the screen. Hard contact keyboards consist of a series of switches that complete the circuit each time a key is pressed. The information is then sent into the computer as the capacitive keyboard.

The keyboard is not always the most efficient way of entering data into a computer. The simple fact that it sometimes requires the user to be familiar with the keys can be a problem. For the new user, data inputting can be a slow process and often daunting. Computer programmers and hardware manufacturers have recognised this drawback. Accordingly, the modern microcomputer comes equipped with a 'pointing' device. Pointing devices have made computer use very easy for most people. The pointing device places a **cursor** on the computer screen. Taking the cursor to any active area of the screen and clicking issues a command to the computer. In the early days of its development, to be a successful computer user required knowing the computer command language that would enable the user to give instructions to the computer through the keyboard. This job has been made easier with the advent of the **mouse**. The mouse has in it an electronic circuit that communicates with the processor. Through this communication, the user merely has to move the mouse cursor to an item, and a click of the mouse activates the command. Other pointing devices on the market include the **track ball**, and the **joystick**, both employing the same principles of operation.

There are other input devices outside the above. Very popular in sales outlets is the bar code reader. Using an infrared light, this gadget reads the price coding on an item, and feeds the price information directly to the computer. With the bar code reader, the supermarket checkout assistant does not need to manually enter the prices, but these are done automatically as the light beam passes over the code. Also becoming popular is the electronic pen, used for inputting information from the pen directly to the computer.

16.5.3 Output devices

The term **output** refers to the information that is sent from the computer. Computer output can be either temporary or permanent, or can take the form of an action performed by a mechanical device. The most common of output are the **monitor**, sometimes called the **VDU** (acronym for **Visual Display Unit**), and the **printer**.

Characters typed at the keyboard input appear on the screen of the VDU. Output on the screen is temporary. It can easily be erased; and disappears when the computer is switched off. Consequently, the VDU gives a **temporary output**. Another temporary output is a **voice response system**. Computer systems equipped with this system give out speech output when certain conditions are met. Warning systems in modern automobiles informing the driver of the state of the fuel, and elevators announcing the floor level are common examples of the voice response outputs. Output from such devices are described as **soft copy**.

Fig. 16.5 The most popular output devices: (a) the monitor, (b) the printer.

By contrast, output from the printer stays even when the computer is switched off. This output is often on paper, and referred to as **permanent output**, or **hard copy**. **Fig. 16.5** shows the monitor and the printer as examples of output devices.

16.5.4 Data storage

Computers also do have the ability to store data. This storage can be statement of program instructions, the raw data before processing, and processed data that would be required at a later date. The program statements or data can be saved on a **storage device**. Then, when it is time to use the program or data again, the stored program or data is transferred from the storage unit to the **main memory** of the computer.

The main memory of the computer is the RAM, or random access memory. The computer stores data in RAM during program execution. Data stored in RAM constitutes only a small part of the computer's memory. Random access memory is contained on **memory modules**, called single inline memory module (or SIMM), or double inline memory modules (or DIMM). These are often colloquially called RAM sticks. Data stored on RAM is only available during program execution while the computer is turned on. This data disappears when the computer is turned off. Consequently, the RAM is described as **volatile memory**.

In computers today, there is an important type of RAM, known as CMOS – an acronym for **Complementary Metal Oxide Semiconductor**. This semiconductor is powered by a little battery, typically about 3.0 V. The CMOS chip holds the configuration information about the system. For instance, it holds information on the type of monitor, the specifications (i.e. cylinders, sectors, tracks, clusters, etc.) of the hard disk drive, and the RAM size and allocation. This CMOS chip can be found on the motherboard, usually near its battery. Being battery powered, CMOS does not lose its contents even when the computer is turned off. With use over a long period, the CMOS battery can run low and need replacing. In such circumstances, the computer would not **boot** up, and a message would appear on the screen to the effect when the computer is turned on. The battery is replaced, and life gets back to normal.

At any given time only a small percentage of data need to be available to the computer for use. The rest of the data is stored in external memory devices. This memory is called **secondary memory**. Also called **auxiliary storage**, this memory is cheaper and takes longer to access than RAM. Data stored on secondary devices is permanent – hence it is called **non-volatile memory**. Although secondary memory takes longer to access than primary memory, it is still quicker to access data on secondary storage than to input the data all over again.

There are many different types of secondary **storage media**. Storage media are the material on which data is recorded. Special **drives** are used to read information on the storage medium into the computer. **Fig. 16.6** shows some secondary memory media. As we studied earlier in chapter 10, suitable magnetic media can be made to store computer data. These magnetic media can either be in the form of a disk, or tape.

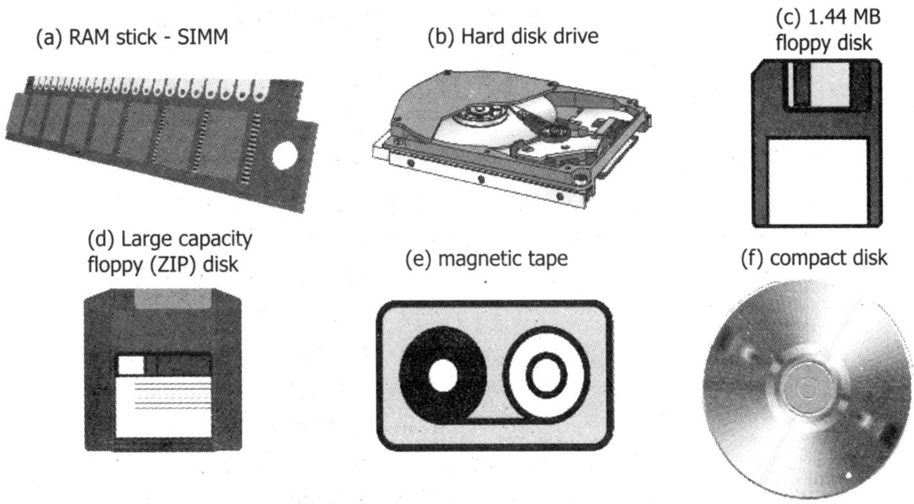

Fig. 16.6 Some memory media: (a) single in-line memory module, (b) hard disk, (c) 1.44 MB floppy disk, (d) high capacity floppy (ZIP) disk, (e) magnetic tape cassette, and (f) compact disk (or CD ROM).

Examples of the magnetic storage disk are the **hard disk**, the 1.44 MB floppy disk, and larger capacity floppy disks, like the 250 MB ZIP disk. In addition to the above memory media, computer technology has recently introduced little what is known as "false memory". These come as little electronic cards or sticks, and are becoming very common.

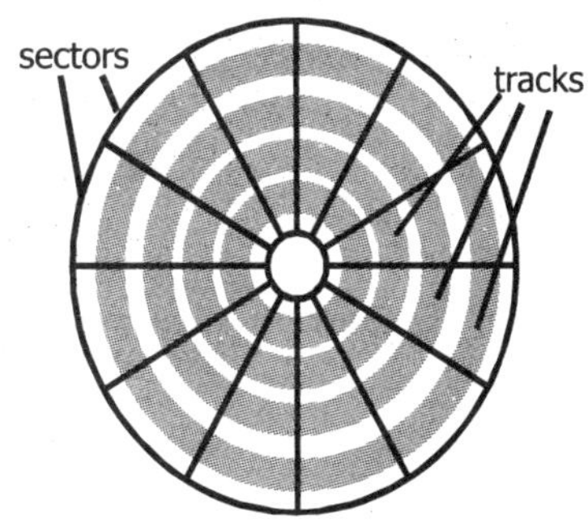

Fig. 16.7 Tracks and sectors on magnetic storage disk

Data on magnetic disks are arranged in **tracks** and **sectors** – a few are shown schematically in **Fig. 16.7**. Sometimes data in a file can extend from one sector to another to form a **cluster**.

Prior to recording data, the disk must be electronically prepared to accept data. This process is called **formatting**. Special computer programs carry out this procedure. Formatting erases any existing data on the disk, and prepares it to accept new data. Newly manufactured disks are **unformatted**. However, most of the floppy disks we buy in the stores have been formatted already.

Magnetic tape storage is the least expensive of the high capacity storage media. These media are very much similar to the videocassette or the audiotape cassette. As a matter of fact they all use the same principles of operation. Most microcomputers today store data on hard disks, but many data backup systems use magnetic tape. The tape is a long strip of plastic material, with a coating of iron oxide compound that can be magnetised. Data is stored along the length of the tape at a specified density, measured in **bytes per centimetre**. One byte stores one character. The rate at which data can be retrieved from the tape depends on the tape speed. A disadvantage of the magnetic tape is that data access can be very slow, as data is accessed sequentially.

The **compact disk** has today become part of the basic microcomputer hardware. They are also called **CDROM**, and acronym for **compact disk read-only memory**. They are not used the same way as floppy disks, or hard disks for storing and accessing data. The main advantage of the compact disk is that the disk can hold very large amount of data. Today large graphics and application programs are sold on compact disks. Up to 700 MB of data can be stored on these media. A special drive – the CD ROM drive – is used to access data stored on the disk.

Becoming common is the DVD – short for **digital video disk**. These hold much larger data than the CD-ROM.

16.5.5 Other peripheral devices

The computer market today has a wide range of peripheral devices. One may consider a peripheral device as an add-on to the basic computer system. These devices have greatly added to the capabilities of the computer. Each device enables the computer to perform some special function. **Fig. 16.8** shows some common peripheral devices.

The **modem** – short for **modulator-demodulator** – enables the computer to communicate with other systems over long distances using the telephone line. We use the modem to send faxes, and communicate with other computers on the **Internet**. We shall discuss this communication in greater detail later. The modem can be either external (as shown in the diagram) or internal. Internal modems come on computer interface cards that can be inserted into one of the expansion or PCI slots on the motherboard. Many computers are equipped with the modem as part of the motherboard.

The **scanner** is used to transfer documents into the internal memory of the computer. It works in a manner similar to the photocopier, with one difference that the scanner converts the document into electronic form, which can be saved on an external memory device. The scanner has proved very useful in transferring copies of many documents (e.g. maps, paintings, legal documents) into the computer. Prior to this, photocopying was the only way in which such documents can be reproduced. A basic scanner can be bought today for as little as a few hundred dollars.

The basic microcomputer system comes equipped with a 1.44 MB floppy drive. For work requiring the use of large capacity drives to transfer files between computers,

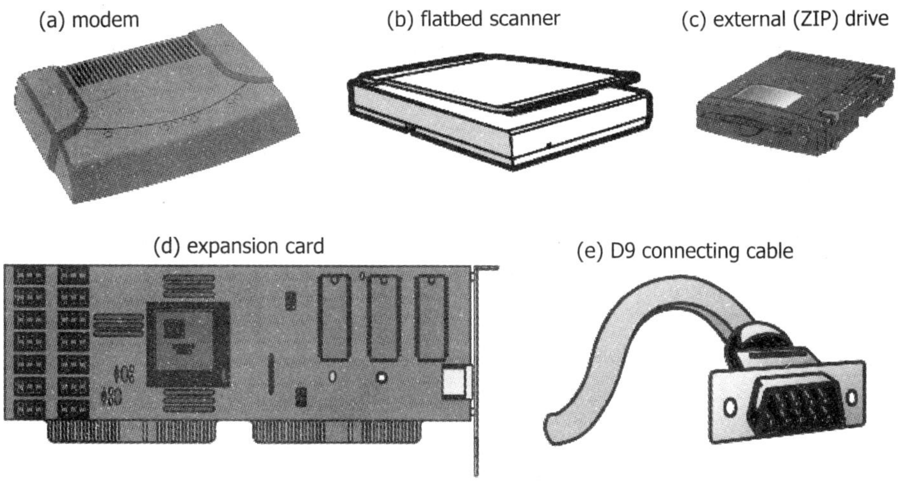

Fig. 16.8 Peripheral devices and their connections; (a) modem, (b) flatbed scanner, (c) ZIP drive, (d) adaptor card, (e) D9 cable.

and making document backups, the ZIP drive offers a good alternative. ZIP drives come in a range of sizes depending on the storage capacity. Flash memory devices are replacing ZIP drives, particularly since the introduction of the USB interface. These come as flash compact cards, typically used in digital cameras, and as flash drives (or memory sticks), most times used instead of floppy disks.

There is a wide range of peripheral devices, most not mentioned here, and the list is ever growing. High capacity hard disk drives, CD Writers, digital cameras, etc., all come as peripheral devices Peripheral devices always come with the following:

1. A means of connecting them to the host computer. Some devices use **adaptor cards**, while others use a direct cable connection. **Fig. 16.8** (d) shows what an interface looks like. This goes into one of the expansion slots in the unit case. The interface card provides the platform for two-way communication between the peripheral device and the host computer. There is also the direct cable connection, which uses the output **ports** at the back of the unit case. An example is the D9 cable, having nine connecting pins.
2. A device driver for operating the device. Usually a program on a floppy disk or CD-ROM, this device driver alerts the computer of the presence of the peripheral, and allows communication between the device and the computer.

16.6 ORGANISATION OF DATA

It is important at this juncture to spend a few moments to understand how data is stored and accessed by the computer. The fundamental unit of data is the **bit**, and eight bits are grouped together to form a **byte.** A byte represents a character, which can be a single letter, or symbol.

Information in secondary storage is organized into files. The file may be a word processing file, a database file, or the output of any software program. Different files have different structures and different characteristics. These differences enable the computer to recognise exactly the file being requested by the user.

The computer can be commanded to perform certain basic operations on data. These operations include **save**, which saves the data onto a secondary memory device, **load**, which loads the data on the memory device for immediate use, the **delete** command which is the electronic equivalent of erasing the data from the file or memory, and the **print** command, which sends the data in a file to a printer.

There are two main ways of accessing data: the **sequential access**, and the **direct access**. **Indexed sequential organisation** represents a hybrid between the two access methods. We discuss these methods of data access in the sections that follow.

16.6.1 Sequential access

In sequential access, the data is retrieved record by record, in a manner practically similar to looking up names on a list that is not arranged in alphabetical order, or looking for a particular song on an audiocassette tape. In sequential access, each record is read. The records are usually stored in the sequence in which they were entered into the file. This access method is also called **First-In-First Out** (or **FIFO**) method. Not surprisingly, this type of data access has the disadvantage of being very slow. In it the system has to scan through all the data in its memory in order to get the one being requested.

16.6.2 Direct access

Direct data access allows access to the certain record or data without sequentially checking every data. In other words, data can be retrieved at random. For this reason, direct data access is also known as **random access**. We can liken direct data access to names arranged in alphabetical order.

In previous generations of computers based on MS-DOS, a **file allocation table** (abbreviated FAT) is used for random data access. This table is held near the outer edge of the disk. The FAT told the computer which sectors are allocated to each file. Corrupting a file allocation table renders the data on the disk inaccessible. No file would be read if the allocation table does not contain an entry of it. Deleting a file simply involved removing the file entry from the FAT. In the **NT** system – short for **New Technology**, there are two file allocation tables, one on the edge, and the other in the middle, making the system less prone to file corruption.

16.6.3 Indexed sequential access

Sequential access is preferred in instances involving massive amounts of information to be processed. The method is used when a large percentage of the data in the computer is to be processed each time the application is run. Direct or random access is used in situations where only a small percentage of the computer data is to he accessed at any one time. The files of interest need to be located quickly.

Indexed sequential accessing is a combination of the two. In indexed sequential accessing, the files are organized such that information in them can be accessed both sequentially and randomly. Files organised for indexed sequential access have a separate index file that lists the key for each record. Some database files run in this way.

16.7 MICROCOMPUTER SOFTWARE

From what we have studied so far, the computer hardware is a very complicated electronic arrangement of parts constituting the basic system, and its peripherals. For all its complexity, however, the hardware by itself is almost useless. This is because by itself the hardware does not know what to do. The various parts of the computer know what to do under the directions of the **software**. The software holds the key to all the marvels of the computer performance. Without the software the computer is just an assortment of complicated logic circuits. Software is a program of instructions written in such a manner that the computer can understand and execute. Once the program is completed and tested to be right, it can be purchased from the shops just as you would buy a piece of hardware. Prices of programs range from a few hundred dollars to thousands of dollars.

Programming is a specialised field that requires a lot of patience, persistence, and logical thinking. Instructing the computer on what to do entails learning the **programming language**. Mastering the language itself takes a number of years of hard work, on the road to becoming a computer programmer and software developer. A computer is able to perform many functions depending on the software that drives it. For instance, software can 'tell' the computer to send its output to the printer in order to produce a hard copy. Similarly, software can instruct the computer to fax a document to a remote location. The software and the hardware must be compatible for the computer to function properly. Software programs are written for specific hardware systems. For instance, the common brands of computers on the market are the **Apple Macintosh** computer (referred to simply as Apple), and the **IBM** compatible computer. IBM is an acronym for **International Business Machines**, an international Company that specialises in the manufacture of computer hardware. Some software are written for the IBM, while others are written for the Mackintosh. Fortunately, many software developers do write the two versions of the program for each hardware system. It is essential to know which hardware system your computer has, so you can buy the right software.

Computer software fall into four main categories: the **firmware**, **operating systems**, **utility programs**, and **applications** software.

16.7.1 Firmware

The lowest level of software is the **firmware**, considered a hybrid between hardware and software. This software is also called **BIOS**, and acronym for **basic input/output system**. This software is coded in the ROM on the motherboard, and is not under any user control. BIOS performs non-user functions such as checking the computer's components when the machine is switched on. This program tests the performance of all components connected on the computer. The testing process is called **POST** – short for **Power On Self Test**. The program also tests the integrity of the communication ports and all input and output devices connected to the computer. The computer manufacturing Company designs the firmware; the user has no control over it.

16.7.2 Operating systems

An **operating system** is a software program that controls the computer and allows other programs to run on the machine. Without the operating system, the computer cannot do anything meaningful. The operating system enables the computer to perform very vital operations such as saving files into memory, reading and writing files, and changing the names (or **renaming**) files. Using **batch files**, the user can have direct access to the operating system; instruct the computer to run certain programs loaded into the computer.

A software Company called Microsoft Corporation in America designed one of the oldest operating systems. This operating system is named **MS-DOS**, short for **Microsoft Disk Operating System**. Over the decades, this operating system turned out to be the most popular for microcomputers. Since the release of its first version, Microsoft has progressively improved on this operating system by widening the range of capabilities. One disadvantage of DOS is that the user needs to learn the commands and type these at the keyboard. It can be seen how a user can be frustrated if he/she forgets the right command, or inputs the wrong command for the desired action. Because of the need to understand and type in these commands, many people got the impression that the computer user necessarily has to be keen on computation and memorising commands. Not so these days.

Apple Macintosh was the first to introduce the **Graphics User Interface**, or **GUI** making computer use very easy. Programs with GUI display the parts of the software as active cells (called **icons**) on the computer screen. Using the mouse pointer, the user selects a desired action by moving the mouse pointer to the cell and clicking on the button of the mouse. This clicking action activates the cell, and tells the computer to execute an action such as saving a file, deleting a file, or renaming a file. This ease in computer use brought a lot of customers to Apple Macintosh computers. In response, Microsoft introduced its own GUI named **MS Windows** operating system. The first version, **MS Windows™ 3.0** appeared on the market in early 1993. The Windows operating system has been modified and greatly improved over the years. Today the operating system is by far the most popular on the market.

16.7.3 Utility programs

Utility programs are software that help the user manage the computer hardware and software. These programs are also known as **utilities**. Utilities perform the electronic analog of housekeeping. Utility programs are designed to improve the performance of the computer, thus making the computer run better. There are two classes of utility programs. To one class belong utilities that the user runs to manage the system, and exits from it once the program has accomplished its task. When the program is exited, the RAM is cleared and made ready for use by other programs. Examples of utility programs falling in this category are hard disk cleanup programs, file defragmentation programs, memory management programs, and **file zipping** programs.

Disk cleanup programs erase files from the hard disk that are no longer needed, or bits of files left by programs that were run previously on the computer. These unwanted data and files make the computer run slower with time if they are not removed – as they are scanned when any file is opened and worked on.

Defragmentation programs reorganize the data area such that materials that belong to the same file are kept as close to each other as possible. The computer would run slow if, for instance one paragraph of a document were in another memory location from the rest of the document.

Memory management programs set areas of the internal memory for specific types of data. Sometimes the amount of data in a file can be too large to save on a small secondary memory device, such as a 1.44 MB floppy disk. A file zipping program can make all the difference in such circumstances. This program condenses the file – a process described in computer language as **zip**. A zipping program can condense a file to less than one-tenth is actual size, so it can fit on a floppy disk. Complementary to this is the unzip program, which works on a zipped file and opens it to its normal size. There are many other utility programs in this category than the ones cited above. A common feature of them is that they are only made to run when needed, after which they are switched off automatically.

The other class is the **terminate and stay resident** (or **TSR**) utility programs. Like the first class above, a TSR starts working on the user's command, and gives access to its functions. However, exiting or terminating does not clear a TSR from the memory. It returns control to the computer to the user, but stays in RAM. While in RAM, a TSR can either perform its actions automatically, or can be easily activated by a **hot key** that activates it. Features like spell checking, calculator functions, calendar control, dates, etc., are provided by TSRs.

16.7.4 Applications software

The vast majority of computer users today use **applications software**. Into this category of software is lumped all programs that allow the user to **do** something with the computer. For all these people, the computer would be useless if this role is not performed. Application programs are what make the computer so powerful in the modern society. We can relate to the software world with reference to **Fig. 16.9**. This diagram depicts the computer system very much like a vertical section through an onion. The centre depicts the hardware, and around it are layers showing the arrangement of the different software programs, ranging from the firmware, which is closest to the hardware, to the application programs.

Fig. 16.9 Computer hardware and layers of software, very much like a vertical section through an onion.

While the rest of society keeps busy doing something else, there is a small group of professionals who spend tireless and painstaking efforts writing programs for use by the computer. We call these people **computer programmers**. Their efforts have brought the computer to the stage at which it is today. There is a wide range of applications software on the computer market. When we group existing software into categories, we find many categories – all thanks to the efforts of the programmers. The major categories are presented below.

(i) Word processing

These programs enable us to prepare documents. A word processing software allows the user to erase, edit, move, and print words and blocks of text. With these programs we prepare letters, presentations, reports, and manuscripts. The forerunner of the word processor is the typewriter. This, however, has the disadvantage that the words are directly typed on paper, and editing the document is a slow and painstaking job. The most popular word processing programs are those from Microsoft Corporation, called MS Word, and Word Perfect.

We are able to use word processing programs quite easily as a result of the in-built **menu** functions. A menu is an on-screen list of options presented to the user to choose from. To activate a menu option we take the mouse pointer to the particular item and click its button. This way the user does not have to remember commands for erasing a word, renaming the file, etc.

(ii) Electronic spreadsheets

A spreadsheet consists of an arrangement of rows and columns. It is used to do calculations on data entered into its **cells**. Each cell is the intersection of a column and a row. A computer spreadsheet is often known as an electronic spreadsheet, to distinguish it from its predecessor, the manual spreadsheet. The big advantage the electronic spreadsheet has over the manual spreadsheet is that calculations are made automatically each time a cell entry is changed. Entries into the spreadsheet can be pure number (e.g. 5.0), text, or formula using cells in the same spreadsheet or different sheet in the same file. **Fig. 16.10** shows a typical spreadsheet, showing some columns and rows. A formula is entered in cell E2 (i.e. cell identified by column E and row 2).

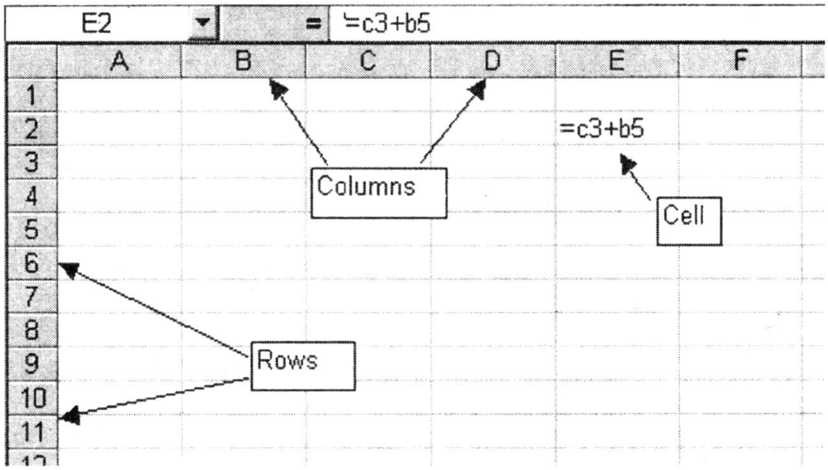

Fig. 16.10 An electronic spreadsheet

(iii) Data management

Data are information we need for an activity. A good example of an everyday database is the address book. In this we can store addresses of our contacts, their telephone numbers, and other information like birthdays, our driver's license numbers, bank account numbers, etc.

In computing, we also have programs that allow us to input and access stored data. Data management programs arrange stored information into files and lists. An example of a data management program can be found in a video CD library. The operator enters information on each CD – the type of film on it, the cost of the CD, the borrower details, the source of the CD, etc. With these entries, a database management program can produce a list of all Documentary CDs, for example, list of all CDs that had been borrowed by a particular user, etc.

Database programs have wide areas of application. They are used extensively by the Police Department in law enforcement. Data on known criminals exist, and serve in the solution of crimes. Big business houses keep databases on their assets. This way the Company keeps track of each capital item. It is all part of good business.

(iv) Graphics programs

Graphics programs allow us to draw diagrams, import pictures from peripheral devices such as digital cameras and scanners. For the most part, however, we use graphics programs to produce pictorial representations of our work. Modern graphics programs are often referred to as **CAD** packages. The acronym CAD stands for **computer-aided design**. Among the areas of application of graphics packages are the following:

1. **In education**. Many educational software consist of a blend of graphics and other software packages. Using graphics programs, teachers turn the task of introducing computers to young children into fun. This way the young minds are able to explore their rudimentary talents, while at the same time getting acquainted to the use of the computer.
2. **In engineering design**. A large number of graphics programs are used by engineers. The civil engineer, for instance, uses graphics packages in the design of buildings and bridges. Getting the ideas presented in the form of a good diagram makes it easier for the engineer to visualise the final output of the work even before starting.

   The electronics engineer and technician use graphics packages in the design of circuit diagrams. Once the schematic design has been done, they set out using another graphics package to design the printed circuit board, or PCB. A quick glance at the inside of an old electronic gadget would reveal the PCB as a card on which the circuit is 'written'. This way the electronics industry has been able to transform hard wire connections (which will look unwieldy particularly for complicated circuits) into a work of art.

   In the automotive design and manufacturing industry, the mechanical graphics design engineers are constantly coming up with new designs of motorcars. Indeed while the computer is the new face of our technology world, graphics programs have played no smaller role in this development.
3. In the textile industry, graphics programs are used in the colour design of new pieces of yarn. Graphics programs have put a new meaning to the concept of fabric design. Thanks to graphics programs, we today have a wide range of choices in fabrics and colour combinations.

(v) Communications

Improvements in information technology has produced methods that enable us to communicate easily using the computer. Programs that allow us to achieve this communication are called **communication programs**. With these programs we can send faxes from one computer to another computer, or fax machine in different locations. The amazing aspect of this communication is that it is very fast and a lot cheaper than other means.

To communicate in this way requires an important piece of hardware called the **modem**. The word 'modem' is an acronym for **MO**dulator **DEM**odulator. This device encodes the message into machine language before sending it. This encoding process is termed 'modulation'. Upon receiving a modulated message, the modem decodes the message into text – this being the 'demodulation' process. Sending electronic mails (or **emails**) also involves this modulation-demodulation. Like any other peripheral device, the modem comes with its driver software that allows the user to set it up on the computer for use.

(vi) Programming languages

Also called **development languages**, these programs enable us to design and run our own software. A programming language allows us to communicate with the computer. It is the common language between man and the machine. The text of any programming language contains coded scripts that the computer translates into a form to make the computer carry out the instructions. This translation process is referred to as **compilation**. Thus the instructions are input at the keyboard, **compiled** by the computer, and stored for use each time the program is called or **executed**. Programming languages fall into two categories: **low-level** languages, and **high-level** languages.

Also referred to as 'machine code', or assembler language, low-level languages address the basic circuitry of the microprocessor. Bearing in mind that the microprocessor contains a large collection of transistor switches and logic gates, a low-level language often addresses these switches directly, closing some and opening others under the instructions of the program. Microprocessor-based appliances (e.g. washing machines, intruder alarms) are programmed in machine code. Computer users concerned with computation seldom use machine code.

The programming languages we use in our everyday computer work are classed as high-level languages. These employ common words that are 'understood' by the microprocessor and converted into machine code in the process of compilation. About the most common of these is called **BASIC,** and acronym for **'Beginners' All-purpose Symbolic Instruction Code**'. Other programming languages include **FORTRAN** – an acronym for **FORMULA TRANSLATION,** used in scientific programming, and **PASCAL**.

(vii) Educational programs

Computer programs have been very useful in guiding students and teachers in almost every subject we find in the school or college curriculum. These programs are called **computer-aided instruction** (or **CAI**) packages. Sophisticated CAI packages would include teaching/learning notes, worked examples, self-test exercises, animated clips, videos and sound clips. Whilst the need for the class teacher still remains, an educational software package often provides an invaluable guide and assistance to both the teacher and the student.

In addition to CAI programs, there are a host of programs designed to help the teacher in various ways. Some of these, for instance, assist the teacher to make teaching notes, and to set exercises and examination questions.

(viii) Financial programs

These are extensions of other programs, notable spreadsheet, and database packages. They are designed to use in the finance and accounting industry. These programs make easy for the businessman to get the financial records in order, and hence keep track of the performance of the business at all times. A typical small business financial package would enable the business to do the following:

1. Keep a record of orders for goods and services, their costs, the variations in the variables that control these costs.

2. Keep data on customers and their needs.
3. Keep data on employees, their salaries, and other benefits.
4. Keep up to date with tax and other financial liabilities.
5. Keep record of running costs of the business, and the overall performance of the business.

(ix) Reference programs

These include a wide range of data that are accessed from a disk. Examples of reference programs are Encyclopaedias, dictionaries, synonym finders (or Thesaurus), maps, atlases, almanacs, telephone and email directories. The materials are all available in a form that can be read by the computer. The reference materials come on standard diskettes or CD-ROMs that are useful in the home or office. Some of them do even include video and sound. A popular computer reference program is 'Encarta', containing basic information on a wide range of subjects, including sports, world history, world geography, science, anthropology, art, and economics. Because they are available in electronic form, these reference programs have some advantages over printed materials.

(x) Desktop publishing

Desktop publishing programs are used to design books, newspapers, magazines, newsletters, and other similar documents. With desktop publishing programs, we are able to create columns of text, change the size and style of text, rotate text, and insert tables and graphics. A quick look at a magazine on the shelf would show the features of desktop publishing software. A desktop publishing program can be regarded as an extension of a word processing program.

(xi) Integrated programs

Many programs we use today are classed as **integrated programs**. These programs combine aspects of two or more software to give the final result. The major word processing programs these days fall into this category of integrated programs. Such word processing programs integrate a spreadsheet program and a graphics program working together as required. The manuscript of this book, for instance was prepared using an integrated program. Hence you find graphics, equations, calculation tables (i.e. aspects of a spreadsheet), and text sometimes all blended together as part of a single document.

(xii) Recreational programs

These are software we use for recreation – i.e. leisure. They are more commonly referred to as **computer games**. They come in many forms; some operating on the resources of only one microprocessor, and do not require the full resources of the computer. These come packaged in boxes, with the controls for playing the game. Some games are played using the computer keyboard, mouse, and a joystick.

Whether it is the stand-alone type, or one used with the computer, the heart of any computer game is the software that sets it up and allows the user to play the game. Many of these software come on CD-ROMs, to be loaded either into the computer, or into the console terminal of the game controller.

With the improvements of the Internet, many games can nowadays be played over the Internet. A player can play against many others from around the world simultaneously.

(xiii) Process control programs

These programs enable the computer to control a mechanical or electronic device, and carry out some external electronic or mechanical action. A process control program works through an electronic circuitry known as **interface**. This interface provides the platform for the computer to communicate with the external device. Examples of process control programs are to be found in science and engineering laboratories.

16.8 SOFTWARE AND HARDWARE COMPATIBILITY

Putting together a computer system to work involves buying the computer hardware on the one hand, and on the other hand, getting software to work. A given hardware can only accept software of a particular type to work on it. Similarly, software can only work on a computer system for which it was designed. Matching software and hardware to work together is known as **compatibility**. Simply put, the software must be compatible with the hardware in order to work.

Another aspect of compatibility involves having a chosen peripheral device to work on the computer system. We learned earlier that each peripheral device comes with its own driver software. This software enables the computer to communicate with the peripheral device. Thus compatibility of a peripheral device with host computer again comes down to software and hardware compatibility.

Compatibility also involves having software to work with other software on the same hardware system. Some software tend to conflict with each other, making it frustrating for a user to work with both software on the same system. Often it becomes necessary to remove one software off the system in order to use the other.

16.8.1 Coordinating software and hardware

Three main factors are to be taken into account when mixing and matching software and hardware to work on a computer system. These are as follows.

(i) the computer's operating system

As we can recall, an operating system is a program that controls the computer and makes it possible for the user to run other programs on the computer. The most common operating systems in the marketplace are those designed by Microsoft Corporation. The earlier ones are called **MS-DOS™**, (or Microsoft Disk Operating System). These have been superseded by the Windows platform, making computing easier through its graphics user interface. These are all designed to work on the **Intel™** family of microprocessors.

The main problem that often arises in software applications has to do with the version of the operating system running on the computer. Periodically, software companies update and improve their operating systems. For example, there are many versions of MS-DOS. Higher numbers designate later versions of the program. Applications programs are usually written with some version of the operating system as target. These programs will work perfectly on computers with higher version of the operating system. On the other hand, software written for a higher version of operating system will not run properly on a computer that has an earlier version. When buying any new software, it is wise to ascertain that you have the operating system for which the software was designed.

(ii) the RAM

Every software package needs minimum RAM requirement to run. This requirement is often indicated in the documentation of the software. The user must make sure there at least enough RAM to satisfy this requirement. It is usually advisable to have more than the minimum RAM. Generally, the larger the RAM, the better the program would run. Programs tend to run very slowly, or sometimes hang up, when the RAM is insufficient.

(iii) the disk drive

The computer industry is going to great lengths these days to making things simple. To this extent most software are now distributed on compatible CD-ROMS, or on 3.5-inch floppy disks. This has reduced the compatibility problems with the disk drive that used to arise with the earlier generation of machines. There still, however, does remain the problem of size of the hard disk, particularly with graphics programs. Graphics programs are large programs, and in operation, they very often need to make frequent **disk calls.** Disk calls are the computer equivalent of 'visits' a program makes to store and retrieve data to and from the hard disk. Apart from graphics programs, any program that uses a graphics user interface needs hard disk space large enough to make these disk calls while running. In addition to RAM, the user therefore needs to have enough hard disk space for such calls. Disk calls make running graphics programs much slower than other software.

Even where the user meets the requirements of the operating system, the RAM and hard disk space, a software program may still not be able to run properly. These are critical requirements, but are by no means the only ones. In the following section we study compatibility requirements between the software and peripheral devices.

16.8.2 Coordinating software and peripherals

Prior to purchasing software, the user must consider what types of peripheral devices are connected to the computer that needs to use the software. All telecommunications programs, for instance, require a modem to operate. The modem has to be compatible with the communication port on the computer, and the software package one plans to use. Graphics programs may need certain types of monitor, while some spreadsheet programs may need a printer with special features. Many entertainment software need joysticks. While some peripherals are connected to the communication ports of the computer, others use adaptors connected into expansion slots inside the computer unit case. Below we discuss the main items in this compatibility jigsaw.

(i) Ports

A port is a connector, much like an electrical socket, that allows the CPU to communicate with peripheral hardware. There are two types of computer ports: **serial, USB** and **parallel** ports. A serial port communicates with a modem or printer through a string of information one bit at a time. A parallel port, on the other hand, sends one byte at a time to the peripheral device. External modems are normally connected to the serial port, while printers communicate through the parallel port. Most computer systems provide two or three ports, one parallel and two serial, or vice versa. Thus the user is limited to no more than two or three peripheral devices at a time. As part of coordinating software with the hardware, the user must ascertain which communication ports exist on the system being used, and install the necessary software.

(ii) Expansion slots

Desktop computers have **expansion slots**, into which extra circuitry can be added for the control of a peripheral device. A computer has two kinds of expansion slots. One of these is used for connecting peripherals. The other, although not normally referred to as expansion slots, are usually a bank of empty slots on the motherboard for increasing the RAM of the computer. The user or computer technician often inserts a **single in-line memory module (SIMM)**, or **double in-line memory module (DIMM)**, to increase the RAM size of the computer. Although many computers these days would 'recognise' the extra RAM added and make the necessary changes to the firmware, it is advisable that RAM changes be referred to qualified personnel.

(iii) Monitors

From the point of view of software compatibility, monitors are described in terms of their display capacity (i.e. the number of characters that can be displayed at one time), and their colour capacity. Most monitors on the market today are of the **VGA (Variable Graphics Array)**, or **Super VGA** type, capable of displaying up to 256 colours or more.

Becoming very popular are the **LCD (Liquid Crystal Display)** monitors. All portable computers have this kind of monitor for easy transport. There are now the desktop versions of these monitors. Their advantage over the older **CRT (Cathode Ray Tube)** type is that they are much smaller – a property often described by the phrase 'small footprint'.

Each software package requires a monitor with specified properties. It is important to ascertain that the software can use the special capabilities of the monitor, or that the monitor is of sufficient quality to display the fine details of the software.

(iv) Modems

In using a modem, it is important to make sure the modem is compatible with the port on the computer as well as the communications program being used. Most modems use the instruction set developed by **Hayes Computer Corporation**. They are hence said to be Hayes compatible.

(v) Printers

There are lots of printers on today's computer market. These range from low quality **dot-matrix** printers to the high-resolution laser or colour printers. Every software works best with a certain type of printer. However, the job of the user has been made easier by the fact that each printer comes with its own **driver software**, which allows the computer to give instructions to the printer. This way, all the user does it to install the printer driver on the computer; the rest of the job is then taken over without the user's intervention.

(vi) Mice and joysticks

Most programs use a graphics user interface. They thus require some sort of pointing device. The usual pointing devices are the mouse and the trackball. The software thus needs to be compatible with the mouse. Again, the computer industry today has made sure that most software on the market are compatible with the existing brands of mouse. The same is true for the trackball and the joystick, the latter being normally used with many entertainment software.

16.8.3 Coordinating software with other software

Software compatibility does not end with mixing and matching software with peripherals. Sometimes compatibility problems arise when mixing one software with another.

Many software today are of the integrated type. They need to be compatible with other software. For example, a desktop publishing software requires to be compatible with other software. The main function of a desktop publishing software is to design page layout for text and graphics that had already been prepared by a word processing software. Equally so, the word processing software has to be compatible with the spreadsheet program, so tables can be imported into a document with ease.

Where one needs to use software that are not completely compatible, the first task is to translate the files in both software into a code that can be read by both programs. This code is called the **ASCII** code. ASCII is an acronym for 'American Standard Code for Information Interchange'. Almost all programs can import and export files that are written in ASCII code.

Software compatibility is a complicated area. To avoid some of these complexities, many people opt for integrated software packages that offer multiple capabilities. An example of such a program is Microsoft Works, which is an integrated package containing word processing, spreadsheet, database, and communications software all rolled into the one program.

16.9 HOW A COMPUTER SYSTEM WORKS

We have so far learnt a lot about parts of the microcomputer, and how these parts are brought to work together. In this process, we have studied some terminologies – computer data, hardware, software, operating systems, applications programs, peripherals, RAM, ROM, etc. We are now ready to learn a bit more on how a computer system really works. Bearing in mind that the computer is a complicated device that puts together electronic circuitry and programs, we shall go about this study by considering how the different parts interact with each other.

Our study here will focus on four broad areas: the power supply and power requirements, data input, data processing, and data output. We shall also discuss some of the reasons why computers fail.

16.9.1 The power supply

The power supply for the computer is provided either by a standard source, or a battery. The power supply unit converts the alternating mains voltage into to +5 V, -5V, +12V, and –12V of direct current that is distributed to the motherboard and disk drives by connecting cables. The motherboard distributes the power to the components mounted on it, and to certain peripheral devices as required.

In normal operation, there is very little the user needs to do with the power supply unit. Problems do arise, however, if the supply to the computer is either too low or too high. Computers expect a voltage supply of around 230 V (in Europe, Asia, Australia, New Zealand) or 115 V (in the Americas). Normal fluctuations on these voltages rarely exceed 5 V. The power supply unit handles such fluctuations. However, there are times when the voltage might rise above, fall below the required amount over prolonged periods of time. On a busy day in a office block, with all electrical appliances (e.g. air conditioners, fans) switched on, the voltage might fall below the normal. This situation is normally referred to as 'brown out', or under-voltage. Under these conditions, the computer system may shut down and restart in quick succession. The power supply might also try to overcompensate for the low voltage and draw too much current, thereby overheating. The opposite situation occurs when there is too much power – a situation known as a 'power surge'. There are a few causes of power surge, listed below.

1. During storm or lightning, a power surge might occur.
2. Power comes on after it had been off for a long time.
3. Unstable power supply. Many countries in the developing world have common occurrence of this.

Power surges can damage any electrical appliance, and the computer is no exception. **Surge protectors**, and **voltage regulators** help protect delicate electronic devices against surges and brown outs.

16.9.2 Input

Data is input to the computer through an input device such as a scanner, the keyboard, the mouse, modem, etc. Each device connected to the device is assigned a data path and memory location in RAM. An **interrupt** signal is sent to the CPU. This interrupt informs the CPU that data is being communicated from an input device. If the signal makes sense, the computer handles it according to instructions in the BIOS (Basic Input/Output system), the operating system, or an application program. Most often, this handling involves storing the input data in a RAM position.

Problems with the data input system might be caused by some mechanical failure. An adaptor not "sitting" properly in an expansion slot, for example, might prevent an input signal from the device not getting to the CPU. Foreign objects caught underneath a key on the keyboard, or piece of dirt on the roller of the mouse might interfere with the input. If the input signal gets through correctly, there might still be problems of handling because of software problems. The rate of data transfer from a modem, for example, has to match that of retrieval by the CPU. This rate of data transfer is called the **baud rate**. If a modem is set at 32,000 baud, and the communication software is set at 32,000 baud, problems of reception will arise with the CPU not making sense of the data.

16.9.3 Processing

Data and instructions for the CPU to process are placed in RAM. When a program is to be started, the CPU 'looks' for the appropriate RAM locations, reads the instructions it finds there, and executes them in the order designated in the program. In order to achieve this, the CPU uses the BIOS and the disk operating system.

If, for example the user wants to load a spreadsheet program called CALC, the word CALC is typed at the keyboard, or a mouse pointer taken to an appropriate cell on the screen and clicked. The CPU then reads each letter of the command (CALC), puts them together as a word, and searches for the word against a list of **executable** files on the hard disk, floppy disk, or CD-ROM. If the name exists, the computer opens the file and loads its contents into RAM. The operating system helps the CPU by opening the file, and BIOS 'tells' the computer where to load the program into RAM.

Once the program is loaded into RAM, the CPU executes its commands. It might, for instance, clear the screen, and substitute what was there before with the user interface, and wait for an input from the keyboard or mouse. Editing a file involves a similar process of reading from the disk, storing the data into RAM, and moving the beginning of the document to the portion of RAM that corresponds to the video screen. Text or figure can then appear on the monitor. Each time a key is pressed, the CPU interprets the key code and stores the data in the appropriate sections in RAM and on the screen.

The above processing cycle sounds very simple, but it involves executing thousands, or millions of commends each second. Moreover, there tare thousands of electrical components involved.

16.9.4 Output

The results of processing are output via a range of devices, including the monitor, printer, modem, and disk drives. The CPU sends data to a memory location dedicated to a specific device. The data is then transferred to the output device through an interface card, serial connector, or parallel connector.

On reaching the device, the data is processed again by the circuitry of the output device, to give the output in the desired form. For the modem, for instance, this entails converting the digital signals from the CPU into analog pulses to travel down the telephone line. For a printer, a set of commands is issued to the printer, thus allowing it to print characters, move the rollers that turn the paper, detect the end of a line, do a carriage return to start on another line, etc. Indeed, the output system is very complex due to the many activities that need to be coordinated simultaneously. Whilst the result is a remarkable piece of production, within the boxes the electronics simply consists of a series of switches coming on and off at precisely the right moments to produce the desired effect – e.g. a box drawn on a piece of paper by the laser printer connected to the computer as the peripheral device. All these switches operate under the command of the CPU, with the BIOS and DOS giving assistance is it is needed.

Coordinating the system unit and output devices is critical to successful computing. The secret to managing output successfully is careful planning, attention to detail, and double-checking your work. All the necessary software must be installed in the computer, and the motherboard components must be sitting firmly in their slots. A dirty or corroded connector would prevent a computer from communicating properly with a printer or other peripheral device.

16.10 COMMUNICATION WITH COMPUTERS

Nowadays, it is possible to link any two computers together using existing telephone lines and the necessary communications equipment and software. This new world in information technology allows computer users to share data and programs, allows one computer to use a peripheral device connected to another computer in a remote location, allows users to send letters to each other by electronic mail, and enables us to search for information from any location in the world. Merging communications with computers is the subject known as **telecommunications**. Telecommunications gives people access to a wide range of information and services. Among the areas that have benefited from this link is the business community. Using telecommunications, we can have access to our funds in financial societies – e.g. banks, building societies, etc. – efficiently and quickly. In this section, we study aspects of computer communications, and how this kind of communication has helped societies. Emphasis will be on the broad aspects of these communications for our understanding, rather than on extensive technical details. We shall learn the hardware elements necessary for office data communications through networking, information accessing using the Internet, and some of the other uses in today's technology world.

16.10.1 Data communications

The components needed for data communication between computers at two locations are few. However, there are many complications that may arise in the data transmission. The basic items used for data transmission are as follows:

1. A sending device. This can be a modem or a straight wire connection between the computers.
2. A communications link. Between remote locations, this usually takes the form of a telephone line or a fibre-optic link.
3. A receiving device. This device is connected to the computer receiving the data. Modem-to-modem communications is normally used for two computers in remote locations.

We now study how the components of a communications system work together, starting with how the data is transmitted.

16.10.2 Data transmission

Computers generate 'digital' data. The data produced by computers are either '1', corresponding to a switch being on, or '0', for the switch being off. It would be easy if all components of the communications systems were of this form. However, the telephone cable network on which the system relies was constructed for voice transmission, and hence it is designed for handling **analog** signals. For the signal to be sent through analog lines, the data has to be converted into analog form. The analog signal is in the form of a transverse wave. Converting this into a digital signal might, for instance, mean designating '1' for the maximum displacement, and '0' for zero displacement. Converting from digital to analog signal is referred to as **modulation**, and the reverse process is called **demodulation**. We use the modem for this inter-conversion.

Like every other component in the computing world today, modems are changing fast in technology and capabilities. Many computers have modem circuits as part of the motherboard. These computers offer a socket for the normal telephone plug. For two modems on remote computers to communicate with each other, they both must be sending data at rates not too fast or too slow for each other, so every bit of data can be received as sent. The word 'handshake' is often used to describe this relationship in a manner that we can easily understand. The technical term **baud rate** is used; this refers to the number of bits of data that is transmitted per second. The old generation modems used to communicate at baud rates of 100, and 300 bits per second (bps). Today's modems transmit at much higher speeds, typically going up to 33,600 bps, 56,000 bps, or more. This is good news especially for business houses where communication times can mean huge sums of money.

16.10.3 Communication links

These refer to both the hardware and the software that go toward making a communication link. Some of the hardware communications links are as follows:

Twisted pairs. These are very similar in form to the telephone cable. Most office computers are linked together using twisted pairs. They are inexpensive, and used for data communication over short distances. They are, however, susceptible to electrical interferences, called **noise**, which causes distortion on the signal.

Coaxial cables. Coaxial cables are more expensive than twisted pair cables, but are less prone to electrical interferences. A coaxial cable consists of bundles of insulated wires within a shielded enclosure. The radio frequency television cable is an example of a coaxial cable. Coaxial cables can transmit data much faster than twisted pairs.

Fibre optic cables. Since its invention by Bell Laboratories, fibre-optic cables have been used extensively in data and telephone communication. In these cables, the electric signal is transformed into light pulses of fixed frequency, and sent down the core of the fibre-optic cable. Being non-conducting, fibre-optic cables are less likely to cause fires, are harder to wiretap, and are not affected by electrical interferences. The broad **bandwidth** (i.e. range of frequencies that the device can handle) of the fibre-optic cable translates into a wide range of multimedia opportunities, since the cable is suited well for handling all types of data. These data might include voice, pictures, music, and video – all at the same time.

Microwave transmission. Another popular medium for transmitting data is by microwaves. This uses line-of-sight transmission of data signals through the atmosphere. Microwave transmission uses antennas situated at convenient distances from the microwave relay station. The microwaves received are converted into digital signals and fed to the computer for data transmission.

Satellite transmission. More advanced and expensive methods of transmission sometimes transmit data between two earth stations via a geostationary satellite. These satellites are in orbit at a height of about 35, 680 km above the earth. At this height, the satellite appears stationary above a given point on the earth.

After the communication links are established, there still remains the question of **protocol** to be addressed before two computers can 'talk' to each other. For data to be transmitted safely from one computer to another, each computer musk know whether the other received the last data sent, and whether it is ready to receive further data. In addition, both systems must agree on how data is to be transferred. With the many manufacturers on the marketplace each producing their own brand of gadgets, the problem of protocols is solved by the use of **standards**.

The **International Standards Organisation** (**ISO**) has defined a set of communications protocols called the **Open Systems Interconnection (OSI) model**. Manufacturers of communication systems make sure they use this or other standard based on the same principles. That way, a modem is made to work on a computer system, and communicate with other modems even though the gadgets come from different manufacturers.

16.10.4 Local area networks

From our study of microcomputer systems, we have learned that it is possible to have a number of peripheral devices connected to a computer. These enhance the capability of the system. Thus, for instance, we can have a scanner connected, enabling us to scan documents. We can also have a printer connected, for document printing, a modem for telecommunication, an external hard disk drive, possibly a high capacity floppy drive, etc. With more computers working in an office, for example, and each trying to get as many peripherals as possible, the end result is a huge array of devices that may cost phenomenal amounts all put together. Unfortunately, there is a limit to the number of peripherals that can be connected to a single machine. Most machines have only two or three external ports.

When many computers are to be used in close proximity, it becomes cheaper and more convenient to **network** them. **Fig. 16.11** is a schematic diagram of a typical networking arrangement. Before going into the details of networking it is worth considering a few advantages of networking computers. The major advantage of networking is to let computers share **resources**. Resources might include printers, modems, directories or files on other computers. In **Fig. 16.11**, for example, computer system 1 has a modem, station 4 had a high capacity ZIP drive connected, whilst station 7 has a printer. In computer language, what is referred to as 'station' here is also referred to as **client**. Each client in the network can use the peripherals and files on another client, if that station allows these facilities to be shared. Networking has made working in groups very easy and affordable, as each computer does not need to have all peripherals connected to it. Perhaps, your school or college offices already have a LAN, and acronym for **Local Area Network**. Before proceeding further, we must study a few basic requirements of networking computers. These fall under the two categories: hardware and software.

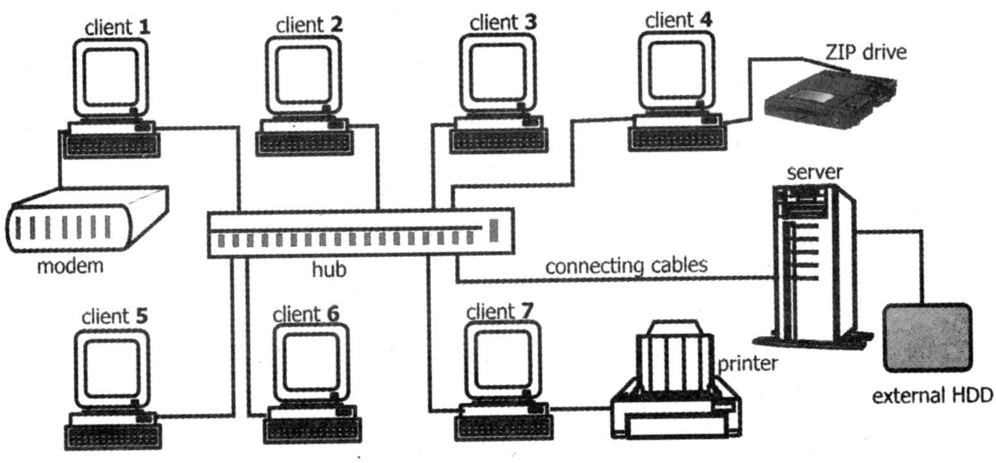

Fig. 16.11 Schematic diagram of a local area network (LAN) system.

On the hardware side, a local area network consists of the following:

1. **Server**. The server is usually another computer with large hard disk space, and a fast CPU.
2. **Clients**. The other computers connected to the server for sharing data and resources are referred to as clients. A client could be a computer with its own hard disk, or be without a hard disk. In the latter case the client is referred to as a **diskless** station, and would use programs installed on the server, on an external drive in the network, or on one of the other clients with hard disks. Clients are also called **workstations**.
3. **Hub**. The hub contains circuitry enabling the clients to communicate with each other through the server. At the back of the hub are sockets for connecting the cables from each computer, and at the front panel are indicators corresponding to each socket being plugged in, and its computer switched on.
4. **Connecting cables**. In most networks, computers are connected to the hub by eight-core cabling, similar to telephone cables. Whilst they appear similar to telephone cables, the two are not to be confused with each other.
5. **Peripherals**. These may include external hard disks, modems, printers, and other devices.

The software control of a network system is in many ways a lot more complicated than the hardware. It starts from installing a **network operating system** on the server. Examples of such operating systems are **Windows NT**, **Netware**, and **Linux**. Operating systems compatible with the network software are loaded on the workstations. Every workstation on the network is identified, so the server 'knows' which client is in which room. Sometimes, this identification might be the officer to whom the computer is assigned, or the room in which the machine is located. All these assignments and many of the details of network operations are in the hands of a **network administrator**. The network administrator sees into the day-to-day smooth running of all the computers on the network. He/she is professionally trained to do this job. Once the network is set up and running everything usually runs fine. Problems are reported to the network administrator who fixes them. The details of how computers share data and use connected resources are in the hands of the administrator, and lie outside the scope of this book. The reader is advised to consult computer books dedicated to the subject.

16.10.5 Wide area networks

A wide area network (or WAN) takes many forms. In a wide area network, communication between computers is sometimes over long distances, using cabling, microwave, or satellite connections. There are two examples of wide area networks. In one, we use a single server, sometimes called the **host computer**, and many workstations over long distances. The computers communicate with each other through the host computer. There is strong similarity between this system and the local area network just studied. The difference between the two lies in the fact that in the wide area network data is sent over much longer distances, whilst distances in the local area network are considerably less. The other type of wide area networking arrangement uses many local area networks, each with its own server, with all the servers linked to a central server. This is the type of wide area network that exists in many colleges and other similar institutions. **Fig. 16.12** is a block diagram of this arrangement.

The system depicted in this figure represents quite a few options of computer brands, networking cables, and peripheral devices. For simplicity, there are four local area networks or **workgroups**, linked via fibre-optic cabling to a central server. The workgroups are named Bumbum, Charlie, Purari, and Markham workgroups. Like in local area network, any identification name would do for a workgroup. Usually the members of the workgroup agree on what name to call their workgroup. This name is then entered in the setting up of the server systems. As can be expected, the software management of the WAN would be a lot more complicated than for the local area network. When managed properly, the advantages of the wide area network are enormous. Suppose, for example, Evelyn, a client of the Bumbum workgroup needs to use a peripheral device connected on to the client named Edmond on the Markham workgroup. Data sent from Evelyn go through the Bumbum server to the main server, through the Markham server before getting to Edmond. For successful communication, however, the two LANs (i.e. Bumbum and Markham) must be set up to share data and resources with each other, and Edmond must have given right to Evelyn to use the software. The 'agreement' between the servers is referred to as **trust relationship**. For Evelyn to use the device on Edmond, Edmond must be setup to allow Evelyn to use its resource. Evelyn must use an agreed **password** in order for Edmond to grant access to use the device. The whole system of interrelations among servers, and workstations can be quite complicated in software terms.

Often, the main server has a communication device such as a modem connected to it. When set up, this allows all the computers in the workgroups to communicate with computers at other locations a long distance away via the telephone link.

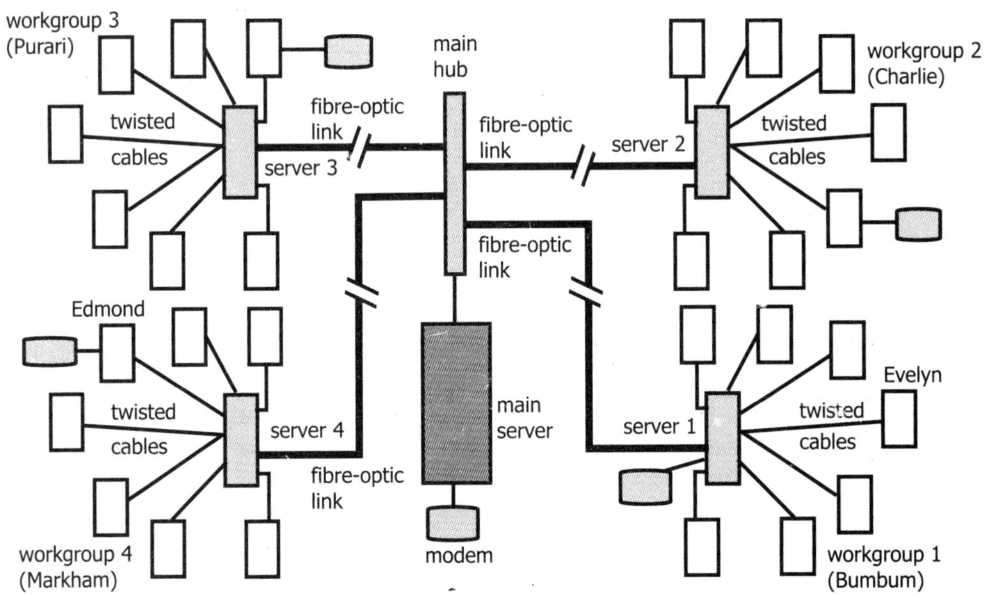

Fig. 16.12 Wide area network (WAN)

Managing a wide area network is a very complicated and highly specialised job. It requires expertise in computer hardware to put together all the necessary components and get them working correctly. It also required skills in software management to ensure the main server is working all the time, and coordinating activities with the other servers. Most important is to make sure the system is working all the time. Fortunately for most users, this is the job of the wide area network manager, and the managers of the workgroups. The rest of us only call on them when things go wrong.

16.10.6 The Internet

In many respects, we can consider the **Internet** as the ultimate wide area network. Originally developed for military and governmental applications, the Internet is now within the domain of the small time computer user. With the Internet, a computer is linked to any other similar system at any location in the world. It therefore gives users the power to search for information anywhere in the world. With millions of computers connected together and sharing information, we can get practically any type of information on almost any subject. The Internet can, for instance, give you the results of last weekend's football matches played in Brazil, provide information on recipes for preparing your favourite dishes, give details on the geography, history, and politics of any country of interest, plus the world news and current affairs, and many more. The list is almost endless. It can be said that the Internet has indeed opened the world, and with its facilities, we are truly in a 'global village', where everyone (with the facilities) is within easy access of every other person for information sharing. Whilst it not intended here to give full technical details of the makeup of the Internet and how it works, we can describe a few features of the Internet that you need to know as a computer user.

Connecting to the Internet. In order to connect a computer to the Internet, the following are needed:

1. A modem to be connected and installed on the computer. This modem connects to the telephone cable.
2. Internet software. Also called the **browser**, this software performs the job of translating the data from one computer into standard form, so any other computer can access it. We can understand the job of the browser from the following example: Jane has a document on her computer that she needs to share with her friend Liz. The document was prepared using software that Liz does not have on her computer. Giving a diskette of the document to Liz will be no use, as she (Liz) would not be able to open the document. However, if both friends are connected to the Internet, and Jane can convert her document to **html** – short for **hypertext markup language –** and send through the Internet, Liz could then open it using her browser software. With the browser software, all computers need not have the same programs installed in order to share data and files. Examples of browser software on today's computer market are **Netscape Navigator™**, and **Microsoft™ Internet Explorer**. These programs can be

bought separately, or sometimes they can be part of the computer's operating system. An example of such an operating system is **MS Windows 98™**.

3. With the hardware and software requirements met, the final step toward accessing the Internet is to have a server, managed in such a way that data can be obtained from it at all times. In any country these days, there are many business houses that specialise in providing this service. Such Companies are called **Internet Service Providers**, with the acronym **ISP**. A computer user normally applies to an ISP for connecting to the Internet. The ISP gives its charges (usually monthly), and issues the applicant with identification details anytime he/she wants to use the Internet. In summary, we regard an ISP as a network server manager whose main job is to enable customers to access the Internet.

Looking up information. Information on the Internet resides on computers, which in turn are attached to servers. Every server has a **website address**. Users **log on** to this address in order to get the desired information. A website address is also known as the **URL,** and acronym for **Uniform Resource Locator**. A typical URL has information on the country in which the server is, the type of server, the name of the server, the Internet service, and the Internet protocol being used. The following is an example of a website: **http://www.daltron.com.pg**. We normally read the URL from the end backwards. The last segment (pg) is the country code. From this we know that the server is in Papua New Guinea. A web address without country identifier refers to the U.S.A. The second segment (com) indicates a commercial service. There are many kinds of internet services, some are commercial, some are set up by government (.gov), some by organisations (.org), etc. In the URL of this example, 'daltron' is the name of the Company. The segment http://www refers to the Internet service; www. stands for the **World Wide Web**, the 'friendly face' of the Internet, and http stands for **hypertext transmission protocol**. With this protocol, a document is transformed into the language of the Internet – the **hypertext markup language**, or **html**. With this common language, Liz's computer in our example can read Jane's document. This method of unscrambling the URL is similar to sorting letters in an international post office. The last line of the address (in the Imperial and American systems) gives the country of destination, the next on top gives the town or city, the next above is the street name, etc., until we get to the top line, which is the name of the person to whom the letter is addressed.

From the above, it may appear that to find information on the Internet one needs to know is the URL. What if one does not know any address holding the information one needs? The answer to this question is provided by the availability of **search engines**. Search engines are special kinds of websites. Their job is to search all the websites for the information one needs. There are few of these engines: **yahoo!.com**, **altavista.com**, **geocities.com**, **google.com**, and many more. By 'visiting' any of these engines a menu screen appears, and all one would be required to do is to type in the space provided a key word (or group of words) that identify the subject one wants to find information on. The search engine then returns with all possible results of the search. Some of these might be useful to the subject of interest, some might not. The user then goes through the listings and select the material of interest. **Surfing** the Internet, as it is called, has been made easy by the use of these search engines, and many search engines are being developed regularly.

16.10.7 Electronic mail

We are all to some extent familiar with the telephone message facility. You telephone your friend John who is not at home. You leave a message on his answering machine saying that you called. John comes home, finds the message, and calls you while for some reason you were out. He also leaves a message on your machine, and so on. This way, it might perhaps take a couple of days before the two of you are simultaneously sitting by your telephones before you can actually talk. Electronic mail, or **email**, for short had been invented to help solve this problem. With this facility on John's computer, you can send a written message that he would get when he turns his computer on. He gets your message, and responds to it, without the two of you having to sit down by the telephone. Electronic mail facility also uses a modem connected to the computer, and the carrier software. Examples of the carrier software in use are **Microsoft™ Outlook, Outlook Express™**, **Eudora™**, and **Netscape™**. The facility has proven very invaluable for people communicating with each other, either to promote business or personal relationship.

16.10.8 Electronic banking

In electronic banking, financial transactions are handled electronically using computer technology. One popular electronic banking scheme is the **EFTs**, short for **Electronics Funds Transfer**. People pay for goods and services by having funds transferred from their bank accounts electronically. The **Automatic Teller Machine** (or **ATM**) is about the most vivid manifestation of the EFTs.

16.10.9 Networks and Security

In any remote communication system, it is necessary to ascertain that communication is made with the right person at the other end of the line. In telephone conversations, for instance, we normally start with identifying ourselves or the telephone number dialed. This way, we get to know whether the wrong number was dialed or not. In computer communication, the issue of user names and corresponding passwords solves this security problem. In some systems (e.g. in banking), the account number and a **personal identification number** (or **PIN**) are used in place of the user name and password. A matching user name and password (or matching account number and PIN) allows the person to use the communication facility to contact other people on the system, to leave messages, to use the banking facility, etc. When set up and working properly, the system is very smooth, identifying those that should be using the facility, and denying access to those who shouldn't. The system is managed in such a manner that all the information on it is safe from access to unauthorised persons.

QUESTIONS

1. Define hardware and software. Give examples.

2. Describe the roles that the central processing unit plays in a computer system.

3. Name three common forms of input devices, and three forms of output devices.

4. State the function of a secondary storage system. Describe the most common types of secondary storage devices.

5. Name and describe three types of computers.

6. When people call software "friendly", what do they usually mean?

7. Explain why the binary system is used to represent computer data.

8. Discuss the methods of data addressing.

9. Define the terms: bit, byte, kilobyte, megabyte, gigabyte, RAM, and ROM.

10. Describe possible ways of increasing the speed of a microcomputer. Explain how each method works.

11. Define the terms hard copy, and soft copy. State devices that produce each.

12. Explain the term batch processing. State one advantage and one disadvantage of it.

13. Explain the term telecommunication. State the pieces of hardware and soft ware necessary for successful computer telecommunication.

14. Describe the advantages of each of the following: twisted pairs, coaxial cables, fibre-optic link, and microwave transmission.

15. What is a protocol? Why are protocols important in computer communications?

16. Describe the differences between a LAN and a WAN.

17. What is the Internet? State the advantages of the Internet over other methods of computer communication.

Answers to Exercises

Chapter 1

1. 5.0 mm, 33.142 m, 800 Hz, 2514 m/s **2.** 3.26 m **3.** 3.8 ms, 263 Hz **4.** 0.17 s **5.** 232 N **6.** 13.72 m/s **7.** $y = 5.1 \sin 2\pi(33t - 0.42x)$ **8.** 860 Hz **9.** 0.275 m/s **10.** 250 m **11.** 330 m/s **12.** 47.1 m/s away from person **13.** (i) 62.5 Hz, 392.75 s^{-1}, 3.93 m^{-1} (ii) $y = -0.8 \sin 2\pi(62.5t - 3.93x)$, (iii) $y = -0.8 \sin 392.75t$, (iv) $y = -0.8 \sin 2\pi(62.5t - 0.75x)$, (v) 31.42 m/s **14.** $y = 2.5 \sin 20\pi(15t - 0.4)$ **15.** 1.33 **16.** (i) 100 Hz, (ii) 1.7 m, (iii) 170 m/s, (iv) $\lambda/2$ **17.** (i) 783.9 Hz, (ii) 558.7 Hz **18.** (i) positive x, (ii) 0.30 m, (iii) 1.0 Hz, 0.5 m, 0.5 m/s, (iv) $y = -0.6 \sin 2\pi t \sin 4\pi x$ **19.** $y = 0.05 \cos 2\pi(50t - 40x)$ **20.** 0.85 **21.** 2.88 m **22.** (i) 10 cm, (ii) 0.8 Hz, 1.57 m, 1.25 m/s **23.** 1.33 **24.** (i) 10 ms, (ii) 3.33 m, (iii) 100 Hz, (iv) 0.04 **25.** $y = 2.50 \sin 2\pi(50t + 1.33)$ mm **26.** (i) 495.9 Hz, (ii) 622.3 Hz **27.** 1.31 km, **28.** 0.0175 m – 17.15 m.

Chapter 2

1. (i) 2.0 cm, (ii) 4.0 cm **2.** const. int. **3.** (a) 0, 11.53°, 23.58°, 36.87°, 53.13°, 90.00° (b) 5.74°, 17.45°, 30.00°, 44.42° , 64.16° **4.** (a) 4.0 mm, 2.1 m, 1.57 s, -ve x-direction, (b) $y = 8 \sin 3x \cos 4t$ 0.17 **5.** 456 m/s **6.** (i) 0.67 cm, (ii) 17.2 cm, (iii) 33.5 cm/s

7. 3.0×10^{11} Hz **8.** 100 Hz, 50 mm, 5.0 m/s, 0 m/s (c) 12.5 cm, 37.5 cm, 62.5 cm, ... (for maximum), 0 cm, 25 cm, 50 cm,... (for minimum) **9.** (i) 1.7 cm, (ii) 1.1 cm, 0.8 cm, 1.1 cm, (iii) 0 cm, (iv) 0.30 cm. **11.** (a) 0.02 m, (b) (i) $y = -0.01 \cos 2\pi(100t + 5xx)$

(ii) $y = -0.02 \sin 200\pi t \sin 10\pi x$

Chapter 3

1. (i) 1.50 km (ii) 0.037 m **2.** 2.071 km **3.** 1103 m **4.** 957 m **5.** 30 db **6.** (a) 25 W/cm^2, (b) 62.9 db **7.** 0.067 μW **8.** 34.3 db **9.** 5.28 **10.** 0.46 m **11.** 3 **12.** 260.8 Hz **13.** 1.96 m

Chapter 4

1. (i) 1.19 mm, (ii) 0.59 mm **2.** 2.36 mm **3.** 37 **4.** 0.44 mm **5.** 303.3 m, 106.7 m, 21 **6.** 7.3% **7.** 1.25 mm **8.** (i) 590 nm, (ii) 0.13° **9.** 9.1 km **10.** 0°, 75.5° **11.** 4044 **12.** 11.5°, 23.6°, 36.9°, 53.1° **13.** 433 **14.** 0.54π **15.** (i) 1.17 mm, (ii) 6.48 mm **16.** 2.28×10^8 m/s, 324 nm; 2.25×10^8 m/s, 319 nm

Chapter 5

1. 2.25×10^8 m/s **2.** 1.50 **3.** 1.65 **4.** 1.03 **5.** (i) 1.12, (ii) 0.89 **6.** (a) 28.1°, (b) 36 28.1° **7.** (i) 35°, (ii) 4.8° **8.** (i) 0.021, 47.7, (ii) 0.08° **9.** 39.96°, 40.54° **10.** 24.8° **11.** (a) 33.3°, (b) 65.6° **12.** 1.316 **13.** 9.05° **14.** (i) 17.53°, (ii) 56.10° **15.** 18.49 cm **16.** (i) 53.13°, (ii) 46.26° **17.** 51.4° **18.** (i) 61°, (ii) 51° **19.** 26.8°

Chapter 6

1. real, inverted, 9.6 cm in front **2.** 24 cm **3.** (i) concave, (ii) 3.0 **4.** 30.0 cm, 0.5 cm **5.** (i) 35.0 cm, (ii) 33.8 cm **6.** 5.45 cm, upright, 0.45 **8.** (i) 3.28 cm, (ii) 4.93 cm **9.** 8.0 **10.** 12 cm, 36 cm, 3, 0.33 **12.** 28.57 cm **13.** 3.85 cm **14.** 23.3 cm **15.** 60 cm, 30 cr **17.** (i) 3.25 cm, 36.75 cm, (ii) 22.6 cm, 0.18 cm. **18.** 0.11 cm

Chapter 7

1. (i) 6.0, (ii) 6.0 **2.** 3.33 **3.** (a) 1.08 cm, (b) 1162.5 **4.** 3.06 cm **5.** (i) 50, (ii) 0.45 r **6.** 360 **7.** 50 **9.** (i) 6.67, (ii) 0.57 cm **10.** 18.0 cm, 3.0 cm **11.** (i) 25, (ii) 6.8 **12.** (i) 12.24 cm, (ii) 1.18m x 1.76 m **13.** 0.038 rad **14.** 5.5 cm, **15.** 13.7 m **16.** 1.13

Chapter 8

1. 28.8 μN **2.** 2.27×10^{39} **3.** 1.38 cm **4.** 39.15 μN **5.** (i) 3.0×10^{-7} N (ii) 4.8×10^{-6} N (iii) 1.3×10^{-7} N **6.** 4.14×10^{-3} N **7.** 3.82×10^{-2} N **8.** 1.31 μC **9.** 5.6×10^{-11} N/C **10.** 1.02×10^{-7} N/C **11.** 1178 N/C, 105.9 V **12.** (i) 3.3 ns, (ii) 1.58×10^8 m/s **13.** 102.4 **14.** 1.65 J **15.** 540 N toward the -20 μC charge **16.** 2.75×10^6 N/C **17.** (i) 02 m, 0.3 m (ii) 0.15 m, 0.35 m **18.** (a) $V_o > V_1$, (b) V_o/V_1 **19.** (i) 1.44×10^{-15} J, (ii) 1.31×10^6 m/s **20.** (i) 3.0×10^4 J, (ii) 0 **21.** (i) 708 pF, (ii) 155 nC, (iii) 17 μJ **22.** 11.4 μJ **23.** (i) 40000 (ii) 9.6 °C **24.** (b) 200 μC, 66.67 V, 20 V, 13.33 V (c) 0.015 J, 0.05 J, 0.075 J **25.** 1000 V **26.** 1.33 μF **27.** (i) 2.3×10^4 V, (ii) 226 pF, (iii) 12.23 cm^2, (iv) 598 μJ **28.** (i) 66.6 μF, (ii) 233.3 μF.

Chapter 9

1. 0.58 mm **2.** 45 cm **3.** 6.0 A **4.** (i) 14 Ω, (ii) 1.7 A **5.** (i) 2.9 Ω (ii) 2.4 A, 1.2 A, 0.6 A (iii) 4.2 A (iv) 28.8 W, 14.4 W, 7.2 W **6.** (i) 9.8 Ω (ii) 2.45 A **7.** (i) 0.6 A, (ii) 3.3 A, (iii) 0.

8. \$17.28 **9.** (i) 2.0 A, 2.0 A, 1.0 A, 5.0 A **10.** (i) 1.21×10^8 J (ii) \$5,400.00 **11.** 1.2 Ω **12.** (i) 2.2 A, (ii) 30.4 W (iii) 4.3 W (iv) 1 hr 9 min **13.** (a) 3.0 W (b) 3.0 W **14.** (b) 1.50 m **15.** 24.6 Ω **16.** (i) 2.29 A, 1.43 A, 0.86 A (ii) 5.13 V **18.** (i) 3.7 Ω, (ii) 24 W, (iii) 14.8 W **19.** (i) 6.0 A (ii) 3.0 A (iii) 12.0 A, (iv) 6.0 A **20.** 144 W, 72 W, 288 W, 144 W **21.** -6.8°C **22.** $4.7 \times 10^{-4} K^{-1}$ **23.** (i) 1.84 mA, 1.08 V, (ii) 1.08 V, (iii) 0.29 V **24.** (i) 5 Ω, (ii) 36 mA, (iii) 720 mV **25.** (a) 0.5 Ω, (b) 8 Ω, 32.5 Ω, (c) parallel **26.** 12.65 V, 0.03 Ω **27.** 0.41 A **28.** (i) 71.4 mA, (ii) 0.2 A, 0.1 A **29.** (i) 13 Ω, (ii) 15.0 V, (iii) 0.87 A, **30.** 4.7 A, 13.13 V

Chapter 10

2. 0.025 Wb **3.** 4.0×10^{-5} T **4.** (i) 20 μT, (ii) E-W **6.** 1.43 A **7.** (c) 4.0×10^{-5} T **8.** 1.5×10^{-5} T **9.** (i) 50 mT, (ii) 2.0×10^{-5} Wb **10.** 15.7 mT **11.** 4.97 A **12.** 100 turns **13.** 1989 turns **14.** (b) 0.3 T **15.** 9.81 A **16.** 63 mT, 18.9 T **17.** (a) 36 mT, (b) 0.18 N/m **18.** (a) +ve x-direction, upwards (b) 0 Nm, 31.4 Nm (c) 14.5° **19.** (b) 8.32 mm **20.** (b) 0 N, 3 N, 0 N, 3 N **21.** 6.13 mm/s **22.** 50 cm **23.** 50.2° **24.** 12.5 mm **25.** 1.29×10^7 C/kg **26.** (i) 9.6×10^6 C/kg, (ii) 1.67×10^{-27} kg **27.** 25.0 cm.
30. 4.7 A, 13.13 V

Chapter 11

1. 0.1 V **2.** 0.09 T **3.** (i) 2.6 V, (ii) 1.3 V **4.** 13.3 H **5.** 0.5 ms **6.** 1.6 μV **7.** (i) 1.57 V, (ii) 3.14 A **8.** (i) 15 rad/s, (ii) 12 mA **9.** 1.13 V **10.** (i) 57, (ii) 109

Chapter 12

1. (i) 2.12 A, (ii) $90 \sin 100\pi t$, (iii) 63.6 V **2.** 45.3 min **3.** 12.7 A **4.** (i) 33.9 Ω, (ii) 194.2 V, 101.4 V, (iii) 27.6° **5.** 20.7 min **6.** 4.36 nF **7.** (i) 2.8 A, (ii) 1.98 A, (iii) 314 W **8.** (i) 225 μH, (ii) 1 μA, (iii) 0.47 μV, (iv) 16 **9.** (i) 100 mA, (ii) 100 Ω, (iii) 1000 Hz **10.** (i) 118 Ω, (ii) 212 V, 133 V, (iii) 32°, (iv) 450 W **11.** (i) 2.19 A, (ii) 139 μF, (iii) 174 mH, (iv) 220 W **12.** (i) 67.3 V, (ii) 156.2 W **13.** (i) 190 Ω, (ii) 113 V, 30 V, 213 V, 215 V, **14.** (i) 58.4°, (ii) $304 \sin(120\pi t + 1.02)$ V, (iii) 129 W **15.** (i) 11.3 Ω, 53 Ω, (ii) 48 Ω, (iii) 60°, (iv) 2.49 A, (v) 149 W **16.** (i) 202.5 μF, (ii) 17.7 V, -17.7 V, 113.0 V, 0 V, (iii) 113 mA, (iv) 12.8 W, (v) 0.16 **17.** (i) 15.7 Ω, 63.6 Ω, (ii) 111.8 Ω, (iii) 25.6°, (iv) 1.98 A, (v) 392 W.

Chapter 13

1. (a) 25 ms, (b) 20 ms, (c) 2.5 s, (d) 2.0 s, (e) 600 μs, (f) 10 ms, (g) 250 μs, (h) 2.0 ms **2.** (i) 1.0 mA, (ii) 200 ms **3.** (i) 200 μs, (ii) 36.8 V **4.** 75 kΩ **5.** (i) 10 ms, (ii) 73 V

Chapter 14

1. (i) 0.01 Ω shunt, (ii) 0.10 Ω shunt, (iii) 1995 Ω bobbing, (iv) 495 Ω bobbing **2.** (i) 2.4 mV (ii) 40 mV, (iii) 91 Hz, (iv) 77 Hz, (v) 2 ms

Appendices

CONTENTS

Appendix 1
Constants and Symbols

Physical constants

Quantity	Usual symbol	Value with units
Acceleration due to gravity	g	9.81 m/s^2
Boltzmann's constant	k_B	1.38×10^{-23} J/K
Velocity of electromagnetic waves in vacuo	c	3.0×10^{8} m/s
Permitivity of free space	ε_o	8.85×10^{-12} F/m
Permeability of free space	μ_o	$4\pi \times 10^{-7}$ H/m
Speed of sound in air at 20 °C		334 m/s
Density of air at S.T.P.		1.29 kg/m^3
Charge on the electron	e	-1.6021×10^{-19} C
Mass of the electron	m_e	9.11×10^{-31} kg
Charge on the proton	p	$+1.6021 \times 10^{-19}$ C
Coulomb's law constant	$k\left(= \frac{1}{4\pi\varepsilon_o}\right)$	9.0×10^{9} Nm^2/C^2
Mass of proton	m_p	1.667×10^{-27} kg
Universal gas constant	R	8.32 J/K/mol
Avogadro's number	N_A	6.023×10^{23} mol^{-1}
Universal gravitational constant	G	6.67×10^{-11} Nm^2 / kg^2
Planck's constant	h	6.62×10^{-34} Js

Some frequently used Greek characters

Character Upper case	Character Lower case	Name	Character Upper case	Character Lower case	Name
A	α	Alpha	Λ	λ	Lambda
B	β	Beta	M	μ	Mu
Δ	δ	Delta	N	ν	Nu
E	ε	Epsilon	Π	π	Pi
Φ	ϕ	Phi	Θ	θ	Theta
Γ	γ	Gamma	P	ρ	Rho
H	η	Eta	Σ	σ	Sigma
K	κ	Kappa	T	τ	Tau
			Ω	ω	Omega

Appendix 2
Some Useful Formulae

Waves and Optics

Wave velocity

$$v = f\lambda \;;\quad f = \frac{v}{\lambda} \;;\quad \lambda = \frac{v}{f}$$

$$v = \sqrt{\frac{\text{Elastic modulus}}{\text{density}}}$$

$$v = \sqrt{\frac{\text{Tension in string}}{\text{linear density}}}$$

Relative refractive index

$$_1n_2 = \frac{\text{speed in medium 1}}{\text{speed in medium 2}} = \frac{v_1}{v_2}$$

Progressive wave function

$$y = A\sin(\omega t - kx)$$

Stationary wave equation

$$y = 2A\sin kx\cos\omega t$$

Doppler effect

$$f' = \frac{V \pm v_o}{V \mp v_s}.f$$ (apparent frequency formula)

Sound intensity

$$1\text{ db} = 10\log_{10}\left(\frac{I}{I_o}\right)$$

Inverse-square law

$$\frac{I_1}{I_o} = \frac{d_o^2}{d_1^2}$$

Interference and diffraction

$$\Delta = m\lambda, \quad m = 0, 1, 2, 3, \ldots$$ (condition for constructive interference)

$$\Delta = \left(m + \tfrac{1}{2}\right), \quad m = 0, 1, 2, 3, \ldots$$ (condition for destructive interference)

$$a\frac{y_m}{D} = m\lambda, \quad m = 0, 1, 2, 3, \ldots$$ (Young's bright fringe formula)

$$y = \frac{\lambda D}{a}, \quad m = 0, 1, 2, 3, \ldots$$ (Young's fringe separation formula)

$$2t = \left(m + \tfrac{1}{2}\right)\lambda, \quad m = 0, 1, 2, 3, \ldots$$ (bright fringe in thin wedge)

$$2nt\cos r = \left(m + \tfrac{1}{2}\right)\lambda, \quad m = 1, 2, \ldots$$ (bright fringe in thin liquid film)

$$d\sin\theta = m\lambda, \quad m = 0, 1, 2, 3, \ldots$$ (diffraction maxima)

Refraction of light

$$n\sin i = \text{constant}$$ (Snell's law)

$$\sin c = \frac{n_1}{n_2}, \text{ with } n_2 > n_1$$ (total internal reflection)

$$\frac{1}{u} + \frac{1}{v} = \frac{1}{f}$$ (passage of light through lens)

$$m = -\frac{v}{u}$$ (magnification by a thin lens)

$$\frac{1}{f_1} + \frac{1}{f_2} = \frac{1}{f}$$ (two thin lenses in contact)

$$\frac{1}{u} + \frac{1}{v} = \frac{1}{f} = (n-1)\left(\frac{1}{r_1} + \frac{1}{r_2}\right)$$ (lens makers' formula)

Optical instruments

$$M = \frac{\alpha'}{\alpha}$$ (angular magnification)

$$M = \left(\frac{D}{f_e} - 1\right)\left(\frac{v}{f_o} - 1\right)$$ (compound microscope)

$$M = \frac{f_o}{f_e}$$ (astronomical telescope)

Electricity and Magnetism

Electrostatics

$$F = \frac{1}{4\pi\varepsilon_o} \cdot \frac{q_1 q_2}{r^2}$$ (magnitude of force in Coulomb's law)

$$F = qE$$ (E = electric field)

$$C = \frac{Q}{V}$$ (capacitance)

$$C_T = C_1 + C_2 + C_3 + \dots\dots\dots$$ (capacitors in parallel)

$$\frac{1}{C_T} = \frac{1}{C_1} + \frac{1}{C_2} + \frac{1}{C_3} + \dots\dots$$ (capacitors in series)

$$E = \frac{1}{2}QV = \frac{1}{2}CV^2 = \frac{1}{2}\frac{Q^2}{C}$$ (energy stored in a capacitor)

Direct current electricity

$$I = nqvA \; ; \quad J = \frac{I}{A} = nqv$$ (Current density)

$$J = \sigma E = \sigma \frac{V}{L}$$ (Ohm's law)

$$V = IR$$ (Ohm's law)

$$R = \rho \frac{L}{A}$$ (resistance of wire)

$$R_T = R_1 + R_2 + R_3 + \dots\dots\dots$$ (resistances in series)

$$\frac{1}{R_T} = \frac{1}{R_1} + \frac{1}{R_2} + \frac{1}{R_3} + \dots\dots$$ (resistances in parallel)

$$P = IV = I^2R = \frac{V^2}{R}$$ (power in d.c. circuit)

$$E_T = E_1 + E_2 + E_3 + \dots\dots\dots$$ (cells in series)

The magnetic field

$$dB = \frac{\mu_o}{4\pi} \frac{Idl\sin\theta}{r^2}$$ (Biot-Savart law)

$$B = \frac{\mu_o I}{2\pi a}$$ (field due to long straight wire)

$$B = \frac{\mu_o NI}{2a}$$ (field at centre of a circular coil)

$$B = \frac{\mu_o NIx^2}{2(a^2 + x^2)^{3/2}}$$ (field along axis of circular coil)

$B = \mu_0 nI$	(field at centre of long solenoid)
$H = nI$	(magnetising force due to current loop)
$B = \mu H$	(flux density dependence on magnetising force)
$F = BIl\sin\theta$	(force on current wire in magnetic field)

Electromagnetic induction

$\varepsilon = -\frac{d\Phi}{dt} = -\frac{Nd\phi}{dt}$	(Law of electromagnetic induction)
$\varepsilon = -L\frac{dI}{dt}$	(self inductance)
$\varepsilon = -M\frac{dI}{dt}$	(mutual inductance)
$L = L_1 + L_2 + L_3 + \ldots$	(non-interacting inductances in series)
$\frac{1}{L} = \frac{1}{L_1} + \frac{1}{L_2} + \frac{1}{L_3} + \ldots$	(non-interacting inductances in parallel)
$\frac{\varepsilon_S}{\varepsilon_P} = \frac{N_S}{N_P}$	(transformer equation)

Alternating currents

$I = I_0 \sin\omega t, \quad V = V_0 \sin(\omega t + \phi)$	(Alternating current and voltage)
$I_{rms} = \frac{I_0}{\sqrt{2}}, \quad V_{rms} = \frac{V_0}{\sqrt{2}},$	(rms values of current and voltage)
$\tan\phi = \frac{X}{R} = \frac{X_L - X_C}{R}$	(phase angle between current and voltage)
$X_L = \omega L, \quad X_C = \frac{1}{\omega C}$	(inductive and capacitive reactances)
$Z = \sqrt{R^2 + X^2}$	(impedance in a.c. circuit)
$P = \frac{1}{2} I_0 V_0 \cos\phi$	(power in circuit)

Transient currents

$I = I_0 e^{-t/RC}, \quad \tau = RC$	(current and time constant in RC circuit)
$I = I_0\left(1 - e^{-Lt/R}\right), \quad \tau = \frac{R}{L}$	(current and time constant in RL circuit)

Appendix 3
Mathematical Notes

A3.1 GEOMETRY

A3.1.1 The straight line

The straight line is defined as the shortest distance between two points. **Fig. A3.1** represents a general straight line. Using Cartesian coordinates, the straight line can be represented by

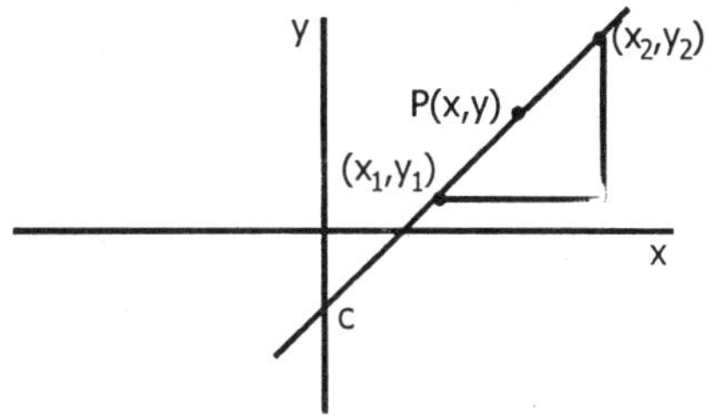

Fig. A3.1 The Straight line

$$y = mx + c, \quad (A3.1)$$

where c is the intercept on the y-axis, and m is the slope of the graph, given by the expression $m = \frac{y_2 - y_1}{x_2 - x_1}$, where (x_1,y_1) and (x_2,y_2) are any two points on the line. Alternatively, (A3.1) can be written in the form

$$\frac{y - y_1}{x - x_1} = \frac{y_2 - y_1}{x_2 - x_1}. \quad (A3.2)$$

A3.1.2 Other curves

Table A3.1 gives the equations of other curves we deal with in physics. In each of these Cartesian equations, x is the independent variable, whilst y is the dependent variable.

Table A3.1 Cartesian equations of some curves

Curve	General equation	Special forms
Quadratic (or parabola)	$y = Ax^2 + Bx + C$	$y - k = 4a(x - h)^2$
Circle	$(x - h)^2 + (y - k)^2 = r^2$	$x^2 + y^2 = r^2$
Cubic	$y = Ax^3 + Bx^2 + Cx + D$	
Ellipse	$\sqrt{(x - x_1)^2 + (y - y_1)^2} + \sqrt{(x - x_2)^2 + (y - y_2)^2} = \text{constant}$	$\frac{x^2}{a^2} + \frac{y^2}{b^2} = 1$
Hyperbola	$\sqrt{(x - x_1)^2 + (y - y_1)^2} - \sqrt{(x - x_2)^2 + (y - y_2)^2} = \text{constant}$	$xy = c^2$
Exponential decay	$y = y_o e^{-kx}$	
Exponential growth	$y = y_o e^{kx}$; also $y = y_s(1 - e^{-kx})$	

A3.2 TRIGONOMETRY

A3.2.1 Trigonometric ratios of a right-angled triangle

Fig. A3.2 (a) shows the general right-angled triangle, with sides of lengths x, y and z. In terms of these, we can define the basic trigonometric ratios of sine, cosine, and tangent of an angle A. The reciprocal trigonometric ratios are also given. Shown in **Fig. A3.2** (b) are the special angles of 45°, 30°, and 60°. **Table A3.2** gives the trigonometric and reciprocal trigonometric ratios of these angles.

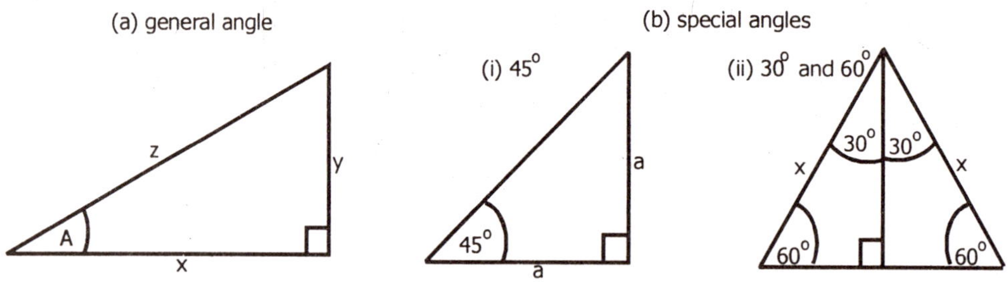

Fig. A3.2 Trigonometric ratios; (a) for general angle, (b) for special angles of 30° and 60°.

Table A3.2 Trigonometric and reciprocal trigonometric ratios

General angle A	Special angles: 45°, 30°, and 60°		
	(i) 45°	(ii) 30°	(iii) 60°
$\sin A = \frac{y}{z}$	$\sin 45^o = \frac{a}{a\sqrt{2}} = \frac{1}{\sqrt{2}}$	$\sin 30^o = \frac{x/2}{x} = \frac{1}{2}$	$\sin 60^o = \frac{x\sqrt{3}/2}{x} = \frac{\sqrt{3}}{2}$
$\cos A = \frac{x}{z}$	$\cos 45^o = \frac{a}{a\sqrt{2}} = \frac{1}{\sqrt{2}}$	$\cos 30^o = \frac{x\sqrt{3}/2}{x} = \frac{\sqrt{3}}{2}$	$\cos 60^o = \frac{x/2}{x} = \frac{1}{2}$
$\tan A = \frac{y}{x}$	$\tan 45^o = \frac{a}{a} = 1.000$	$\tan 30^o = \frac{x/2}{x\sqrt{3}/2} = \frac{1}{\sqrt{3}}$	$\tan 60^o = \frac{x\sqrt{3}/2}{x/2} = \sqrt{3}$
$\text{cosec } A = \frac{1}{\sin A} = \frac{z}{y}$	$\text{cosec } 45^o = \sqrt{2}$	$\text{cosec } 30^o = \frac{1}{\sin 30^o} = 2$	$\text{cosec } 60^o = \frac{1}{\sin 60^o} = \frac{2}{\sqrt{3}}$
$\sec A = \frac{1}{\cos A} = \frac{z}{x}$	$\sec 45^o = \frac{a\sqrt{2}}{a} = \sqrt{2}$	$\sec 30^o = \frac{1}{\cos 30^o} = \frac{2}{\sqrt{3}}$	$\sec 60^o = \frac{1}{\cos 60^o} = 2$
$\cot A = \frac{1}{\tan A} = \frac{x}{y}$	$\cot 45^o = \frac{a}{a} = 1.000$	$\cot 30^o = \frac{1}{\tan 30^o} = \sqrt{3}$	$\cot 60^o = \frac{1}{\tan 60^o} = \frac{1}{\sqrt{3}}$

For any angle A, the following hold:

$$\tan A = \frac{\sin A}{\cos A}; \qquad \sin A^{o} = \cos(90 - A)^{o},$$

and

$$\cos^2 A + \sin^2 A = 1.$$

This last equation is an alternative statement of Pythagoras' theorem.

A3.2.2 Graphs of trigonometric ratios

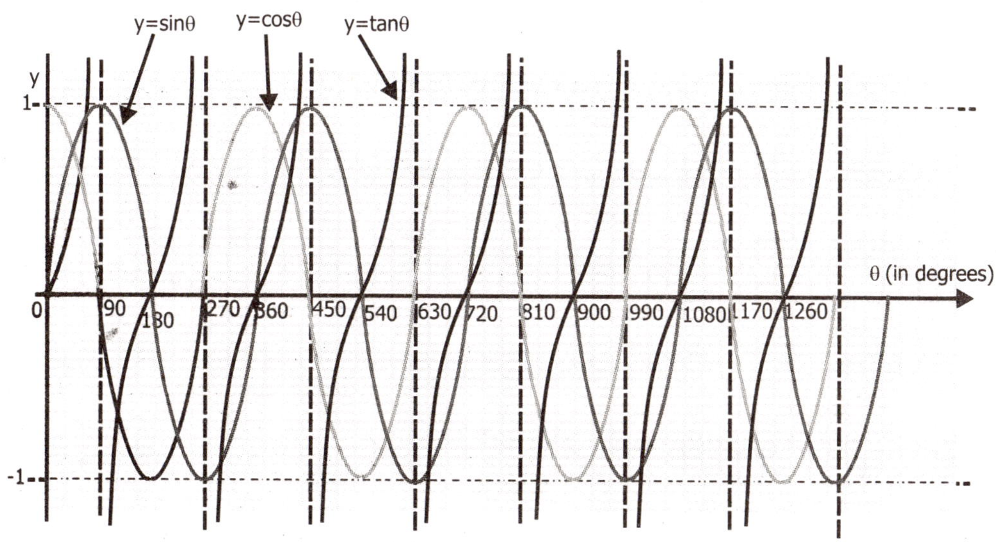

Fig. A3.3 Plots of the values of the basic trigonometric ratios as functions of the angle θ.

A3.2.3 Compound and multiple angles

$\sin(A + B) = \sin A \cos B + \cos A \sin B$; $\quad \sin(A - B) = \sin A \cos B - \cos A \sin B$

$\cos(A + B) = \cos A \cos B - \sin A \sin B$; $\quad \cos(A - B) = \cos A \cos B + \sin A \sin B$

$$\tan(A + B) = \frac{\tan A + \tan B}{1 - \tan A \tan B}; \qquad \tan(A - B) = \frac{\tan A - \tan B}{1 + \tan A \tan B}$$

$$\cos 2A = 2\cos^2 A - 1 = 1 - 2\sin^2 A \qquad \tan 2A = \frac{2\tan A}{1 - \tan^2 A}$$

$\sin 2A = 2\sin A \cos A$; $\quad \sin 3A = 3\sin A - 4\sin^3 A$

$\cos 3A = 4\cos^3 A - 3\cos A$

A3.2.4 Addition formulae

$$\sin P + \sin Q = 2\sin\left(\frac{P+Q}{2}\right)\cos\left(\frac{P-Q}{2}\right); \qquad \sin P - \sin Q = 2\cos\left(\frac{P+Q}{2}\right)\sin\left(\frac{P-Q}{2}\right)$$

$$\cos P + \cos Q = 2\cos\left(\frac{P+Q}{2}\right)\cos\left(\frac{P-Q}{2}\right); \qquad \cos P - \cos Q = -2\sin\left(\frac{P+Q}{2}\right)\sin\left(\frac{P-Q}{2}\right)$$

A3.2.5 Radian measure of an angle

$$\theta\,(\text{in radians}) = \frac{\pi\theta^{o}}{180}; \quad \text{for small angles, } \sin\theta \approx \tan\theta \approx \theta\,(\text{in radians}).$$

A3.3 EXPONENTIAL AND LOGARITHMIC FUNCTIONS

A3.4.1 The exponential function

Exponential functions exist in many calculations, in both the physical and natural sciences. In all these functions, the change in the quantity at any one time depends on the number of items present. A few examples of exponential functions are as follows:

(i) Biological growth: The growth of a population size depends on how large the population is. The greater the population (of insects, people, cultured bacteria, etc.) the greater the number of offspring that are going to be added to the population.

(ii) Financial interest: Interests to deposits in banks and similar financial institutions depends on the amount of money the customer has in his/her account.

(iii) Radioactive decay: the number of radioactive nuclei decaying at any instant depends on the number of radioactive nuclei present at the time.

(iv) In a crate of fruits, the number of fruits going bad at any time depends on the number of good fruits present in the crate.

The above examples give two types of exponential functions: the growth function, and the decay function. Exponential growth functions are represented by one of two equations:

$$y = Ae^{kx}, \tag{A3.3}$$

and

$$y = y_s\left(1 - e^{-kx}\right). \tag{A3.4}$$

Exponential decay functions are represented by

$$y = Ae^{-kx}. \tag{A3.5}$$

In (A3.3), (A3.4), and (A3.5), the quantities, A, k and y_s are constants, and e is the exponential, with value 2.71838282.... In the growth curve represented by (A3.3), y grows to infinity as x increases to infinitely large values. In (A3.4), however, y only grows to a **saturation** value y_s, when x tends to infinity. In the decay relationship of (A3.5), y tends to zero value when x goes to infinity. **Fig. A3.4** are graphs of these functions.

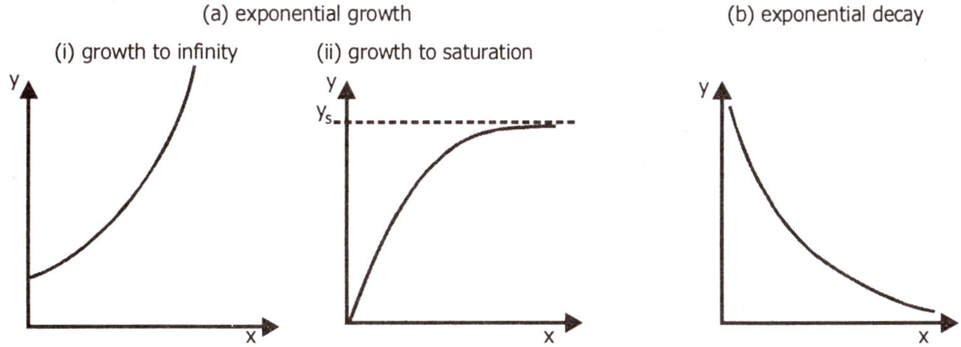

Fig. A3.4 Graphs of exponential growth and decay.

For the graph $y = e^x$, the intercept on the y-axis is 1.0, and the slope at this point is also 1.0. This latter property becomes very useful in calculus.

A3.4.2 Logarithms

The logarithm of a number is defined as the power to which the base must be raised to get the number. Using this definition, if $y = a^x$, then $x = \log_a y$. A few basic theorems of logarithms are stated as follows:

$$\log_a b + \log_a c = \log_a bc\,; \tag{A3.6}$$

$$\log_a b - \log_a c = \log_a\left(\frac{b}{c}\right); \tag{A3.7}$$

$$\log_a y = \frac{\log_b y}{\log_b a}\,; \tag{A3.8}$$

$$\log_a y^n = n \log_a y\,; \tag{A3.9}$$

$$\log_a a^n = n\,; \tag{A3.10}$$

$$\log_a ka^n = n + \log_a k\,. \tag{A3.11}$$

For many everyday calculations using logarithms, the base of 10 is usually selected. For these, tabulated values are readily available. A more useful base when using calculus and when dealing with questions on exponential functions is the exponent e. Logarithms quoted to this base are referred to as **natural logarithms**. The logarithm of a number A to base e is written as $\log_e A$, or $\ln A$. Conversion between logarithms to base 10 to logarithm to base e, and vice versa, is obtained through the relation

$$\log_e A = 2.3026 \times \log_{10} A. \tag{A3.12}$$

A3.4 CONVERTING NON-LINEAR EQUATIONS TO LINEAR FORM

Linear equations are very useful in experimental science. In these relations, we can deduce properties of the functions from the slope of the line and the intercept on the y-axis. These quantities can be used to study the variation of one parameter against another. Below are examples of how this can be done.

A3.4.1 Graph of $y = \frac{a}{x} + b$

For this relationship, we assume a and b are constants, and x is the independent variable. If we let $z = \frac{1}{x}$, the above equation is reduced to the form $y = az + b$. Thus a plot of y as ordinate against z (or 1/x) as abscissa is a straight line with slope a, and intercept b on the y-axis.

A3.4.2 Graph of $y = ax^2 + b$

Here, we let $z = x^2$, and plot a graph of y against z (or x^2). This produces a straight-line graph with slope a, and intercept b on the y-axis.

A3.4.3 Graph of $y = \frac{a}{x^2} + b$

In this, we let $z = \frac{1}{x^2}$, and plot a graph of y against z (or $1/x^2$). A straight line results with slope a, and intercept b on the y-axis.

A3.4.4 Graph of $y = a\sqrt{x} + b$

Following the above methods, it is seen that a plot of y as ordinate against $\sqrt{x}$ as abscissa gives straight line with slope a and intercept b.

A3.4.5 Graph of $y = \frac{a}{\sqrt{x}} + b$

Here, we plot y against $\frac{1}{\sqrt{x}}$. The resulting straight line has slope a and intercept b.

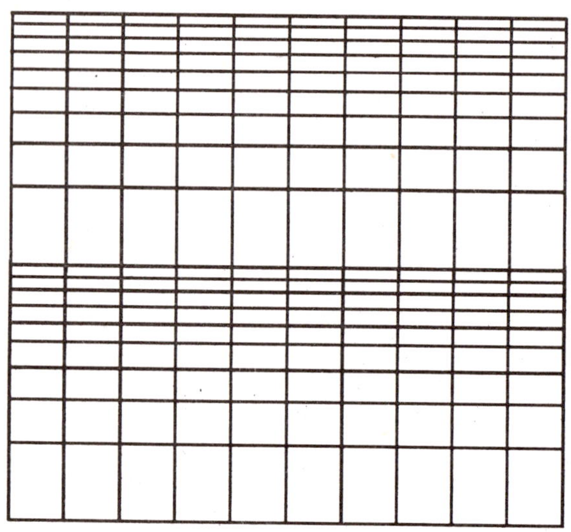

Fig. A3.5 Two cycles of semilog graph paper.

A3.4.6 Graph of $y = ae^{bx}$

In this relation, e is the exponential. To put this into linear form, we take the logarithm of both sides. This gives $\log y = \log a + \log(bx) = bx + \log a$.

Thus a plot of log y as ordinate against x as abscissa yields a linear graph with slope b, and intercept of log a. Such a graph is normally plotted on a paper that has a linear scale on one axis (the x-axis), and a logarithmic scale on the other axis. This graph paper is also called **semilog** paper. **Fig. A3.5** shows two cycles of semilog paper.

A3.5 DIFFERENTIATION

The process of differentiation is a method of finding the rate of change of a function. The rate of change of a function at a point can be done in one of two ways: one of these involves plotting the graph of the function and finding the slope of the tangent to the curve at the point in question. This method we can describe as the practical method. In the theoretical method, we use the equation representing the function, and find the **differential coefficient** at the point of interest. The practical method is already covered extensively in the previous section. In this section, we concentrate on the rules for finding the differential coefficient for certain functions.

A3.5.1 The differential coefficient

Table A3.3 gives the differential coefficients $\frac{dy}{dx}$ for some common functions

Table A3.3 Differential coefficients of some common functions

Function	Equation	Differential coefficient
Polynomial	$y = f(x) = Ax^n$; A, n, constants	nAx^{n-1}
Trigonometric	$y = f(x) = \sin kx$; k constant	$k \cos kx$
	$y = f(x) = \cos kx$	$-k \sin kx$
	$y = f(x) = \tan kx$	$k \sec^2 kx$
Exponential	$y = f(x) = Ae^{kx}$	kAe^{kx}
Logarithmic	$y = \log_e kx$	$k \cdot 1/x$

Using these and some rules, we can get the differential coefficients for other more complicated functions. These rules are stated in the following section.

A3.5.2 Methods of differentiation

(a) Addition and subtraction

The differential coefficient of the sum or difference of two functions is the sum or difference of the coefficients. This rule is stated mathematically as

$$\frac{d}{dx}[f(x) \pm g(x)] = \frac{d}{dx}[f(x)] \pm \frac{d}{dx}[g(x)], \qquad \text{(A3.13)}$$

where $f(x)$ and $g(x)$ are functions of the variable x.

Example:

$$\frac{d}{dx}\left[3x^5 + \sin 4x\right] = 15x^4 + 4\cos 4x$$

(b) The product rule

The rule for differentiating the product of two functions is as follows:

$$\frac{d}{dx}[f(x)\times g(x)] = g(x)\frac{d}{dx}[f(x)] + f(x)\frac{d}{dx}[g(x)], \tag{A3.14}$$

where f(x) and g(x) are functions of the variable x.

Example:

$$\frac{d}{dx}\left[3x^5 \sin 4x\right] = 15x^4 \sin 4x + 3x^5 \times 4\cos 4x = 15x^4 \sin 4x + 12x^5 \cos 4x.$$

(c) The quotient rule

For division, we have

$$\frac{d}{dx}\left[\frac{f(x)}{g(x)}\right] = \frac{g(x)\frac{d}{dx}[f(x)] - f(x)\frac{d}{dx}[g(x)]}{[g(x)]^2}, \tag{A3.15}$$

where f(x) and g(x) are functions of the variable x.

Example: $\frac{d}{dx}\left[\frac{3x^5}{\sin 4x}\right] = \frac{15x^4 \sin 4x - 3x^5 \times 4\cos 4x}{\sin^2 4x}.$

(d) Function of a function

If z = f(u), and u = f(x), then

$$\frac{dz}{dx} = \frac{dz}{du} \times \frac{du}{dx}. \tag{A3.16}$$

Example: To find $\frac{d}{dx}\left[\left(3x^5 + \sin 4x\right)^2\right]$, we let $u(x) = 3x^5 + \sin 4x$. With this substitution, we write $z = f(u) = u^2$. Thus $\frac{dz}{du} = 2u$; $\frac{du}{dx} = 15x^4 + 4\cos 4x$. We hence have $\frac{d}{dx}\left[\left(3x^5 + \sin 4x\right)^2\right] = 2\left(3x^5 + \sin 4x\right)\left(15x^4 + 4\cos 4x\right).$

A3.5.3 The second derivative

Sometimes, it is useful to find the **second** derivative of a function. This second derivative has many practical applications in physics, some of which we have already come across in the two volumes of these texts. For any function, we obtain the second derivative by differentiating the first derivative. The second derivative of the function y with respect to x is written as

$$\frac{d^2y}{dx^2} = \frac{d}{dx}\left(\frac{dy}{dx}\right).$$

To get the second derivative, we apply the differentiation rules described above but this time using the first derivative as the function. For example, we get the second derivative on the function $y = 3x^5 + \sin 4x$.

From the above, we have $\frac{d}{dx}\left[3x^5 + \sin 4x\right] = 15x^4 + 4\cos 4x$.

Thus $\frac{d^2y}{dx^2} = \frac{d}{dx}\left(\frac{dy}{dx}\right) = 60x^3 - 16\sin 4x$. The reader can easily verify this.

A3.6 INTEGRATION

Integration is the reverse of differentiation. Thus, the integral of a function is that one whose differential coefficient gives the function in question. With this connection between the two, we can rewrite **Table A3.3** to give instead the integrals of the functions. This is presented in **Table A3.4**. Notice that differentiating the integral gives the equation of the function. The symbol for integral is an elongated "S". $\int y dx$ is interpreted as the integral of the function y with respect to the variable x.

A3.6.1 Indefinite and definite integrals

An indefinite integral is one that has no limits. The solution of an indefinite integral has a constant of integration. The value of this constant can be found if we know at least one set of corresponding values of the variables.

Table A3.4 Integrals of some common functions

Function	Equation	Integral
Polynomial	$y = f(x) = Ax^n$; A, n, constants	$\frac{A}{n+1}x^{n+1}$; for $n \neq -1$
Trigonometric	$y = f(x) = \sin kx$; k constant	$-1/k \cos kx$
	$y = f(x) = \cos kx$	$1/k \sin kx$
	$y = f(x) = \sec^2 kx$	$1/k \tan kx$
Exponential	$y = f(x) = Ae^{kx}$	$A/k\ e^{kx}$
Logarithmic	$y = 1/x$	$\log_e x$

Example: $\int 4x^2dx = 4\frac{x^3}{3} + C = \frac{4}{3}x^3 + C,$ where C is the constant of integration.

A definite integral is calculated between limits of integration. The solution of an indefinite integral does not have a constant of integration. We find the value of the definite integral by substituting the limits into the integral, and subtracting.

Example: $\int_1^4 4x^2dx = \left[\frac{4}{3}x^3\right]_1^4 = \frac{4}{3}(64-1) = 84.$

A3.6.2 Some technical applications of integration

There are many computational applications of integration. These include
(i) areas between curves,
(ii) volumes of solids of revolution,
(iii) moments of inertia,
(iv) centre-of-mass calculations,
(v) curvature and radius of curvature.
The reader is advised to refer to standard mathematics texts for the formulae for these quantities. Below we briefly survey some applications useful in the technical field.

(a) Finding mean values

The mean value of a quantity y, written as $\bar{y}$, within the region from a to b of the independent variable is defined by the equation

$$\bar{y} = \frac{\int_a^b y dx}{b-a}. \qquad \text{(A3.17)}$$

Example: Find the mean values of the output voltages from
(i) sinusoidal voltage,
(ii) full sinusoidal voltage.
In both cases the function we need to integrate is given by $V = V_0 \sin 2\pi ft$, where f is the frequency of the sinusoidal voltage, and V_o is the peak voltage.

(i) **Fig. A3.6** shows the half-wave. For this, the limits of integration are from $t = 0$ to $t = \frac{1}{2f}$. Using the definition for mean value from (A3.7), we have

$$\bar{V} = \frac{\int_0^{1/2f} V_0 \sin 2\pi ft dt}{1/2f} = 0.637V_0.$$

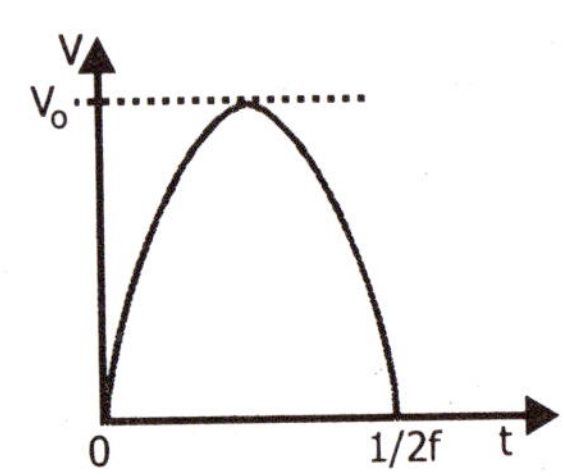

Fig. A3.7 Half-sinusoidal wave

(ii) For the full-wave, the limits of integration are from $t = 0$ to $t = \frac{1}{f}$. The full wave is shown in **Fig. A3.8**.

Over one cycle, we have

$$\overline{V} = \frac{\int_0^{1/f} V_0 \sin 2\pi ft\,dt}{1/f} = 0$$

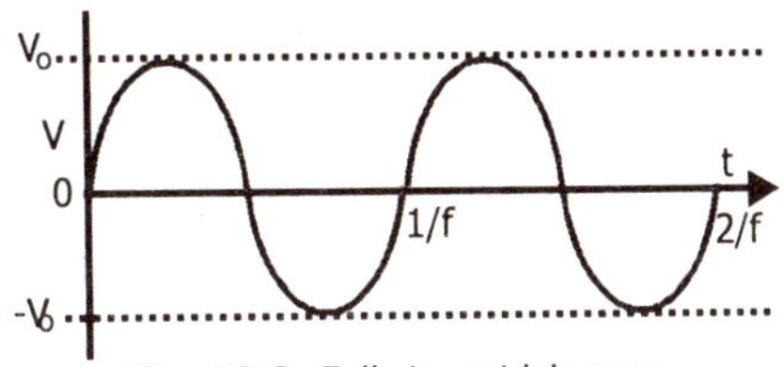

Fig. A3.8 Full sinusoidal wave

(b) Finding root-mean-square values

Our study of alternating currents in chapter 12 led us to an important parameter when a function varies sinusoidally over a cycle. We defined a quantity known as the root-mean-square value of the function. In general, we proceed via the mean square value to the root-mean square, and the method described above for the mean value can be applied to the mean square value. Using the same sinusoidal function as before, we have $V^2 = V_0^2 \sin^2 2\pi ft$. Thus the mean-square value over one cycle is given by

$$\overline{V^2} = \frac{\int_0^{1/f} V_0^2 \sin^2(2\pi ft)dt}{1/f} = \frac{V_0^2}{1/f}\int_0^{1/f}\frac{1}{2}(1-\cos 4\pi ft)dt = \frac{V_0^2}{1/f}\cdot\frac{1}{2}\left[t + \frac{1}{4\pi f}\right]_0^{1/f} = \frac{V_0^2}{2}.$$

The root-mean-square value is hence given by $V_{rms} = \sqrt{\overline{V^2}} = \frac{V_0}{\sqrt{2}}$. This is the same value obtained earlier using analytical argument.

A3.7 DIFFERENTIAL EQUATIONS

A differential equation is defined as one containing differential coefficients. Examples of differential equations are

(i) $\frac{dy}{dx} = 3,$

(ii) $\frac{dy}{dx} = 2x + 1,$

(iii) $\frac{dy}{dx} = 3y,$

(iv) $\frac{d^2y}{dx^2} + 3x\frac{dy}{dx} + 5y = 3,$

and many more. The solution to a differential equation gives general properties of a family of curves. The first three examples above are called **first order differential equations**, whilst (iv) gives an example of the **second order differential equation**. The highest differential coefficient defines the order of the differential equation. We shall study here some methods of solving first order differential equations. Second order equations are more complicated. For the solution of these, the reader can reference standard mathematics texts.

A3.7.1 Solution of first-order differential equations

(a) Basic equations

The most basic differential equation is one with the differential coefficient on one side, and a constant or the other variable on the other side. We solve these by straightforward integration. In example (i) above, we have

$\frac{dy}{dx} = 3$. Thus $dy = 3dx$, or $y = \int 3dx = 3x + C$, where C is a constant of integration. This equation describes a family of curves with a slope of 3. To get a specific equation, we must know one set of corresponding values of the variables x and y. Using the same principles, we can solve the differential equation given in (ii).

(b) By separation of variables

Differential equations of the type $\frac{dy}{dx} = 3y$ are solved by separating the variables. In this example we transform the given equation to give $\frac{dy}{y} = 3dx$. Solving for y gives

$\int \frac{dy}{y} = 3\int dx$. From this we get $\log_e y = 3x + C$, where C is the constant of integration.

A more useful form of this equation is $y = Ae^{3x}$, where A is a constant. We have already come across some physical applications of this type of equation in connection with transient currents.

Index

A

B

C